T0305990

Model Risk Management

This book provides the first systematic treatment of model risk, outlining the tools needed to quantify model uncertainty, to study its effects, and, in particular, to determine the best upper and lower risk bounds for various risk aggregation functionals of interest.

Drawing on both numerical and analytical examples, this is a thorough reference work for actuaries, risk managers, and regulators. Supervisory authorities can use the methods discussed to challenge the models used by banks and insurers, and banks and insurers can use them to prioritize the activities on model development, identifying which ones require more attention than others.

In sum, it is essential reading for all those working in portfolio theory and the theory of financial and engineering risk, as well as for practitioners in these areas. It can also be used as a textbook for graduate courses on risk bounds and model uncertainty.

Ludger Rüschendorf is Professor of Mathematics at the University of Freiburg. He is author of more than 200 research papers and a number of textbooks, in a variety of subjects in probability, statistics, and analysis of algorithms, as well as in risk analysis and mathematical finance. A main topic in his research is the modeling and analysis of dependence structures.

Steven Vanduffel is Professor in Risk Management at the Solvay Business School at Vrije Universiteit Brussel. He has authored papers for leading journals including *Journal of Risk and Insurance, Finance and Stochastics, Mathematical Finance*, and *Journal of Econometrics*. He has won prizes including the Robert I. Mehr Award (2022), the Robert C. Witt Award (2018), and the Redington Prize (2015).

Carole Bernard is Professor in Finance at Grenoble Ecole de Management and Vrije Universiteit Brussel. She has published articles in leading journals in finance, insurance, operations research, and risk management, including *Management Science, Journal of Risk and Insurance, Journal of Banking and Finance*, and *Mathematical Finance*.

Model Risk Management

Risk Bounds under Uncertainty

LUDGER RÜSCHENDORF
University of Freiburg

STEVEN VANDUFFEL
Vrije Universiteit Brussel

CAROLE BERNARD
Grenoble Ecole de Management

Shaftesbury Road, Cambridge CB2 8EA, United Kingdom

One Liberty Plaza, 20th Floor, New York, NY 10006, USA

477 Williamstown Road, Port Melbourne, VIC 3207, Australia

314–321, 3rd Floor, Plot 3, Splendor Forum, Jasola District Centre, New Delhi – 110025, India

103 Penang Road,#05–06/07, Visioncrest Commercial, Singapore 238467

Cambridge University Press is part of the University of Cambridge.

Cambridge University Press is part of Cambridge University Press & Assessment, a department of the University of Cambridge.

We share the University's mission to contribute to society through the pursuit of education, learning and research at the highest international levels of excellence

www.cambridge.org
Information on this title: www.cambridge.org/9781009367165

DOI: 10.1017/9781009367189

First published 2024

A catalogue record for this publication is available from the British Library

Library of Congress Cataloging-in-Publication Data
Names: Rüschendorf, Ludger, 1948- author. | Vanduffel, Steven, author. | Bernard, Carole, author.
Title: Model risk management : risk bounds under uncertainty / Ludger Rüschendorf, Steven Vanduffel, Carole Bernard.
Description: New York, NY : Cambridge University Press, [2023] | Includes bibliographical references and index.
Identifiers: LCCN 2023032170 (print) | LCCN 2023032171 (ebook) | ISBN 9781009367165 (hardback) | ISBN 9781009367189 (epub)
Subjects: LCSH: Risk management. | Management. | Computer simulation.
Classification: LCC HD61 .R867 2023 (print) | LCC HD61 (ebook) |
DDC 658.15/52–dc23/eng/20230810
LC record available at https://lccn.loc.gov/2023032170
LC ebook record available at https://lccn.loc.gov/2023032171

ISBN 978-1-009-36716-5 Hardback

Contents

Preface

What is this book about? This book deals with the ubiquitous problem of model uncertainty, which is one of the most relevant topics in quantitative risk management. It identifies a series of relevant instances of model uncertainty, such as departures from assumed independence, incomplete dependence information, factor models that are only partially specified, or portfolio information that is only available on an aggregate level (e.g., mean and variance of the portfolio loss). It provides the necessary tools to quantify this model uncertainty and in particular to determine the best upper and lower risk bounds for various risk aggregation functionals of interest.

Why did we write this book? While there are good textbooks available dealing with basic methods, concepts, and models in quantitative risk management, this book is the first systematic treatment of the topic of model risk. Not only does it elaborate on the necessary theoretical results for the determination of risk bounds, but it also provides numerical procedures for the effective evaluation of these bounds.

For whom is this book? The book is a relevant reference text on the topic of model risk assessment for actuaries, risk managers, and regulators. It also serves as a textbook for graduate courses on the topic of risk bounds and model uncertainty within the general subject area of risk management in quantitative finance and insurance. The methodology draws on diverse quantitative disciplines ranging from mathematical finance, probability, and statistics to actuarial mathematics.

Acknowledgments: The book is a culmination of a long series of research papers on the topic of model risk. We are indebted to numerous colleagues and former Ph.D. students who either coauthored some of these research papers or who have helped us in our understanding of model risk and the mathematics underlying it. In particular, we would like to mention the substantial contributions of Paul Embrechts and Giovanni Puccetti, who in fact initiated this type of research, as well as the early basic contributions of Etienne de Vylder. Our cooperation includes joint work with Jonathan Ansari, Valeria Bignozzi, Kris Boudt, Andrew Chernih, Ka Chun Cheung, Dries Cornilly, Michel Denuit, Corrado De Vecchi, Paul Embrechts, Luc Henrard, Edgars Jakobsons, Rodrigue Kazzi, Thibaut Lux, Dennis Manko, Silvana Pesenti, Giovanni Puccetti, Daniel Small, Jan Tuitman, Ruodu Wang, Julian Witting, and Jing Yao. We are also grateful to the many anonymous referees whose comments helped us to improve the quality of these research papers and thus of the book. We thank

David Tranah and the team at Cambridge University Press for all their help in the production of this book. Special thanks go to Monika Hattenbach for her dedicated work in preparing the manuscript. Finally, we are indebted to our partners, Gabi, Sonja, and Christophe, for their patience and for forgiving our absentmindedness.

Introduction

This book deals with the problem of model uncertainty, which is a most relevant topic in quantitative risk management. It identifies a series of relevant instances of model uncertainty, such as departures from assumed independence, incomplete dependence information, factor models that are only partially specified, or portfolio information that is only available on an aggregate level (e.g., mean and variance of the portfolio loss). It provides necessary tools to quantify this model uncertainty and in particular to determine the best upper and lower risk bounds for various risk aggregation functionals of interest.

Making sound decisions under uncertainty generally requires quantitative analysis and the use of models. However, a "perfect" model does not exist since some divergence between the model and the reality it attempts to describe cannot be avoided. In a broad sense, model risk is about the extent to which the quality of model-based decisions is sensitive to underlying model deviations and data issues. Quantifying model risk is a key problem in nearly all applied disciplines, including epidemiology, engineering, finance, and insurance. For instance, sums of variables (portfolios) are at the core of the insurance business, as the insurer counts on diversification effects to control the risk of the entire portfolio. For an insurance portfolio, the assumption of independence between the policies is sometimes realistic, in which case the insurer can, for instance, resort to the central limit theorem or to Monte Carlo methods to quantify the maximum loss in a given period of time at a certain probability level (i.e., the VaR). In the majority of cases, however, the individual risks are influenced by one or more common factors, such as geography or economic environment, and it is difficult to specify the joint distribution. Another example concerns the establishment of the capital buffers banks need to put aside to absorb unforeseen losses for a portfolio of risky loans. Doing so requires accurate estimates of the likelihood that various obligors default together, which is very difficult due to a scarcity of data. In this book we focus on model risk in a financial and insurance context.

Model risk may have a real impact on society, such as damage to an institution's reputation or even systemic risk implications. For example, Long-Term Capital Management (LTCM) was a hedge fund that used quantitative models based on normality assumptions but neglected the importance of stress-testing. In 1998 it lost 4.5 billion and required the financial intervention by the Federal Reserve (Lowenstein, 2008). In fact, one of the drivers of the worldwide 2008–2009 financial crisis was the almost blind reliance on certain model assumptions (Salmon, 2009).

While the problem of model uncertainty in statistics has a long history – being developed there under the notion of robust statistics – at the time of writing this book, measuring and managing model risk in a financial context is a relatively new activity. Model risk has, however, already taken a prominent position in the agenda of regulators and supervisors. For instance, in February 2017, the European Central Bank published a guide to the targeted review of internal models (TRIM) in which it is declared that every institution "should have a model risk management framework in place that allows it to identify, understand, and manage its model risk" (ECB, 2017). However, in a status update on TRIM published in June 2018, more than a quarter of the companies supervised had no model risk management framework in place. In their discussion paper on the review of specific items in the Solvency II Regulation, the Actuarial Association of Europe is insisting on focusing more on model risk assessment (AAE, 2017). In the UK, the prudential regulation authority of the Bank of England published some notes on stress test model management principles in which they insist on the necessity of understanding and accounting for the assessment of model uncertainties (PRA, 2018).

In addition to the inherent necessity for insurers and banks to continuously monitor and challenge the models they use in their operations (pricing, product design, risk management), the need for model risk management is further strengthened by the fact that the insurance and banking market is innovating at a very fast pace. New products emerge with unique characteristics, such as driverless car liability insurance, artificial intelligence/robotics liability insurance, and nanotechnology liability insurance. The advent of these new products calls for new pricing and reserving models. In response to this, and in particular driven by the lessons learnt from the 2008–2009 financial crisis, regulators and rating agencies are thus increasing pressure on the financial industry to measure the risk they run and to demonstrate that enough capital is available for absorbing adverse shocks. In order to mitigate this model risk efficiently without restraining too much the arrival of new products and models, the model risk management function (MRM) is emerging in the financial industry. The well-established Professional Risk Managers' International Association (PRMIA) follows this evolution in that it is giving more and more attention to the MRM function in its seminars and other activities.

In the remainder of this introductory chapter, we explain some basic notions of risk assessment and risk models. Furthermore, we describe some basic tools of how to measure aggregate risk, and we conclude by providing an overview of the various subjects dealt with in this book.

A. Risk Assessment and Risk Models

The risk assessment of a multidimensional portfolio $(X_1, X_2, \ldots, X_n)$ is a core issue in risk management of financial institutions. In particular, this problem appears naturally for an insurance company. Any insurer is exposed to different risk factors (e.g., non-life risk, longevity risk, credit risk, market risk, operational risk), has different business lines, or has an exposure to several portfolios of clients. In this regard, one typically attempts to measure the risk of a sum, $S = \sum_{i=1}^{n} X_i$ or of another aggregation function of the risk vector X, in which the individual risks X_i depict losses (claims of the different customers, changes in the different market risk factors, . . .) using a risk measure such as the variance, the VaR, or the tail value-at-risk[1] (TVaR). It is clear that solving this problem is mainly a numerical issue once the joint distribution of $(X_1, X_2, \ldots, X_n)$ is completely specified. Estimating a multivariate distribution or testing its adequacy is in general a difficult task. In many cases, the actuary will be able to use mathematical and statistical techniques to describe the marginal risks X_i, but the dependence among the risks is not available, or only partially known. In other words, the assessment of portfolio risk based on specific models is prone to model misspecification (model risk).

From a mathematical point of view, it is often convenient to assume that the random variables X_i are mutually independent, because powerful and accurate computation methods such as Panjer's recursion and the technique of convolution can then be applied. In this case, one can also take advantage of the central limit theorem, which states that the sum of risks, S, is approximately normally distributed if the number of risks is sufficiently high. In fact, the mere existence of insurance is based on the assumption of mutual independence between the insured risks, and sometimes this complies, approximately, with reality. In the majority of cases, however, the different risks will be interrelated to a certain extent. For example, a sum S of dependent risks occurs when considering the aggregate claims amount of a non-life insurance portfolio because the insured risks are subject to some common factors such as geography, climate, or economic environment. The cumulative distribution function of S can no longer be easily specified.

Standard approaches to estimating a multivariate distribution of a portfolio $(X_1, X_2, \ldots, X_n)$ consist in using a multivariate Gaussian distribution or a multivariate Student t distribution, but there is ample evidence that these models are not always adequate. More precisely, while the multivariate Gaussian distribution can be suitable as a fit to a dataset "on the whole," it is usually a poor choice if one wants to use it to obtain accurate estimates of the probability of simultaneous extreme ("tail") events or, equivalently, if one wants to estimate the VaR of the aggregate portfolio $S = \sum_{i=1}^{n} X_i$ at a given high confidence interval; see McNeil et al. (2015). The use of the multivariate Gaussian model is also based on the (wrong) intuition that correlations are enough to model dependence (Embrechts et al., 1999, 2002). This fallacy also underpins the variance-covariance standard approach that is used for capital aggregation regulatory frameworks such as Basel III and Solvency II, and which also appears in many risk management frameworks in the industry. Furthermore, in practice, there are not enough observations that can be considered as tail events. In fact, there is always a level beyond which there is no observation. Therefore, if one makes a choice for modeling tail dependence, it has to be somewhat arbitrary.

In the literature, one can find flexible multivariate models that allow a much better fit to the data, for example using pair copula constructions and vines (see, e.g., Aas et al., 2009 or Czado, 2010 for an overview). While these models have theoretical and intuitive appeal, their successful use in practice requires a dataset that is sufficiently

---

[1] In the literature it is also called the expected shortfall and the conditional value-at-risk, among others.

rich. However, no model is perfect, and while such developments are clearly needed for an accurate assessment of portfolio risk, they are only useful to regulators and risk managers if they are able to significantly reduce the model risk that is inherent in risk assessments.

B. Measuring Aggregate Risk

Insurance companies essentially exchange premiums against (future) random claims. Consider a portfolio containing n policies, and let X_i, $i = 1, 2, \ldots, n$, denote the loss, defined as the random claim net of the premium, of the ith policy. In order to protect policyholders and other debtholders against insolvency, the regulator will require the portfolio loss $S = X_1 + X_2 + \cdots + X_n$ to be "low enough" as compared to the available resources, say a capital requirement K, which means that the available capital K has to be such that $S - K$ is a "safe bet" for the debtholders, i.e., one is "reasonably sure" that the event "$S > K$" is of minor importance (Tsanakas and Desli, 2005; Dhaene et al., 2012).

It is clear that measuring the riskiness of $S = X_1 + X_2 + \cdots + X_n$ is of key importance for setting capital requirements. However, there are several other reasons for studying the properties of the aggregate loss S. Indeed, an important task of an enterprise risk management (ERM) framework concerns capital (risk) allocation, i.e., the allocation of total capital held by the insurer across its various constituents (subgroups), such as business lines, risk types, and geographical areas, among others. Indeed, doing so makes it possible to redistribute the cost of holding capital across the various constituents so that it can be transferred back to the depositors or policyholders in the form of charges (premiums). Risk allocation also makes it possible to assess the performance of the different business lines by determining the return on allocated capital for each line. Finally, the exercise of risk aggregation and allocation may help to identify areas of risk consumption within a given organization and thus to support the decision-making concerning business expansions, reductions, or even eliminations; see Panjer (2001) and Tsanakas (2009).

When measuring the aggregate risk S, it is also important to consider the context at hand. In particular, different stakeholders may have different perceptions of riskiness. For example, depositors and policyholders mainly care only about the probability that the company will meet its obligations. Regulators primarily share the interest of depositors and policyholders and establish rules to determine the required capital to be held by the company. However, they also care about the magnitude of the loss given that it exceeds the capital held, as this is the amount that needs to be funded by society when a bailout is needed. Formally, they care about the *shortfall* of the portfolio loss S with solvency capital requirement $\varrho(S)$; that is,

$$(S - \varrho(S))_+ := \max(0, S - \varrho(S)). \tag{0.1}$$

The shortfall is thus part of the total loss that cannot be covered by the insurer. It is also referred to as the *loss to society* or the *policyholders' deficit*. In view of their limited liability, shareholders do not really have to care about the residual risk but rather focus on the properties of the variable $S - (S - \varrho(S))_+ = \min(S, \varrho(S))$. In summary, various stakeholders may have different perceptions and sensitivities with respect to the meaning of the risk they run, and they may employ different paradigms for defining and measuring it.

As for measuring the risk, the two most influential risk measures are the *value-at-risk* (VaR) and the *tail value-at-risk* (TVaR).[2] For a given confidence level α, they are denoted by VaR_α and TVaR_α, respectively, and are defined as

$$\text{VaR}_\alpha(S) = \min\{x \mid P(S \le x) \ge \alpha\}, \quad 0 < \alpha < 1 \tag{0.2}$$

and

$$\text{TVaR}_\alpha(S) = \frac{1}{1-\alpha} \int_\alpha^1 \text{VaR}_q(S) \, dq, \quad 0 < \alpha < 1. \tag{0.3}$$

So, VaR_α is the minimum loss one observes with probability $1 - \alpha$, whereas TVaR_α is the average of all upper VaRs. A general class of risk measures that includes VaR and TVaR is the distortion risk measures, which are essentially weighted averages of VaRs.

Another interesting risk measure that is not in this class is the "tail risk," which is the probability that the aggregate risk exceeds some level K, i.e., $P(S \ge K)$.

C. Regulatory Frameworks

The report of the Basel Committee on Banking Supervision (2010) describes the modeling methods used by financial firms and regulators in various countries to aggregate risk. It also aims to identify the conditions under which these aggregation techniques perform as anticipated in the model and to suggest potential improvements. The report expresses doubts about the reliability of internal risk aggregation results that incorporate diversification benefits: "Model results should be reviewed carefully and treated with caution, to determine whether claimed diversification benefits are reliable and robust." In this subsection, we very briefly summarize how the main "regulatory frameworks" deal with risk aggregation and diversification.

Basel III Regulation for Banks

One calculates a bank's overall minimum capital requirement as the sum of capital requirements for the credit risk, operational risk, and market risk, without recognizing possible diversification benefits between the three risk types. As diversification is not acknowledged, the practice of adding up capital is seen as a conservative approach. However, this also depends on the risk measure that is used for establishing capital. If VaR is used, it might indeed occur that the VaR of the aggregate risk is higher than the sums of the individual VaRs. By contract, if a coherent risk measure ϱ is used, then it always holds that $\varrho(X_1 + \cdots + X_n) \le \varrho(X_1) + \cdots + \varrho(X_n)$.

---

[2] VaR appears to be the most popular risk measure in practice, among both regulators and risk managers; see, for example, Jorion (2006).

As for market risk, banks have the choice between two methods. They may benefit from diversification if they use an internal model approach (IMA). With the standardized measurement method (SMM), the minimum capital requirement for market risk is the sum of the capital charges calculated for each individual risk type (interest rate risk, equity risk, foreign exchange risk, commodities risk, and price risk in options).

Solvency II

The Solvency Capital Requirement (SCR) under Solvency II is defined as the VaR at 99.5 % over a time horizon of one year. When aggregating risks, insurers may benefit from diversification: They have the option to use an internal model (without any particular method prescribed) or a standard formula. The standard formula aggregates risks using a correlation matrix (variance-covariance approach) to take dependencies into account.

Comparison and Comments on Regulatory Frameworks

Generally, regulatory rules incorporate diversification by taking into account some correlation effect to reduce the total capital (at least in some subcategories). Overall, we observe that regulators all implicitly assume that the sum of the risk numbers is the worst possible situation. "No diversification benefits" is then synonymous to "adding up risk numbers (VaRs)."

The easiest method to aggregate risks is the variance-covariance approach (which is explicitly mentioned in Solvency II above and is also used by the Australian regulator OSFI (2014)). It builds on the assumption that the correlation matrix is enough to describe the dependence and that it is possible to aggregate risks based on this correlation matrix. Its strength is in its being a simple approach, but it is only a correct approach for elliptical multivariate distributions, such as the Gaussian multivariate distribution. Furthermore, correlation is a linear measure of dependence and does not capture tail dependence adequately. Using such a method to aggregate risk may perhaps be fine for having some idea on the distribution "globally" but fails when it comes to assess the risk in the tail; also note that capital requirements are typically based on tail risk measures, such as VaR at 99.5 %, which essentially reflects the outcome of a 1-in-200-year scenario.

Instead of using the variance-covariance approach, one may use copulas to aggregate the individual risks. This approach is rather flexible and allows one to separate the risk assessment of the marginal distribution of individual risks and their dependence. By specifying a given copula to model some dependence, it is then possible to recognize tail dependence among some risks. However, determining the "right" copula to use is a very hard task that is prone to significant model risk, as we will see later in this book. Statistical methods to fit a multivariate model involve large numbers of parameters and copula families. In addition, understanding the outputs of the model will then require good expertise in the copula approach to understand the impact of each assumption made on the dependence. This is a concern and a challenge among institutions.

Another way to capture tail risks and tail dependence is to understand "where the dependence comes from" and to model the real risk drivers of the dependence between individual risks of the portfolio and understand their interactions. The report of the Basel Committee on Banking Supervision (2010) suggests using "scenario-based aggregation." Aggregation through scenarios boils down to determining the state of the firm under specific events and summing profits and losses for the various positions under the specific event. In other words, it means that one needs to incorporate information that one knows about the dependence in some specific states.

As observed in the report of the Basel Committee on Banking Supervision (2010), the results of scenario-based aggregation are easier to interpret, with more meaningful economic and financial implications, but use of this method again requires deep expertise to identify risk drivers, derive meaningful sets of scenarios with relevant statistical properties, and then use them to obtain a full loss distribution. A lot of the inputs in these kinds of model come from experts' judgements. Overall, there is no clear unique solution to the problem of risk aggregation. Each method has its pros and cons and may be helpful in some situations and useless in others.

D. Overview of the Contents

As described in the introduction, the main topic of the book is to establish risk bounds for aggregate risk functionals under various forms of model uncertainty for the underlying model. The model uncertainty is described by different types of (partial) knowledge on the underlying model: knowledge of the marginals, various forms of partial knowledge of the dependence structure, or knowledge of the structure of the model, as given for example by (partially specified) factor models or by (partially specified) subgroups of the model. An important part of the book addresses the possible improvement of risk bounds due to information on various functionals of the risk vector, in particular given by moment bounds for the aggregate portfolio. Combination with a neighborhood model assumption yields further improvement of the risk bounds.

The book consists of four main parts. The basic assumption in Parts I–III in order to derive risk bounds is that the marginal distributions of the risk vector $X = (X_1, \ldots, X_n)$ are known, say $X_i \sim F_i$, $1 \leq i \leq n$. This assumption can be realistically made in many applications since it is much easier to model and test simple hypotheses compared to the task of modeling and testing the joint distribution of X. Throughout the book, we present focus on risk bounds for the aggregated portfolio $S = \sum_{i=1}^{i=n}$. Nevertheless in Parts I–III the assumed knowledge of the marginals makes it possible to also derive various risk bounds for other risk functionals, for example, for the maximal risk. In Part IV, we do not assume knowledge of the marginals but only use information that pertains to a portfolio sum S.

In **Part I**, an introduction to the problem of risk bounds with information on the marginal distributions is given. Also, a basic algorithm to determine these bounds – the rearrangement algorithm – and a basic solution method – the dual bounds – are introduced.

In Chapter 1, we introduce some basic notions of risk measures (like VaR, TVaR, and convex risk measures) and describe some corresponding worst case VaR and TVaR portfolios. We also give a rearrangement formulation of various forms of determining worst or best case risk bounds. We describe the connection of upper and lower risk bounds to convex ordering properties, discuss comonotonicity and countermonotonicity (antimonotonicity), and give some basic results to obtain worst case VaR portfolios resp. portfolios with maximal tail risk.

The "standard bounds" for tail risks go back to classical sources, like Sklar (1973) or Moynihan et al. (1978). For $n = 2$ they are shown in Makarov (1981) and Rüschendorf (1982) to be sharp, but for $n > 2$ they typically only deliver rough (i.e., not necessarily sharp) bounds.

The conditional moment method gives an upper bound on the tail risk of the portfolio sum in terms of conditional moments of the marginals. This type of upper risk bound produces sharp bounds under a mixing condition on the upper tail. In the final subsection of this part, we discuss in more detail the notion of mixability, describe some basic results due to Wang and Wang (2011, 2016) on mixability, and explain its role in the determination of convex minima of portfolio sums and similarly for best and worst case portfolios.

Chapter 2 is devoted to the motivation and introduction of the rearrangement algorithm (RA) as introduced in Puccetti and Rüschendorf (2012a), which is a fundamental tool to determine sharp upper and lower bounds for the tail risks resp. for the VaR. It is basically motivated by the formulation of the problem of determining risk bounds as a rearrangement problem and in a second step by a further reduction to an assignment problem. Also, a variation of the RA, the block rearrangement algorithm (BRA), is introduced, which improves some aspects of the RA. Several interesting applications of these algorithms are indicated.

Chapter 3 gives an introduction to Hoeffding–Fréchet functionals, which describe the largest resp. the smallest value of a risk functional over all possible dependence structures, when fixing the marginals. The main result is a duality theorem for these functionals that is a far reaching extension of the dual representation in the Monge–Kantorovich mass transportation problem. A reduction of the admissible dual functions to a simple class of admissible dual functions introduced in Embrechts and Puccetti (2006b) leads to good dual bounds for the tail risk. Sharpness of these bounds is established under a mixing condition in the homogeneous case in Puccetti and Rüschendorf (2013).

An easy-to-calculate upper bound for the worst case VaR of an aggregated portfolio is given by the TVaR of a corresponding comonotonic sum. In Chapter 4 we derive asymptotic sharpness of this upper TVaR bound under an asymptotic mixing condition. A version of this result also holds in the infinite mean case and in the inhomogeneous case.

In contrast to Part I, we assume in **Part II** not only that we know the marginal distributions of the risk vector $X = (X_1, \ldots, X_n)$, say $X_i \sim F_i$, $1 \leq i \leq n$, but that we have additional information on the dependence among these n variables.

In Chapter 5 we present the method of improved standard bounds. We review bounds on the distribution function that improve upon the Hoeffding–Fréchet bounds by including, in addition to the marginal information, some positive or negative dependence information on the distribution functions of the form $F \leq G$ or $F \geq G$. These improved Hoeffding–Fréchet bounds lead to improved upper and lower VaR bounds. To do so, we build on results of Williamson and Downs (1990), Denuit et al. (1999), Embrechts et al. (2003), Rüschendorf (2005), Embrechts and Puccetti (2006a,b), and Puccetti et al. (2016), among others. We present some examples in which the information significantly reduces the standard bounds. Further examples of this type are discussed in Rüschendorf (2017a,b). The last section of this chapter presents an extension of the method of improved standard bounds by Lux and Rüschendorf (2018) to include two-sided dependence information of the form $G \leq F \leq H$ or related bounds for the copula C or for the survival functions $\overline{F}$.

Chapter 6 deals with bounds on VaR under the additional constraint that a bound on the variance of the sum is known (as in Bernard et al., 2017c) or that information on higher moments on the sum (as in Bernard et al., 2017a) is available. As compared to assuming only knowledge of the marginal means or variances (so-called moment bounds), knowledge of the marginal distributions may make it possible to improve the VaR bounds. However, when supplementing the information on the marginal distributions with information on the portfolio variance, it turns out that, when the variance constraint is "low" enough, the analytical VaR-bounds coincide with moment bounds (studied in full detail in Part IV). We then propose a numerical method based on the RA-algorithm to approximate sharp VaR bounds given the marginal constraints and additionally bounds on the variance of the sum. A corresponding rearrangement algorithm is no longer available if only moment bounds are assumed. We show that for large portfolios, our analytical bounds are nearly sharp. However, in the context of smaller portfolios (or when the portfolio depicts significant concentration), one cannot expect the bounds to be sharp. The RA-algorithm can accommodate this situation and makes it possible to approximate sharp bounds. As a by-product, this algorithm also sheds light on the composition of extreme portfolios.

Chapter 7 is devoted to studying the case when information on the joint distribution of the risk vector $X = (X_1, \ldots, X_n)$ is partially available, by knowing the conditional distribution of X on a subset S in R^n the marginal distributions as well as the probability of the subset, or alternatively, knowing bounds for the distribution function F_X resp. the copula C_X on the corresponding subsets. For example, it may be the case that C_X of X is known on a central domain due to availability of sufficient statistical data, allowing for a precise model within this domain. Alternatively, some positive or negative dependence information may be available in the upper or lower tail area.

We derive VaR bounds in several situations of interest in which the conditional distribution of X is known on a central domain. The corresponding improvements of the Hoeffding–Fréchet bounds are derived in Rachev and Rüschendorf (1994), Nelsen et al. (2004), Tankov (2011) and Bernard and Vanduffel (2014) in the two-dimensional case and by Puccetti et al. (2016) and Lux and Papapantoleon (2019) in the n-dimensional case.

We solve several cases exactly or numerically when we know the distribution function on a subset or we know the conditional distribution on a subset (Bernard and

Vanduffel, 2015; Puccetti et al., 2016). While knowing the distribution function F on an open domain A implies knowledge of P^X on A, the converse direction is not valid, even for rectangle domains A. So both types of information are typically different. Based on a numerical approach using the RA-algorithm, the value of this type of information for improving VaR bounds for the sum is determined

In **Part III** the risk bounds and range of dependence uncertainty obtained in the unconstrained case is essentially improved by including relevant structural information on the underlying class of models. This additional structural information may lead to a positive or a negative dependence restriction and thus induce improvements of the upper resp. the lower risk bounds, as shown in Part II.

We deal in this part with several types of structural information, investigate their impact, and show in several applications their potential in reducing dependence uncertainty. In some applications it is possible to combine these structural assumptions with dependence assumptions, as in Part II, leading to particular useful reduction results.

In Chapter 8 we consider first in Section 8.1 the case that besides the one-dimensional marginals, some higher dimensional marginals are also known. This assumption allows us to derive improvements of the classical Fréchet bounds by using a duality theorem. For general higher-dimensional marginal systems, the "reduction method" allows us to derive good upper and lower bounds by an associated reduced risk model with simple marginals. This reduced problem can be solved by the RA. If the higher-dimensional marginals exhibit strong positive dependence, this leads to improvements of the lower bounds in comparison to the unconstrained case; if they exhibit negative dependence, this leads to improvements of the upper bounds.

In Section 8.2 we consider a general case of additional constraints given by the distribution or the expectations of a class of functionals. This includes in principle the constraints due to higher-dimensional marginals (infinite set of restrictions) but also the case of variance constraints as in Part II. We give several improved lower and upper dual bounds for such classes of constraints. Using martingale constraints (infinite set of restrictions), this method can also be used to derive improved price bounds for options.

Chapter 9 gives a detailed discussion on partially specified factor models (PSFM). This model assumption is based on an underlying factor model $X_i = f_i(Z, \varepsilon_i)$ with a systematic risk factor Z and individual risks $\varepsilon_1, \ldots, \varepsilon_n$. In comparison to the usual assumption of completely specified factor models, which is in general a hard to verify assumption, in PSFMs only the joint distributions of (ε_i, Z) are specified, which is a much simpler to verify model assumption. We show that the assumption of PSFM may lead to strongly reduced risk bounds and that it can be combined in a particular effective and flexible way with other dependence assumptions, like variance bounds for the aggregated portfolio.

Chapter 10 deals with a systematic investigation of the assumption that the risk vector X is split into k subgroups. For a comparison vector Y, conditions are given for the comparison of the subgroup sums of X and Y and further for the comparison of the copulas of the vectors of subgroup sums to imply a relevant comparison theorem between the aggregated portfolios $\sum_{i=1}^{n} X_i$ and $\sum_{i=1}^{n} Y_i$. Also, this criterion allows flexible and effective applications and can be combined in a useful way with further constraints, for example, with the assumption of PSFMs within the subgroups.

In **Part IV** risk bounds are studied under the assumption that moment bounds of the portfolio sums are given. It is shown in Chapter 6 that adding a variance constraint on the portfolio loss $S = \sum X_i$ in addition to knowledge of the marginal distribution of the portfolio components X_i may lead to an improved VaR upper bound that corresponds to the classic Cantelli moment bound (involving mean and variance). This observation motivates us to study risk bounds for a portfolio sum $S = \sum X_i$ when the mean of S (but *not* the marginal distributions of its components) is given as well as some of its higher order moments. This approach to assessing model risk is, for instance, useful when loss statistics are only available at an aggregate level.

In Chapter 11 we study moment bounds for the risk measures VaR, TVaR, and the range value-at-risk (RVaR), which can be seen as a generalization of VaR and TVaR. Most of our results assume that only the first two moments are known (but no other higher-order moment). For the important case of VaR, however, we propose under an additional domain restriction a method that can in principle deal with any number of known moments. As a main tool for deriving the bounds, we use convex ordering and its generalization, s-convex order.

In Chapter 12 we provide a detailed study on moment bounds for distortion risk measures. In Section 12.1 we build on results in Cornilly et al. (2018) to derive moment bounds under an additional domain restriction. In Section 12.2 we dispense with domain restrictions and derive bounds using the tool of isotonic projections combined with the use of the Cauchy–Schwarz inequality.

In Chapters 13 and 14 we study the influence of structural information on the moment bounds for VaR, TVaR, and RVaR. First, in Chapter 13, we study bounds when, in addition to moment information, it is also known that the distribution is unimodal. Second, in Chapter 14 we study bounds when the distribution is assumed to stay in the neighborhood of a reference distribution. We use the Wasserstein distance as a metric to determine the neighborhood and use the tool of isotonic projections to obtain sharp bounds on VaR, TVaR, and RVaR.

Part I

Risk Bounds for Portfolios Based on Marginal Information

The basic assumption in Parts I–III of this book in order to derive risk bounds is that the marginal distributions of the risk vector $X = (X_1, \ldots, X_n)$ are known, say $X_i \sim F_i$, $1 \leq i \leq n$. This can be realistically assumed in many applications since it is much easier to model and test simple hypotheses compared to the task of modeling and testing the joint distribution of X. We often deal with risk bounds for the aggregated portfolio $S = \sum_{i=1}^{i=n}(X_i)$, but we also present risk bounds for other risk functionals, for example for the maximal risk, for the variation $\max |X_i - X_j|$ of the risks, or for the risk exposure in certain domains.

In Chapter 1, we introduce some basic notions of risk measures (like VaR, TVaR, and convex risk measures) and describe some corresponding worst case VaR and TVaR portfolios. We also give a rearrangement formulation to determine worst or best case risk bounds. We describe the connection of upper and lower risk bounds to convex ordering properties, discuss comonotonicity and countermonotonicity (or antimonotonicity), and give some basic results to obtain worst case VaR portfolios resp. portfolios with maximal tail risk.

The "standard bounds" for tail risk go back to classical sources like Sklar (1973) or Moynihan et al. (1978). For $n = 2$ they are shown in Makarov (1981) and Rüschendorf (1982) to be sharp, but for $n > 2$ they typically only deliver rough bounds.

The conditional moment method gives an upper bound on the tail risk of the portfolio in terms of conditional moments of the marginals. This type of upper risk bound produces sharp bounds under a mixing condition on the upper tail. In the final subsection of this part, we discuss in more detail the notion of mixability, describe some basic results due to Wang and Wang (2011, 2016) on mixability, and explain its role in the determination of convex minima of portfolio sums and similarly for best and worst case portfolios.

Chapter 2 is devoted to the motivation and introduction of the rearrangement algorithm (RA) as introduced in Puccetti and Rüschendorf (2012a). This is a fundamental tool to determine sharp upper and lower bounds for the tail risk resp. for the VaR. It is basically motivated by the formulation of the problem of determining risk bounds as a rearrangement problem and in a second step by a further reduction to an assignment problem.

Chapter 3 gives an introduction to Hoeffding–Fréchet functionals, which describe the largest and the smallest value of a risk functional over all possible dependence structures, when fixing the marginals. The main result is a duality theorem for these functionals, which is a far reaching extension of the dual representation in the Monge–Kantorovich mass transportation problem. A reduction of the admissible dual functions to a simple class of admissible dual functions introduced in Embrechts and Puccetti (2006b) leads to good dual bounds for the tail risk. Sharpness of these bounds was established under a mixing condition in the homogeneous case in Puccetti and Rüschendorf (2013).

A simple upper bound for the worst case VaR risk is given by the TVaR bound, i.e., the TVaR of the comonotonic risk vector. In Chapter 4 we derive asymptotic sharpness of this upper TVaR bound under an asymptotic mixing condition. A version of this result also holds in the infinite mean case and in the inhomogeneous case.

1 Risk Bounds with Known Marginal Distributions

As described in the introduction, a key problem of risk analysis is to derive (sharp) risk bounds on a portfolio $S = X_1 + \cdots + X_n$ under the given distributional information on a risk vector $X = (X_1, \ldots, X_n)$. In this chapter, we derive several explicit results for this problem under the assumption that only the marginal distributions F_j of X_j are known, but the dependence structure of X is completely unknown. In particular we introduce some basic notions of risk theory, such as worst case value-at-risk and tail value-at-risk portfolios, comonotonic risk vectors, the connection of upper risk bounds to convex ordering, and some basic results to obtain worst case value-at-risk portfolios. A more detailed presentation and extension of these results is given in Rüschendorf (2013, Chapters 2–4). Some detailed mixing results in Section 1.4 are due to several papers of Wang and coauthors (see Wang and Wang, 2011).

1.1 Some Basic Notions and Results of Risk Analysis: VaR, TVaR, Comonotonicity, and Convex Order

There are several risk measures of interest, like the value-at-risk (VaR), the tail value-at-risk (TVaR), and the classes of convex risk measures or of distortion risk measures. The VaR risk measure at level α, VaR_α, $\alpha \in (0, 1)$ of the portfolio S is defined as the α-quantile of the distribution of S, i.e.,

$$\text{VaR}_\alpha(S) = F_S^{-1}(\alpha) = \inf\{x \in \mathbb{R}; \ F_S(x) \geq \alpha\}; \tag{1.1}$$

we also make use of the upper VaR as an upper α-quantile, i.e.,

$$\text{VaR}_\alpha^+(S) = \sup\{x \in \mathbb{R}; \ F_S(x) \leq \alpha\}. \tag{1.2}$$

The TVaR risk measure at level α, TVaR_α, takes into account also the magnitude of the risk above the α-quantile and is defined as

$$\text{TVaR}_\alpha(S) = \frac{1}{1 - \alpha} \int_\alpha^1 \text{VaR}_u^+(S) \, du = \frac{1}{1 - \alpha} \int_\alpha^1 \text{VaR}_u(S) \, du. \tag{1.3}$$

From the definition it follows that TVaR is an upper bound of VaR, i.e., for $\alpha < 1$ it holds that

$$\text{VaR}_\alpha(S) \leq \text{VaR}_\alpha^+(S) \leq \text{TVaR}_\alpha(S). \tag{1.4}$$

In comparison to VaR, TVaR has the important property of being a convex risk measure. A risk measure ϱ is said to be **convex** (Föllmer and Schied, 2004) if it is monotone, translation invariant, and satisfies the important convexity condition,

$$\mathrm{TVaR}_\alpha(aX + (1-a)Y) \le a\,\mathrm{TVaR}_\alpha(X) + (1-a)\,\mathrm{TVaR}_\alpha(Y). \qquad (1.5)$$

If the risk measure ϱ is also positive homogeneous, then it is called **coherent**.

Thus, using TVaR as a risk measure, a diversified portfolio is preferred concerning the magnitude of risk in comparison to an undiversified portfolio. The left TVaR measure at level α, LTVaR_α is similarly defined and considers the left tails (best case) of risks:

$$\mathrm{LTVaR}_\alpha(S) = \frac{1}{\alpha} \int_0^\alpha \mathrm{VaR}_u(S)\,\mathrm{d}s. \qquad (1.6)$$

An important property of a risk measure that is convex **law invariant**, i.e., one that only depends on the marginal distribution, is its consistency with respect to convex order $\le_{\mathrm{cx}}$.

Definition 1.1 (Convex order) Let X and Y be two random variables with finite means. X is smaller than Y in convex order, denoted by $X \le_{\mathrm{cx}} Y$, if for all convex functions f,

$$Ef(X) \le Ef(Y), \qquad (1.7)$$

whenever both sides of (1.7) are well defined.

A law-invariant convex risk measure ϱ (e.g., TVaR) is consistent with respect to convex order on proper probability spaces such as L^1 (integrable random variables) and L^∞ (bounded random variables). In consequence it holds that $X \le_{\mathrm{cx}} Y$ implies

$$\mathrm{TVaR}_\alpha(X) \le \mathrm{TVaR}_\alpha(Y), \qquad (1.8)$$

see Chapter 4 of Föllmer and Schied (2004), Jouini et al. (2006), Bäuerle and Müller (2006), and Burgert and Rüschendorf (2006). From this section on, we consider as the basic space of risks $\mathcal{X} = L^1$ and assume that all marginal distributions of a risk vector X have finite first moments when dealing with TVaR. For given distribution functions $F_1, \ldots, F_n$, let $\mathcal{F}(F_1, \ldots, F_n)$ denote the **Fréchet class** of all n-dimensional distribution functions F with marginal distribution functions $F_1, \ldots, F_n$. The classical Fréchet bounds characterize the Fréchet class $\mathcal{F}(F_1, \ldots, F_n)$.

Theorem 1.2 (Fréchet bounds)
a) For $F \in \mathcal{F}(F_1, \ldots, F_n)$ it holds that

$$F_-(x) := \left(\sum_{i=1}^n F_i(x_i) - (n-1) \right)_+ \le F(x)$$

$$\le F_+(x) := \min_{1 \le i \le n} F_i(x_i), \quad x \in \mathbb{R}^n. \qquad (1.9)$$

F_-, F_+ are called lower resp. upper Fréchet bounds (also called Hoeffding–Fréchet bounds).

b) $F_+ \in \mathcal{F}(F_1, \ldots, F_n)$; if $n = 2$ then $F_- \in \mathcal{F}(F_1, F_2)$.

c) For a distribution function F on $\mathbb{R}^n$, it holds that

$$F \in \mathcal{F}(F_1, \ldots, F_n) \iff F_- \le F \le F_+.$$

In particular, there exists for any n a largest distribution function with marginals F_i, the upper Fréchet bound F_+. For $n = 2$ there exists a smallest distribution function with marginals F_i, the lower Fréchet bound. In general, the upper and lower bounds in (1.9) are sharp. The upper bound F_+ is attained by the comonotonic risk vector.

Definition 1.3 (Comonotonicity, countermonotonicity)
Let $F_1, \ldots, F_n$ be one-dimensional distribution functions, and let $U \sim U(0, 1)$ be uniformly distributed on $[0, 1]$. Then:

a)
$$X^c := (F_1^{-1}(U), \ldots, F_n^{-1}(U)) \qquad (1.10)$$

with $F_i^{-1}(\alpha) = \inf\{x \in \mathbb{R}: F_i(x) \ge \alpha\}$ is called a **comonotonic** risk vector.

b) For $n = 2$,

$$X_c := (F_1^{-1}(U), F_2^{-1}(1 - U)) \qquad (1.11)$$

is called a **countermonotonic** (antimonotonic) risk vector.

Comonotonic risk vectors X are characterized by the fact that the components of X are ordered in the same way.

The co- resp. countermonotonic risk vectors realize the upper resp. lower Fréchet bounds F_+, F_-, i.e.,

$$X^c \sim F_+ \qquad \text{and for } n = 2, \quad X_c \sim F_-. \qquad (1.12)$$

In terms of the lower orthant order $\le_{\text{lo}}$ defined by the pointwise ordering of the distribution functions, therefore, for any vector X with marginal distributions F_i it holds by the Fréchet bounds that

$$X \le_{\text{lo}} X^c \qquad (1.13)$$

and for $n = 2$,

$$X_c \le_{\text{lo}} X. \qquad (1.14)$$

The following basic result due to Meilijson and Nadas (1979) describes the role of the comonotonic vector as a worst case model for the portfolio $S = \sum_{i=1}^{n} X_i$ with respect to all law-invariant convex risk measures.

Theorem 1.4 (Comonotonic risk vector and convex order)
Let $X = (X_1, \ldots, X_n)$ be a risk vector with marginal distributions F_i. Then

a)
$$\sum_{i=1}^{n} X_i \le_{\text{cx}} \sum_{i=1}^{n} X_i^c, \qquad (1.15)$$

i.e., the portfolio of comonotonic risks is the worst case portfolio with respect to convex order.

b)
$$E\left(\sum_{i=1}^{n} X_i - t\right)_+ \leq E\left(\sum_{i=1}^{n} X_i^c - t\right)_+ \qquad (1.16)$$

for all t. Moreover, $E\left(\sum_{i=1}^{n} X_i^c - t\right)_+ =: \Psi_+(t)$, *where*

$$\Psi_+(t) = \inf_{\sum v_i = t} \sum_{i=1}^{n} E(X_i - v_i)_+. \qquad (1.17)$$

The statement in b) says that the excess of loss risk functional of the portfolio is maximized by the comonotonic risk vector.

For $n = 2$, a countermonotonic risk vector $X_c = (F_1^{-1}(U), F_2^{-1}(1 - U))$ realizes the convex minimum of portfolio sums of variables X_i with distribution functions F_i.

Proposition 1.5 (Countermonotonic risk vector and convex order) *Let* $X = (X_1, X_2)$ *be a risk vector of size* $n = 2$ *with marginal distribution functions* F_i. *Then for all* $X_i \sim F_i$ *it holds that*

$$F_1^{-1}(U) + F_2^{-1}(1 - U) \leq_{cx} X_1 + X_2. \qquad (1.18)$$

In consequence, for $n = 2$ we have for all $X_i \sim F_i$,

$$X_{1,c} + X_{2,c} \leq_{cx} X_1 + X_2 \leq_{cx} X_1^c + X_2^c, \qquad (1.19)$$

where $X_c = (X_{1,c}, X_{2,c})$.

We define the worst case risks of the portfolio $S = \sum_{i=1}^{n} X_i$, where X_i have marginal distribution functions F_i with respect to VaR and TVaR by

$$\overline{\text{VaR}}_\alpha := \sup\left\{ \text{VaR}_\alpha(S); S = \sum_{i=1}^{n} X_i, X_i \sim F_i, 1 \leq i \leq n \right\}$$
$$\text{and} \quad \overline{\text{TVaR}}_\alpha := \sup\left\{ \text{TVaR}_\alpha(S); S = \sum_{i=1}^{n} X_i, X_i \sim F_i, 1 \leq i \leq n \right\}. \qquad (1.20)$$

Similarly, the best case of risks at level α is defined as

$$\underline{\text{VaR}}_\alpha := \inf\left\{ \text{VaR}_\alpha(S); S = \sum_{i=1}^{n} X_i, X_i \sim F_i, 1 \leq i \leq n \right\}$$
$$\text{and} \quad \underline{\text{TVaR}}_\alpha := \inf\left\{ \text{LTVaR}_\alpha(S); S = \sum_{i=1}^{n} X_i, X_i \sim F_i, 1 \leq i \leq n \right\}. \qquad (1.21)$$

Then we get by means of Theorem 1.4 the following important connections between these notions. For a risk vector X, let $S = \sum_{i=1}^{n} X_i$ be the portfolio sum and $S^c = \sum_{i=1}^{n} X_i^c$ be the corresponding portfolio sum of the comonotonic risk vector X.

Theorem 1.6 *Let X be a risk vector with distribution function $F \in \mathcal{F}(F_1, \ldots, F_n)$. Then for the portfolio $S = \sum_{i=1}^{n} X_i$, it holds that*

a)
$$\text{VaR}_\alpha(S) \leq \text{TVaR}_\alpha(S) \leq \text{TVaR}_\alpha(S^c) = \sum_{i=1}^{n} \text{TVaR}_\alpha(X_i), \qquad (1.22)$$

b)
$$\sum_{i=1}^{n} \text{LTVaR}_\alpha(X_i) = \text{LTVaR}_\alpha(S^c) \leq \text{LTVaR}_\alpha(S) \leq \text{VaR}_\alpha(S), \qquad (1.23)$$

$$c) \qquad \overline{\text{VaR}}_\alpha \leq \overline{\text{TVaR}}_\alpha = \sum_{i=1}^n \text{TVaR}_\alpha(X_i) \tag{1.24}$$

$$and \quad \underline{\text{LTVaR}}_\alpha = \sum_{i=1}^n \text{LTVaR}_\alpha(X_i) \leq \underline{\text{VaR}}_\alpha,$$

$$d) \qquad \text{VaR}_\alpha(S^c) = \sum_{i=1}^n \text{VaR}_\alpha(X_i). \tag{1.25}$$

Proof The inequality $\text{VaR}_\alpha(S) \leq \text{TVaR}_\alpha(S)$ is immediate from the definition of $\text{TVaR}_\alpha(S)$. Since TVaR_α is a convex law-invariant risk measure, we obtain the inequality $\text{TVaR}_\alpha(S) \leq \text{TVaR}_\alpha(S^c)$ by the consistency with respect to convex order from Theorem 1.4.

Using that

$$\alpha \, \text{LTVaR}_\alpha(S) + (1-\alpha) \, \text{TVaR}_\alpha(S) = ES, \tag{1.26}$$

we obtain

$$\text{LTVaR}_\alpha(S^c) = \sum_{i=1}^n \text{LTVaR}_\alpha(X_i) \leq \text{LTVaR}_\alpha(S) \leq \text{VaR}_\alpha(S).$$

Finally for $S^c = \sum_{i=1}^n F_i^{-1}(U)$, it holds by the comonotonicity of the summands:

$$S^c \geq \text{VaR}_\alpha(S^c)$$

if and only if for all i, $X_i = F_i^{-1}(U) \geq \text{VaR}_\alpha(X_i)$, i.e.,

$$\text{VaR}_\alpha(S^c) = \sum_{i=1}^n \text{VaR}_\alpha(X_i), \tag{1.27}$$

$$and \quad \text{TVaR}_\alpha(S^c) = \frac{1}{1-\alpha} \int_\alpha^1 \text{VaR}_u(S^c) \, du \tag{1.28}$$

$$= \frac{1}{1-\alpha} \int_\alpha^1 \sum_{i=1}^n \text{VaR}_u(X_i) \, du = \sum_{i=1}^n \text{TVaR}_\alpha(X_i). \qquad \square$$

Remark 1.7 The inequalities (1.22) and (1.24) give a simple way to calculate an upper bound for the worst case VaR, whereas inequality (1.23) gives a lower bound for the best case VaR. The VaR of the comonotonic risk portfolio is easy to calculate, but it turns out that it is not a worst case with respect to VaR. The comonotonic portfolio is, however, a worst case portfolio with respect to TVaR, and hence the worst case TVaR bound is easy to determine. ◊

1.2   Standard Bounds, VaR Bounds, and Worst Case Distributions

It is an important task to describe good upper bounds for the value-at-risk and to determine worst case portfolios. The insight that the comonotonic portfolio is not the worst case VaR portfolio was a surprise in the practice of risk analysis and led to a rethinking of basic recommendations in risk regulation.

The standard bounds for the distribution function of the sum

$$M_n^{\leq}(t) = \sup \left\{ P\left(\sum_{i=1}^{n} X_i \leq t \right); X_i \sim F_i, \ 1 \leq i \leq n \right\},$$

$$m_n^{\leq}(t) = \inf \left\{ P\left(\sum_{i=1}^{n} X_i \leq t \right); X_i \sim F_i, \ 1 \leq i \leq n \right\},$$

resp. for the corresponding tail risks

$$M_n(t) = \sup \left\{ P\left(\sum_{i=1}^{n} X_i \geq t \right); X_i \sim F_i, \ 1 \leq i \leq n \right\},$$

$$m_n(t) = \inf \left\{ P\left(\sum_{i=1}^{n} X_i \geq t \right); X_i \sim F_i, \ 1 \leq i \leq n \right\},$$

have been known in the literature for a long time, see Sklar (1973), Moynihan et al. (1978), Denuit et al. (1999), and Rüschendorf (2005).

Theorem 1.8 (Standard bounds) *Let $X = (X_1, \ldots, X_n)$ be a random vector with marginal distribution functions $F_1, \ldots, F_n$. Then for any $t \in \mathbb{R}$, it holds that*

$$\left(\bigvee_{i=1}^{n} F_i(t) - (n-1) \right)_+ \leq P\left(\sum_{i=1}^{n} X_i \leq t \right)$$

$$\leq \min \left(\bigwedge_{i=1}^{n} F_i(t), 1 \right), \tag{1.29}$$

where $\bigwedge_{i=1}^{n} F_i(t) = \inf\{\sum_{i=1}^{n} F_i(u_i); \sum_{i=1}^{n} u_i = t\}$ is the "infimal convolution" of the (F_i), and $\bigvee_{i=1}^{n} F_i(t) = \sup\{\sum_{i=1}^{n} F_i(u_i); \sum_{i=1}^{n} u_i = t\}$ is the "supremal convolution" of the (F_i).

Proof For any $u_1, \ldots, u_n$ with $\sum_{i=1}^{n} u_i = t$, it holds that

$$P\left(\sum_{i=1}^{n} X_i \leq t \right) \geq P\left(\bigcup_{i=1}^{n} (X_i \leq u_i) \right),$$

$$\geq \sum_{i=1}^{n} F_i(u_i), \tag{1.30}$$

which implies the upper bound in (1.29). Similarly, using the Fréchet lower bound in (1.9) we obtain

$$P\left(\sum_{i=1}^{n} X_i \leq t \right) \geq P\left(X_1 \leq u_1, \ldots, X_n \leq u_n \right)$$

$$\geq \left(\sum_{i=1}^{n} F_i(u_i) - (n-1) \right)_+. \tag{1.31}$$

$\square$

In general, the standard bounds in Theorem 1.8 are not sharp and can be considerably improved. Define for general n,

$$A_n(t) := \left\{ (x_1, \ldots, x_n); \sum_{i=1}^{n} x_i \le t \right\},$$

$$A_n^+(t) := \left\{ (x_1, \ldots, x_n); \sum_{i=1}^{n} x_i < t \right\}, \quad t \in \mathbb{R}^1,$$

and let

$$(F_1 \wedge F_2)^-(t) = \inf\{ F_1(x-) + F_2(t-x); \quad x \in \mathbb{R}^1 \}$$

denote the left continuous version of $F_1 \wedge F_2$; similarly, let $(F_1 \vee F_2)^-(t)$ be the left continuous version of $F_1 \vee F_2$. In the case $n = 2$, it was proved independently in Makarov (1981) and Rüschendorf (1982) that the standard bounds are sharp.

Theorem 1.9 (Sharpness of standard bounds, $n = 2$) *If X_i have distribution functions F_i, $1 = 1, 2$, then*

$$P(X_1 + X_2 \le t) \le M_2^{\le}(t) = (F_1 \wedge F_2)^-(t) \tag{1.32}$$

and

$$P(X_1 + X_2 < t) \ge m_2^{\le}(t) = ((F_1 \vee F_2)^-(t) - 1)_+. \tag{1.33}$$

The proof of Theorem 1.9 given in Makarov (1981) uses direct arguments on the copulas, while the proof in Rüschendorf (1982) is based on duality theory. This latter proof allows us also to determine the worst case dependence structure.

On the unit interval $[0, 1]$ supplied with the Lebesgue measure λ, define the random variables

$$Y_1(s) = F_1^{-1}(s), \quad Y_2(s) = F_2^{-1}(\varphi(S)), \tag{1.34}$$

with $\varphi(s) = 1 - s$, $0 \le s \le h(t)$, and $\varphi(s) = s$, $h(t) \le s \le 1$. Then the random variables Y_1, Y_2 maximize the distribution function of the sum at point t, i.e., they maximize $P(X_1 + X_2 < t)$. This means that they minimize the tail risk $P(X_1 + X_2 \ge t)$.

Proposition 1.10 (Maximizing (best case) pairs) *The random variables defined in (1.34) satisfy:*

a) $$Y_1 \sim F_1, \quad Y_2 \sim F_2,$$

b) $$P(Y_1 + Y_2 \le t) = M_2^{\le}(t) = (F_1 \wedge F_2)^-(t). \tag{1.35}$$

Proof The Lebesgue measure λ is invariant with respect to φ, i.e., $\lambda^\varphi = \lambda$. Therefore, $\lambda^{Y_i} = \lambda^{F_i^{-1} \circ \varphi} = \lambda^{F_i^{-1}}$, and thus $Y_i \sim F_i$, $i = 1, 2$. Since $F_i^{-1} \circ F_i(x) \le x$, we obtain for $s = F_1(u)$,

$$F_1^{-1}(s) + F_2^{-1}(h(t) - s) = F_1^{-1} \circ F_1(u) + F_2^{-1}(h(t) - F_1(u))$$
$$= u + F_2^{-1}(F_2(t - u)) \le u + (t - u) = t.$$

For the sup in the definition of $g(t) = F_1^{-1} \vee F_2^{-1}(t)$, it is enough to consider s of the form $F_1(u)$. This implies $g \circ h(t) \leq t$, and it follows that

$$\lambda(\{Y_1 + Y_2 \leq t\}) \geq h(t) = (F_1 \wedge F_2)^-(t).$$

This implies b) and moreover

$$h(t) = \lambda(\{Y_1 + Y_2 \leq t\}) = F_1^{-1} \vee F_2^{-1}(t) = g(t). \qquad \square$$

A similar construction yields a worst case pair of risks minimizing the probability $P(X_1 + X_2 < t)$ or equivalently maximizing $P(X_1 + X_2 \geq t)$.
Define $Y_1(s) = F_1^{-1}(s)$, $Y_2(s) = F_2^{-1}(\varphi(s))$, $s \in [0, 1]$ with

$$\varphi(s) = s, \quad 0 \leq s \leq h(t) \quad \text{and} \quad \varphi(s) = 1 - s, \quad h(t) \leq s \leq 1. \tag{1.36}$$

Then Y_1, Y_2 are obtained by a countermonotonic coupling in the upper part of the distributions, and we obtain in a similar way as in Proposition 1.10 that the risk variables Y_1, Y_2 determine a worst case pair of risks.

Proposition 1.11 (Worst case risks)   *The random variables Y_1, Y_2 defined in (1.36) satisfy*

a) $Y_1 \sim F_1$, $\quad Y_2 \sim F_2$
and
b)
$$P(Y_1 + Y_2 \geq t) = M_2(t) = 1 - m_2^<(t)$$
$$= \min(2 - (F_1 \vee F_2)^-(t), 1) \tag{1.37}$$
$$= \sup\{P(X_1 + X_2 \geq t); X_i \sim F_i\}.$$

Definition 1.12 (Rearrangements)   Let f, g be measurable, real functions on $[0, 1]$. Then f is a rearrangement of g (with respect to the Lebesgue measure λ), notation $f \sim_r g$, if $\lambda^f = \lambda^g$, i.e., f, g have the same distribution with respect to λ.

The best resp. worst case couplings with respect to the tail risk can also be described by rearrangements.

Corollary 1.13 (Best and worst case risks by rearrangements)   *For any $t \in \mathbb{R}$, it holds that*

a)
$$\alpha^* = M_2^<(t) = \sup\{\alpha \in [0, 1]: \exists f_j^\alpha \sim_r F_j^{-1} \text{ on } [0, \alpha], \tag{1.38}$$
$$\text{such that } f_1^\alpha(s) + f_2^\alpha(s) \leq t, \text{ for all } s \in [0, \alpha]\}$$
$$= \inf\{\alpha \in [0, 1]: \exists f_j^\alpha \sim_r F_j^{-1} \text{ on } [\alpha, 1],$$
$$\text{such that } f_1^\alpha(s) + f_2^\alpha(s) > t, \text{ for all } s \in [\alpha, 1]\}$$

b)
$$\beta^* = M_2(t) = \sup\{P(X_1 + X_2 \geq t), X_i \sim F_i, i = 1, 2\} \tag{1.39}$$
$$= \inf\{\alpha \in [0, 1]: \exists f_j^{-1} \sim_r F_j^{-1} \text{ on } [\alpha, 1],$$
$$\text{such that } f_1^\alpha(s) + f_2^\alpha(s) \geq t, \text{ for all } s \in [\alpha, 1]\}.$$

This rearrangement description also characterizes worst and best case couplings in the general case $n \geq 2$ (see Rüschendorf, 1983a; Puccetti and Rüschendorf, 2012a).

Theorem 1.14 (Structure of worst and best case couplings)   *For all distribution functions $F_1, \ldots, F_n$ and $t \in \mathbb{R}$, it holds that*

$$M_n^{\leq}(t) = \sup \left\{ \alpha \in [0, 1]; \text{ there exist } f_j^\alpha \sim_r F_j^{-1} \text{ on } [0, \alpha], \right. \tag{1.40}$$

$$\left. 0 \leq j \leq n, \text{ such that } \sum_{j=1}^{n} f_j^\alpha \leq t \text{ on } [0, \alpha] \right\}.$$

Similarly,

$$M_n(t) = 1 - m_n(t) = \inf \left\{ \alpha \in [0, 1]; \text{ there exist } f_j^\alpha \sim F_j^{-1} \text{ on } [\alpha, 1], \right.$$

$$\left. \text{ such that } \sum_{j=1}^{n} f_j^\alpha \geq t \text{ on } [\alpha, 1] \right\}. \tag{1.41}$$

By Theorem 1.14, the problem of getting sharp bounds on the distribution function of the sum is reduced to a rearrangement problem. This rearrangement formulation motivates the construction of a fast algorithm to approximate the sharp bounds numerically – the rearrangement algorithm (RA) (see Puccetti and Rüschendorf, 2012b and Chapter 3). The proposed algorithm works well for general inhomogeneous portfolios, also those with high dimensions (i.e., d in the thousands).

The results of Propositions 1.10 and 1.11 imply that for $n = 2$, the worst case distribution maximizing $M_n(t)$ resp. maximizing $\mathrm{VaR}_\alpha(X_1 + X_2)$ is obtained by the countermonotonic coupling in the corresponding upper part of the distributions (see (1.36) and (1.37)), and the best case distribution minimizing $M_n(t)$ resp. minimizing $\mathrm{VaR}_\alpha(X_1 + X_2)$ is obtained by the countermonotonic coupling in the lower part of the distributions (see (1.34)).

Let F_i^α ($F_{i,\alpha}$) denote the distribution F_i restricted to the upper (lower) α-part of F_i, i.e., formally F_i^α ($F_{i,\alpha}$) is the distribution of $F_i^{-1}(U)$, where U is uniformly distributed on $[\alpha, 1]$ ($[0, \alpha]$). Then by Propositions (1.4) and (1.5), the worst case distribution (resp. best case distribution) minimizes (resp. maximizes) the distribution function of the portfolio sum in the upper part (resp. in the lower part) of the distribution. In other words, the upper resp. the lower parts of the distributions are flattened as much as possible. This principle also extends to $n \geq 2$: see Bernard et al. (2017c, Theorem 2.5).

Theorem 1.15 (VaR-bounds and convex order)   *Let F_i^α denote the upper α-part of F_i. Then*

a)
$$\overline{\mathrm{VaR}}_\alpha^+ = \sup_{X_i \sim F_i} \mathrm{VaR}_\alpha^+ \left(\sum_{i=1}^{n} X_i \right) = \sup_{Y_i^\alpha \sim F_i^\alpha} \mathrm{VaR}_0^+ \left(\sum_{i=1}^{n} Y_i^\alpha \right). \tag{1.42}$$

b) If $X_i^\alpha, Y_i^\alpha \sim F_i^\alpha$ and

$$S^\alpha = \sum_{i=1}^{n} Y_i^\alpha \leq_{\mathrm{cx}} \sum_{i=1}^{n} X_i^\alpha, \text{ then} \tag{1.43}$$

$$\mathrm{VaR}_0^+ \left(\sum_{i=1}^{n} X_i^\alpha \right) \leq \mathrm{VaR}_0^+ \left(\sum_{i=1}^{n} Y_i^\alpha \right).$$

If it is possible to minimize the sum in the upper part of the distributions, i.e., for F_i^α in convex order, then one obtains as in the case $n = 2$ a worst case joint distribution, maximizing the VaR. This minimization of the convex order in the upper part is achieved in particular in the mixing case.

Definition 1.16 (Mixability)

a) Distribution functions $F_1, \ldots, F_n$ on $\mathbb{R}$ are called **mixable** if there exist $X_i \sim F_i$ such that $\sum_{i=1}^n X_i = \mu$ a.s. for some $\mu \in \mathbb{R}^1$.

b) A distribution function F on $\mathbb{R}$ is called **n-mixable** (with center μ) if $F_1 = F, \ldots, F_n = F$ are mixable.

Since for n-mixable distributions the mixing variables realize the convex minimum, we obtain as an immediate consequence of Theorem 1.15:

Corollary 1.17 *If $Y_i^\alpha \sim F_i^\alpha$, $1 \le i \le n$ exist, such that $S^\alpha = \sum_{i=1}^n Y_i^\alpha = c$, i.e., (F_i^α), $1 \le i \le n$ are mixable, then for all $X_i \sim F_i$, $1 \le i \le n$, it holds that*

$$\mathrm{VaR}_\alpha^+ \left(\sum_{i=1}^n X_i \right) \le \mathrm{VaR}_0^+(S^\alpha) = c. \tag{1.44}$$

Remark 1.18

a) As stated in (1.43), the worst value for $\mathrm{VaR}_\alpha(S)$ is attained by the lower support point of some minimal element in convex order in this class $\mathcal{F}^\alpha = \{ \sum_{i=1}^n Y_i; Y_i \sim F_i^\alpha \}$. For $d = 2$, a smallest element in this class exists and is given by the countermonotonic pair $Y_1^\alpha = (F_1^\alpha)^{-1}(U)$, $Y_2^\alpha = (F_2^\alpha)^{-1}(1 - U)$, where U is uniformly distributed on $[\alpha, 1]$. The resulting $\mathrm{VaR}_0^+(Y_1^\alpha + Y_2^\alpha)$ is by Proposition 1.11 a sharp upper bound and is identical to the solution of this case in Rüschendorf (1982).

b) Similarly to Theorem 1.15, we obtain a corresponding result for the lower bound for $\mathrm{VaR}_\alpha \left(\sum_{i=1}^n X_i \right)$. Let $F_{i,\alpha}$ denote the distributions F_i restricted to the lower α-part. For $X_{i,\alpha}, Y_{i,\alpha} \sim F_{i,\alpha}$ we get: If $S_\alpha = \sum_{i=1}^n Y_{i,\alpha} \le_{cx} \sum_{i=1}^n X_{i,\alpha} = S'_\alpha$, then

$$\mathrm{VaR}_1(S_\alpha) \le \mathrm{VaR}_1(S'_\alpha). \tag{1.45}$$

In consequence we obtain: If S_α is a smallest sum with respect to $(F_{i,\alpha})$ in convex order, then

$$\mathrm{VaR}_1(S_\alpha) = \inf \left\{ \mathrm{VaR}_\alpha \left(\sum_{i=1}^n X_i \right); X_i \sim F_i \right\}. \tag{1.46}$$

c) n-mixability has been established for uniform $U(0, 1)$-distributions and binomial distributions in Gaffke and Rüschendorf (1981) and Rüschendorf (1982, 1983b), and for symmetric unimodal distributions in Rüschendorf and Uckelmann (2002). Wang and Wang (2011) established n-mixability for distributions with a decreasing density on $[0, 1]$ under a moderate moment condition. Mixing in the case of concave densities on an interval (a, b) was established in Puccetti et al. (2012). For n small, say $n = 2, 3$, mixing is a rare property while for large enough n, by the abovementioned results mixing typically holds on bounded domains under monotonicity or concavity conditions. For more details on mixing, see Section 1.4.   ◇

1.3   **Worst Case Risk Vectors: The Conditional Moment Method and the Mixing Method**

The conditional moment method gives an upper bound on the tail risk of the portfolio in terms of conditional moments of the marginals. Combined with a mixing condition on the marginal distributions, the upper bound is attained. This method was introduced in the case of homogeneous portfolios with monotone densities on $[0, 1]$ in Wang and Wang (2011). It was extended to general inhomogeneous distributions $F_1, \ldots, F_n$ in Puccetti and Rüschendorf (2012b) and in Wang et al. (2013). In fact, it turns out that the conditional moment bounds have to be improved to be attainable. This improvement is given in a direct constructive way in Wang et al. (2013), while it is obtained in Puccetti and Rüschendorf (2012b) based on duality theory (see Chapter 3).

Let $X_i \sim F_i$, $G_i = F_i^{-1}$, the generalized inverse of F_i, $G = \sum_{i=1}^n G_i$, and assume that $\mu_i = EX_i$ exists. For $a \in [0, 1]$ define $\Psi(a)$ as the sum of the conditional first moments, given $X_i \geq G_i(a)$, i.e.,

$$\Psi(a) = \frac{1}{1-a} \int_a^1 G(t)\, dt = \sum_{i=1}^n E(X_i \mid X_i \geq G_i(a)). \qquad (1.47)$$

Then Ψ is monotonically non-decreasing and $\Psi(0) = \mu = \sum_{i=1}^n \mu_i$.

Theorem 1.19 (Method of conditional moments)   *Let $X_i \sim F_i$ have first moments μ_i, $1 \leq i \leq n$. Then, for $s \geq \mu$, we have*

$$M_n(s) = \sup \left\{ P\left(\sum_{i=1}^n X_i \geq s \right); \; X_i \sim F_i, 1 \leq i \leq n \right\} \leq 1 - \Psi^-(s), \qquad (1.48)$$

where $\Psi^-(s) = \sup\{t \in [0, 1]; \Psi(t) \leq s\}$ is the left-continuous generalized inverse of Ψ.

Proof   With $X_i \sim F_i$ and $S = \sum_{i=1}^n X_i$, we have

$$\mu = \sum_{i=1}^n \mu_i = E[S] \geq E\left[S \mathbf{1}_{\{S<s\}}\right] + sP(S \geq s) \qquad (1.49)$$

$$= \int_0^{P(S<s)} G(t)\, dt + sP(S \geq s) = \mu - \int_{P(S<s)}^1 G(t)\, dt + sP(S \geq s).$$

If $P(S \geq s) > 0$, this implies that $\Psi(P(S < s)) \geq s$ and thus

$$P(S < s) \geq \Psi^-(s).$$

As a consequence, we obtain

$$P(S \geq s) \leq 1 - \Psi^-(s). \qquad \square$$

Remark 1.20

a) **Sharpness of conditional moment bounds.** The conditional bound in (1.48) is sharp if and only if the estimate in (1.49) is an equality, that is, if for the optimal coupling it holds true that $\{S \geq s\} = \{S = s\}$ a.s. This means, by Theorem 1.14, that the corresponding optimal rearrangements f_i^α on $[\alpha, 1]$ satisfy

$$\sum_{i=1}^{n} f_i^\alpha(u) = s \quad \text{for all } u \in [\alpha, 1],$$

with $1 - \alpha = M(s)$, i.e., the random variables are mixing on the upper part of the distribution.

b) For unbounded domains, the bound in (1.48) typically fails to be sharp. To be a good bound it is indeed necessary that

$$\sum_{i=1}^{n} E\left(X_i \mid X_i \geq G_i(\alpha)\right) \approx s.$$

c) The method to get upper bounds for $M(s)$ implies directly also a lower bound for $P(S > s)$. Denoting by H the conditional moment function associated with the random variable $-X_i$, we obtain

$$P(S > s) = 1 - P((-S) \geq (-s)) \geq H^{-1}(-s).$$

In fact the conditional moment bound in Theorem 1.19 for the tail risk is equivalent to the TVaR bound for VaR in (1.22). The sharpness statement in Remark 1.20 a) corresponds to the sharpness of the bounds in Corollary 1.17,

$$\text{VaR}_\alpha^+\left(\sum_{i=1}^{n} X_i\right) \leq \text{VaR}_0^+(S^\alpha) = c, \tag{1.50}$$

under the mixing condition on the $\{F_i^\alpha\}$, $1 \leq i \leq n$. Note that under this condition $\text{VaR}_0^+(S^\alpha) = c = \text{TVaR}_\alpha(S)$.   ◊

A modification of the method of conditional moments in Theorem 1.19 as in Remark 1.20 allows us to give improved bounds and even sharpness results not only for bounded domains but also for unbounded domains. Define for $s \geq \mu$ and $t \in [0, 1]$ the function $H_t(t_1)$ as the conditional expected moment function on the interval $[t, t_1]$, i.e.,

$$H_t(t_1) = \frac{1}{t_1 - t} \int_t^{t_1} G(u)\, du = EG(U_{[t,t_1]}). \tag{1.51}$$

Here $U_{[t,t_1]}$ denotes a random variable uniformly distributed on $[t, t_1]$. H_t is increasing in t, t_1. Let $H_t(1) \geq s$ and $G(t) \leq s$. This allows us to define

$$t_1 = t_1(t) = H_t^{-1}(s). \tag{1.52}$$

If we assume continuity of the F_i, then we get that the conditional expectation on $[t, t_1(t)]$ is identical to s:

$$H_t(t_1(t)) = s. \tag{1.53}$$

Without continuity we postulate (1.53) for the risk level considered. Next we define the optimal choice of such t's with (1.52) and (1.53):

$$t_0 = t_0(s) = \inf\{t; \ G_i \mid [t, t_1(t)], 1 \leq i \leq n \text{ are mixing with value } s\}, \tag{1.54}$$

that is, t_0 is the infimum of all those t's such that there exist rearrangements $f_i^t \sim_r G_i \mid [t, t_1(t)]$, which satisfy

$$\sum_{i=1}^{n} f_i^t = E \sum_{i=1}^{n} G_i \mid [t, t_1(t)] = s. \tag{1.55}$$

Proposition 1.21 (Extended conditional moment method) *Let $X_i \sim F_i$, $1 \leq i \leq n$, be risk variables, and assume that the mixing condition (1.55) holds for some t. Then for $s \geq \mu$ we obtain the upper tail risk bound*

$$M_n(s) = \sup \left\{ P\left(\sum_{i=1}^{n} X_i \geq s \right); \ X_i \sim F_i \right\} \leq 1 - t_0(s). \tag{1.56}$$

Proof Under the "mixing assumption" (1.55) to $t_0 \in [0, 1]$, we have $t_1(t_0) > t_0$, and the restricted distributions of $(G_i \mid [t_0, t_1(t_0)])$ are mixing. Therefore, there exist $\widetilde{V_i} \sim U_{[t_0, t_1(t_0)]}$ such that $\sum_{i=1}^{n} G_i(\widetilde{V_i}) = s$. Consequently this implies the existence of $V_i \sim U_{[t_0, 1]}$ such that

$$\sum_{i=1}^{n} G_i(V_i) \geq s, \tag{1.57}$$

and the result follows. $\qquad\qquad\qquad\qquad\qquad\qquad\qquad\qquad\qquad\qquad\qquad\square$

In a second step, the bound in (1.56) is further improved. This improvement has been shown in general classes of examples to yield sharp VaR bounds in Puccetti and Rüschendorf (2012a), Wang et al. (2013, Section 2.3), and Wang (2015).

Theorem 1.22 (Improved extended conditional moment bounds) *Assume that the risk variables $X_i \sim F_i$, $1 \leq 1 \leq n$ satisfy the mixing condition (1.55), with $t_0 < 1$ and let $t_1 = t_1(t_0)$. Define*

$$t_2 = t_2(s) = \inf \left\{ t \leq t_0; \text{ there exists a coupling} \tag{1.58} \right.$$
$$\left. U_i \sim U_{[t, t_0] \cup [t_1, 1]} \text{ with } \sum_{i=1}^{n} G_i(U_i) \geq s \right\}.$$

Then

$$M_n(s) \leq 1 - t_2(s). \tag{1.59}$$

Proof The admissible coupling $(\widetilde{V_i})$ on $[t_0, t_1]$ in the proof of Proposition 1.21 can be improved using the admissible coupling (U_i) on $[t_2, t_0] \cup [t_1, 1]$ to an admissible coupling, say (V_i) on $[t_2, 1]$, satisfying $\sum_{i=1}^{n} G_i(V_i) \geq s$ on $[t_2, 1]$; this implies (1.59). $\qquad\square$

Remark 1.23 (Structure of the worst case risks) The structure of the "optimal" coupling yielding worst case portfolios thus has a mixing part on $[t_0, t_1]$ and is admissible on $[t_2, 1]$. In the homogeneous case $F_i = F$, $1 \leq i \leq n$, it turns out that for the admissible part on $[t_2, t_0] \cup [t_1, 1]$ it is often sufficient to couple one "large" observation corresponding to $u \in [t_1, 1]$ with $n - 1$ "small" observations in $[t_2, t_0]$ chosen to be identical (see the following discussion of this structure and the approach via duality in Chapter 3). $\qquad\qquad\qquad\qquad\qquad\qquad\qquad\qquad\qquad\qquad\qquad\qquad\qquad\Diamond$

1.4   Mixability and Convex Minima of Portfolios

As seen in Theorems 1.19 and 1.22, an important ingredient of the structure of worst case portfolios is played by the mixing part of this dependence structure. Next we discuss some basic results and conditions on distributions yielding n-mixing of $F_1, \ldots, F_n$.

Proposition 1.24 (Mixabililty)   *Let F be a distribution function on $\mathbb{R}^1$.*

a) *F is 2-mixable if and only if F is symmetric, i.e., there exists $a \in \mathbb{R}$, such that $X \sim F$ implies $a - X \sim F$.*
b) *If F is n-mixable and T is linear, then F_T is n-mixable.*
c) *The binomial distribution $B(n, \frac{p}{q})$ is q-mixable.*
d) *The uniform distribution $U(a, b)$ on an interval (a, b) is n-mixable for any $n \geq 2$.*
e) *Any continuous distribution function having a symmetric unimodal density is n-mixable for any $n \geq 2$.*

The statements in a), b) are obvious; for c) see Rüschendorf (1983a). The minimal variance $v_k(\vartheta)$ of $\sum_{i=1}^k X_i$, $X_i \sim \mathcal{B}(1, \vartheta)$ has been determined in Snijders (1984):

$$v_k(\vartheta) = a(k, \vartheta)(1 - a(k, \vartheta)), \quad a(k, \vartheta) = k\vartheta \,(\mathrm{mod}\,1). \tag{1.60}$$

d) is established in Gaffke and Rüschendorf (1981), while e) is proved in Rüschendorf and Uckelmann (2002). In particular, several standard distributions, like the normal and Cauchy, are mixing. Some general mixing results are established in Wang and Wang (2011, 2016), Wang et al. (2013), and Puccetti et al. (2012). We highlight some important results from the ample theory developed in these papers.

Theorem 1.25 (Monotone densities)   *Let F be a distribution function on $[a, b]$ with lower (upper) support point a (b) and mean μ.*

a) *If F has an increasing density and $\mu \leq b - \frac{1}{n}(b - a)$, then F is n-mixable.*
b) *If F has a decreasing density and $\mu \geq a + \frac{1}{n}(b - a)$, then F is n-mixable.*

The proof of Theorem 1.25 and of related mixing results is based on combinatorial approximation by discrete uniform distributions on n points, which are n-mixable. In this connection, a useful result states the convexity of the class of n-mixable distributions having the same center μ.

Proposition 1.26 (Convexity of n-mixable distributions)

a) *A countable convex combination of n-mixable distribution functions with the same center μ is n-mixable.*
b) *A discrete distribution F is mixable with center μ if and only if it is a countable convex combination of n-discrete uniform n-mixable distributions.*

This proposition leads to the following result concerning concave densities.

Theorem 1.27 (Concave densities)   *Any distribution on a bounded interval (a, b) with a concave density is n-mixing for any $n \geq 3$.*

For an extension to the inhomogeneous case $F_1, \ldots, F_n$, the following conditions are shown to be necessary in Wang and Wang (2016).

Proposition 1.28 (Necessary conditions for mixability) *If $F_1, \ldots, F_n$ are mixable with support (a_i, b_i) and means $\mu_1, \ldots, \mu_n$, then with the lengths $\ell_1 = b_i - a_i$ it holds that*

a) *mean inequality:*

$$\sum_{i=1}^{n} a_i + \max_{1 \le i \le n} \ell_i \le \sum_{i=1}^{n} \mu_i \le \sum_{i=1}^{n} b_i - \max_{1 \le i \le n} \ell_i; \tag{1.61}$$

b) *norm inequality: For any p-norm, $\| \ \| = \| \ \|_p$, $p \le \infty$ and $X_i \sim F_i$ holds*

$$\sum_{i=1}^{n} \|X_i - \mu_i\| \ge 2 \max_{1 \le i \le n} \|X_i - \mu_i\|; \tag{1.62}$$

c) *length inequality:*

$$\sum_{i=1}^{n} \ell_i \ge 2 \max_{1 \le i \le n} \ell_i; \tag{1.63}$$

d) *variance inequality:*

$$\sum_{i=1}^{n} \sigma_i \ge 2 \max_{1 \le i \le n} \sigma_i, \tag{1.64}$$

where $0 \le \sigma_i^2 \le \infty$ are the variances of F_i.

The convexity property as in Proposition 1.26 also holds in the inhomogeneous case:
If $F_1, \ldots, F_n$ and $G_1, \ldots, G_n$ are mixable, then

$$\alpha F_1 + (1 - \alpha)G_1, \ldots, \alpha F_n + (1 - \alpha)G_n \text{ are also mixable.} \tag{1.65}$$

These properties lead to an extension of Theorem 1.25 to the inhomogeneous case.

Theorem 1.29 (Decreasing densities) *If $F_1, \ldots, F_n$ are distributions with bounded supports and decreasing densities, satisfying the mean inequality (1.61), then $F_1, \ldots, F_n$ are mixable.*

As shown in Theorem 1.15, Corollary 1.17, and Theorem 1.22, mixing allows us to determine sharp bounds for the value-at-risk and for the tail risk of the portfolio. The tail risk bounds in these results depend on the determination of the largest mixing intervals (a, b) for the tail levels allowing the construction of an admissible coupling on these intervals.

Let F be a distribution function on $[0, 1]$ with a decreasing density. For $0 \le c \le \frac{1}{n}$ we define a copula $Q_n^F(c)$ as follows:
The random variables $(U_1, \ldots, U_n)$ have distribution $Q_n^F(c)$ if

1) For each i with $U_i \in [1 - c, 1]$ we have $U_j = (n - 1)(1 - U_i)$ for all $j \ne i$;
2) $F^{-1}(U_1) + \cdots + F^{-1}(U_n)$ is a constant when any of the $U_i \in ((n - 1)c, 1 - c)$,

i.e., a large U_i is coupled with $n-1$ identical small U_j, and if one of the U_i takes values in the intermediate interval then it is mixing with the other U_j and in fact all other U_j lie in the intermediate interval. Denote

$$H(t) = (n-1)F^{-1}((n-1)t) + F^{-1}(1-t).\qquad(1.66)$$

Similarly, in the case of increasing density, define $Q_n^F(c)$ by the properties:

1) For each i with $U_i \in [0, c]$ we have $U_j = 1 - (n-1)U_i$, for all $j \neq i$, and
2) $G(U_1) + \cdots + G(U_n) = \text{const.}$, when any of the $U_i \in (c, 1 - (n-1)c)$, $G = F^{-1}$.

In this case define

$$H(t) = G(t) + (n-1)G(1 - (n-1)t).\qquad(1.67)$$

Proposition 1.30

a) *In the case of monotone densities and $c \in [0, \frac{1}{n}]$, there exists a copula $Q_n^F(c)$ satisfying 1), 2) if*

$$\int_c^{\frac{1}{n}} H(t)\, dt \leq \left(\frac{1}{n} - c\right)H(c).\qquad(1.68)$$

b) *The smallest c such that a copula $Q_n^F(c)$ with 1), 2) exists is given by*

$$c_n = \min\left\{c \in \left[0, \frac{1}{n}\right]; \ \text{inequality (1.68) holds}\right\}.\qquad(1.69)$$

Proof For the proof we use that the assumed mixability for $Q_n^F(c)$ implies the moderate moment condition in Theorem 1.25 on $[c, 1 - (n-1)c]$ in the increasing case resp. on $[(n-1)c, 1-c]$ in the decreasing case.

In consequence, if $Q_n^F(c)$ exists, it follows that in the increasing case the $(G(U_j))$ are mixable on $[c, 1 - (n-1)c]$ when $X_i \in [c, 1 - (n-1)c]$, and thus

$$\sum_{j=1}^n G(X_j) = E\left(\sum_{j=1}^n G(X_j) \mid c \leq U_i \leq 1 - (n-1)c\right)$$

$$= \frac{n}{1-nc}\int_c^{1-(n-1)c} G(t)\, dt;$$

similarly in the decreasing case. By the necessary moderate moment condition in Proposition 1.28, this implies that the conditional mean is less than or equal to $\frac{G(c)}{n+n-1)}\frac{G(1-n-1)c)}{n}$, and thus

$$\int_c^{1-(n-1)c} G(t)\, dt \leq \left(\frac{1}{n} - c\right)(G(c) + (n-1)G(1 - (n-1)c).$$

This implies (1.68).

In the increasing case, $G(t)$ and thus $H(t)$ is concave on $[0, \frac{1}{n}]$. Thus the set of all c satisfying (1.68) is a closed interval $[\widehat{c}_n, \frac{1}{n}]$, and by (1.68), $\widehat{c}_n \leq c \leq \frac{1}{n}$ and thus $c_n \geq \widehat{c}_n$. Since by Proposition 1.30 a) it follows that $Q_n^F(\widehat{c}_n)$ exists, we obtain $c_n = \widehat{c}_n$. The case of decreasing densities is similar. $\square$

An interesting consequence of Proposition 1.30 is the following convex ordering result for the portfolios.

Theorem 1.31 (Convex minima of portfolios) *Let F have a monotone density, and let $(U_1, \ldots, U_n)$ be a copula vector corresponding to $Q_n^F = Q_n^F(c_n)$; then for all $X_i \sim F$, $1 \le i \le n$, it holds that*

a)
$$\sum_{i=1}^{n} X_i \ge_{cx} \sum_{i=1}^{n} G(U_i), \quad G = F^{-1}. \tag{1.70}$$

b) *For any convex function f, it holds that*

$$\min_{\substack{X_i \sim F_i, \\ 1 \le i \le n}} E f\left(\sum_{i=1}^{n} X_i\right) = n \int_0^{c_n} f(H(t))\, dt + (1 - nc_n) f(H(c_n)). \tag{1.71}$$

Proof For the proof of Theorem 1.31 a) it is established that the excess function of the right-hand side dominates that of the left-hand side (see Theorems 3.4, 3.5 in Wang and Wang, 2011), while b) is direct from construction. □

The coupling Q_n^F also solves the following minimization problem, which has an ample historical background.

Proposition 1.32 *For $(U_1, \ldots, U_n) \sim Q_n^F$ it holds that*

$$\Lambda_n = \min_{X_i \sim U(0,1)} E \prod_{i=1}^{n} X_i = E \prod_{i=1}^{n} U_i$$
$$= e^{-n} + \frac{n}{2} e^{-2n} + O(n^4 e^{-3n}). \tag{1.72}$$

In the case of decreasing density on the support $[0, 1]$ with moderate moment condition, it also implies sharpness of the upper bound for the tail risk in the conditional moment method in Theorem 1.19.

Theorem 1.33 *Let F have a decreasing density on its support $[0, 1]$ with mean μ and moderate moment condition $E(X \mid X \ge t) \ge t + \frac{1-t}{n}$ for $X \sim F$ and any $t \in [0, 1]$. Then for $\Psi(t) = E(X \mid X \ge G(t))$, $G = F^{-1}$, it holds that*

$$M_n^+(s) = \sup\left\{ P\left(\sum_{i=1}^{n} X_i \ge s\right); \ X_i \sim F_i, 1 \le i \le n \right\}$$

$$= 1 - m_n(s) = \begin{cases} 1, & s \le n\mu, \\ 1 - \Psi^-\left(\frac{s}{n}\right), & n\mu < s < n, \\ 0, & s \ge n. \end{cases} \tag{1.73}$$

Proof By Theorem 1.19, the right-hand side of (1.73) is an upper bound for the tail risk. By Theorem 1.25, $G(V)$ is n-mixable for $V \sim U[a, 1]$, $a = \Psi^-\left(\frac{s}{n}\right)$. Thus there exist $V_i \sim V$ such that $\sum_{i=1}^{n} G(V_i) = n\Psi(a) = s$. Defining $Y_i = G(U)1_{(U \le a)} + G(V_i)1_{(U > a)}$, $U \sim U(0, 1)$ independent of (V_i), we find $Y_i \sim F$ and

$$P\left(\sum_{i=1}^{n} Y_i \ge s\right) = P(U \ge a) = 1 - a = 1 - \Psi^-\left(\frac{s}{n}\right)$$

for $n\mu < s < n$. □

In the non-mixable case of general support there is the following variant of Theorem 1.33.

For F with decreasing density and $a \in [0, 1]$, define modifications of H, c_n, Φ by

$$H_a(t) = (n-1)F^{-1}(a + (n-1)t) + F^{-1}(1-t), \quad t \in \left[0, \frac{1-a}{n}\right], \tag{1.74}$$

$$c_n(a) = \min \left\{ c \in \left[0, \frac{1}{n}(1-a)\right] : \right. \tag{1.75}$$

$$\int_0^{\frac{1}{n}(1-a)} H_a(t)\, dt \geq \left(\frac{1}{n}(1-a) - c\right) H_a(c) \right\},$$

and $\Phi(a) = \begin{cases} H_a(c_n(a)), & \text{if } c_n(a) > 0, \\ n\Psi(a), & \text{if } c_n(a) = 0. \end{cases}$

Similarly for F with increasing density, $a \in [0, 1]$ define

$$H_a(t) = F^{-1}(a + t) + (n-1)F^{-1}(1 - (n-1)t), \tag{1.76}$$

$$c_n(a) = \min \left\{ c \in \left[0, \frac{1}{n}(1-a)\right] : \right. \tag{1.77}$$

$$\int_c^{\frac{1}{n}(1-a)} H_a(t)\, dt \leq \left(\frac{1}{n}(1-a) - c\right) H_a(c) \right\},$$

and $\Phi(a) = \begin{cases} H_a(0), & \text{if } c_n(a) > 0 \\ n\Psi(a), & \text{if } c_n(a) = 0. \end{cases}$

With $\Phi^-(t) = 0$ if $t < \Phi(0)$ and $\Phi^-(t) = 1$ if $t > \Phi(1)$, then it holds that:

Theorem 1.34 *If F has a monotone density on its support, then the maximal tail risk $M_n^+(s)$ is given by*

$$M_n^+(s) = 1 - m_n(s) = 1 - \Phi^-(s). \tag{1.78}$$

The method of proof is similar to that of the improved extended conditional moment method in Theorem 1.22. That $1 - \Phi^-(s)$ is an upper bound follows from the conditional moment method. The attainment follows by the property of an optimal coupling X that $\{S \geq s\} = \{X_i \geq F^{-1}(a)\}$, $a = m_n(s)$, implying that $m_n(s) = \Phi^-(s)$.

As a result, in the case of tail-decreasing densities, Theorem 1.34 implies a formula for the maximal tail risk for all sufficiently high risk levels s and similarly gives the worst case VaR_α-bounds for all levels $\alpha \geq \alpha_0$ that are sufficiently large.

2   The Rearrangement Algorithm

A basic tool to approximate sharp upper and lower risk bounds in models where only the marginal risks are specified but the dependence structure is completely unspecified is the rearrangement algorithm (RA), along with an extension of it, the block rearrangement algorithm (BRA).

The RA builds upon the formulation of dependence optimization problems as rearrangement problems for real functions on $[0, 1]$. By discretization of the marginals, one obtains in a first step an approximation by a continuous assignment problem – a linear finite (but high-) dimensional optimization problem. Enlarging the discrete support sets then allows this problem to be approximated in a second step by a multidimensional assignment problem – a classical problem in combinatorial optimization.

The RA was suggested for the solution of multidimensional assignment problems in Rüschendorf (1983a). It was then introduced and investigated for the solution of dependence optimization problems in detail in Puccetti and Rüschendorf (2012a,b), Embrechts et al. (2013, 2014), and in a series of further papers on the RA. In particular, Hofert et al. (2017) provide a detailed study of how to efficiently implement the RA. The BRA generalizes the RA and was introduced in Bernard et al. (2017c) and studied in Bernard and McLeish (2016) and Boudt et al. (2018).

Various modifications of the RA and the BRA have been given in the literature to establish sharp risk bounds for models with known marginals and with additional information on the dependence structure – e.g., positive or negative dependence information or some independence information. Some of these developments will be presented in Part II of this book dealing with these kind of models.

Besides the computation of risk bounds and the assessment of model uncertainty, there is a wide range of applications of the (B)RA, including in option pricing (see Bernard et al., 2021) or in operations research for various forms of multivariate assignment problems as described above. An example of the latter is the optimal assembly line crew scheduling problem (ALCS) of allocating n workers sequentially to d jobs to minimize the maximal operation time of the assembly lines; see Coffman and Yannakakis (1984) for the introduction of the problem and Boudt et al. (2018) and Cornilly et al. (2022) for approximate solutions based on the (B)RA.

In Section 2.1, the formulation of dependence problems as rearrangement problems is described, and the basic underlying principles of the RA are explained. In Section 2.2, the RA and its version for determining upper and lower risk bounds is introduced. The high quality of the RA is demonstrated, and several applications are presented. In

Section 2.3, the BRA is introduced, and its quality is investigated. This quality depends on the dynamic of the choice of blocks. A strategy for this component is developed and shown to be an important part in the improvement of the algorithm.

2.1 Formulation of Dependence Problems as Rearrangement Problems

The main motivation for the rearrangement algorithm (RA) is due to the description of the Fréchet class $M(F_1, \ldots, F_n)$, i.e., the class of all distributions with marginals F_i and all possible dependence structures by rearrangements of F_i^{-1}. For $U \sim U(0, 1)$,

$$M(F_1, \ldots, F_n) = \{P^{(f_1(U), \ldots, f_n(U))}; \quad f_i \sim_r F_i^{-1}, 1 \le i \le n\}; \tag{2.1}$$

see Rüschendorf (1983b), Whitt (1976), and Definition 1.12. This representation implies that the determination of worst case risks can be equivalently described by rearrangement problems. In particular, for $\Psi \colon \mathbb{R}^n \longrightarrow \mathbb{R}^1$ measurable, the maximum risk problem is given by:

$$\begin{aligned} M_\Psi &= \sup\{E\Psi(X_1, \ldots, X_n); \quad X_i \sim F_i, 1 \le i \le n\} \\ &= \sup\{E\Psi(f_1(U), \ldots, f_n(U)); \quad f_i \sim_r F_i^{-1}, 1 \le i \le n\}. \end{aligned} \tag{2.2}$$

Similarly, the corresponding inf-problem is stated as

$$\begin{aligned} m_\Psi &= \inf\{E\Psi(X_1, \ldots, X_n); \quad X_i \sim F_i, 1 \le i \le n\} \\ &= \inf\{E\Psi(f_1(U), \ldots, f_n(U)); \quad f_i \sim_r F_i^{-1}, 1 \le i \le n\}. \end{aligned} \tag{2.3}$$

In a similar way, this representation of the Fréchet class allows a rearrangement representation of the maximal resp. minimal tail risk of a function $\Psi(X)$,

$$M_\Psi(s) = \sup\{P(\Psi(X_1, \ldots, X_n) \ge s; \quad X_i \sim F_i, 1 \le i \le n\}$$
$$\text{and} \quad m_\Psi(s) = \inf\{P(\Psi(X_1, \ldots, X_n) \ge s; \quad X_i \sim F_i, 1 \le i \le n\}, \tag{2.4}$$

stated in Puccetti and Rüschendorf (2012a, Theorem 4.1). In particular it gives a representation of the maximum tail risk problem in terms of a minimum rearrangement problem.

Theorem 2.1 *If Ψ is an increasing function then for any $s \in \mathbb{R}^1$, it holds that:*

$$M_\Psi(s) = 1 - \inf\{\alpha \in [0, 1];$$
$$\exists f_j^\alpha \sim_r F_j^{-1} \mid [\alpha, 1], 1 \le j \le n \text{ such that } \Psi(f_1^\alpha, \ldots, f_n^\alpha) \ge s\},$$

$$\tag{2.5}$$

$$m_\Psi(s) = 1 - \sup\{\alpha \in [0, 1];$$
$$\exists f_j^\alpha \sim_r F_j^{-1} \mid [0, \alpha], 1 \le j \le n \text{ such that } \Psi(f_1^\alpha, \ldots, f_n^\alpha) \le s\}.$$

So problems of the determination of the worst, resp. best, possible dependence structure can thus equivalently be formulated as corresponding problems on the rearrangement of functions.

The following exhibition of rearrangement results needed for the introduction and explanation of the RA is based essentially on Rüschendorf (1983a,b). Rearrangements of functions were introduced as generalizations of the finite discrete case in Hardy et al. (1952). They have important applications in many parts of analysis and were intensively studied by Luxemburg (1967), Chong and Rice (1971), and Day (1972). To any measurable function f there exist non-decreasing and non-increasing rearrangements f^*, f_* of f (see Definition 1.12), and in generalization to the discrete case we have:

Proposition 2.2 *For $f, g \in L^2(\lambda)$, $\lambda = \lambda^1$, it holds that:*

$$\int f^* f_* \, d\lambda \leq \int fg \, d\lambda \leq \int f^* g^* \, d\lambda. \tag{2.6}$$

Further define in generalization of the discrete case the Schur order $\prec$ (or majorization order) of functions resp. its weak version $\prec_w$ by

$$f \prec_w g \text{ if } \int_0^x f^*(t) \, d\lambda(t) \leq \int_0^x g^*(t) \, d\lambda(t), x \in (0, 1),$$

$$f \prec g \text{ if } f \prec_w g \text{ and } \int_0^1 f^*(t) \, d\lambda(t) = \int_0^1 g^*(t) \, d\lambda(t).$$

Several important generalizations of the discrete case ordering results were developed in the abovementioned papers. Hardy et al. (1952) and Chong (1974, Theorems 2.3, 2.5) give characterizations of the majorization orderings in terms of convex orderings.

Proposition 2.3 *For $f, g \in L^1(\lambda)$, it holds that:*

$$f \prec_w g \text{ is equivalent to } \int \varphi \circ f \, d\lambda \leq \int \psi \circ g \, d\lambda \tag{2.7}$$

for all convex, non-decreasing φ such that $\varphi \circ f, \varphi \circ g \in L^1(\lambda)$, while

$$f \prec g \text{ is equivalent to } \int \varphi \circ f \, d\lambda \leq \int \varphi \circ g \, d\lambda \tag{2.8}$$

for all convex φ such that $\varphi \circ f, \varphi \circ g \in L^1(\lambda)$.

For the following basic ordering properties, see Day (1972, 6.1 and 6.2).

Proposition 2.4 *For $f, g \in L^1(\lambda)$, it holds that:*

a) $f^* + g_* \prec f + g \prec f^* + g^*.$

b) $f^* - g_* \prec f - g \prec f^* - g_*.$ (2.9)

c) $f^* g_* \prec_w fg \prec_w f^* g^*$ if $fg \in L^1(\lambda).$

Finally, some basic rearrangement results due to Lorentz (1953) and to Fan and Lorentz (1954) imply the following ordering result by making use of the equivalent formulation in terms of dependence structures (for details see Rüschendorf, 1983b):

Proposition 2.5

a) *For distribution functions $F_1, \ldots, F_n$ and a supermodular function $\Psi: \mathbb{R}^n \longrightarrow \mathbb{R}^1$, it holds that*

$$M_\Psi = \sup\{E\Psi(X); \ X_i \sim F_i, 1 \le i \le n\}$$
$$= E\Psi(F_1^{-1}(U), \ldots, F_n^{-1}(U)), \ U \sim U(0, 1). \tag{2.10}$$

b) *If G_i are distribution functions $1 \le i \le d$, with $G_i^{-1} \prec F_i^{-1}$, $1 \le i \le n$, and Ψ is directionally convex, i.e., supermodular and componentwise convex, then*

$$E\Psi(G_1^{-1}(U), \ldots, G_n^{-1}(U)) \le E\Psi(F_1^{-1}(U), \ldots, F_n^{-1}(U)). \tag{2.11}$$

c) *If $n = 2$ and Ψ is supermodular, then*

$$m_\Psi = \inf\{E\Psi(X); \ X_i \sim F_i, i = 1, 2\}$$
$$= E\Psi(F_1^{-1}(U), F_2^{-1}(1 - U)). \tag{2.12}$$

The reformulation of Proposition 2.4 in terms of dependence structures is as follows. Let $\le_{\mathrm{cx}}(\le_{\mathrm{icx}})$ denote the (increasing) convex order as defined in Definition 1.1.

Proposition 2.6 *For integrable random variables X, Y with distribution functions F, G and $U \sim U(0, 1)$, it holds that:*

a) $\quad F^{-1}(U) + G^{-1}(1 - U) \le_{\mathrm{cx}} X + Y \le_{\mathrm{cx}} F^{-1}(U) + G^{-1}(U).$

b) $\quad F^{-1}(U) - G^{-1}(U) \le_{\mathrm{cx}} X - Y \le_{\mathrm{cx}} F^{-1}(U) - G^{-1}(1 - U). \tag{2.13}$

c) $\quad F^{-1}(U)G^{-1}(1 - U) \le_{\mathrm{icx}} XY \le_{\mathrm{icx}} F^{-1}(U)G^{-1}(U) \ if \ X, Y \ge 0.$

d) $\quad$ *If $X_i \sim F_i$ are random variables with distribution functions F_i, $1 \le i \le n$, then*

$$\sum_{i=1}^n X_i \le_{\mathrm{cx}} \sum_{i=1}^n F_i^{-1}(U). \tag{2.14}$$

Proposition 2.5 b) and Proposition 2.6 a), d) imply as a corollary that comonotone pairs of random variables are maximal with respect to the supermodular order $\le_{\mathrm{sm}}$, resp. their sums are maximal with respect to the convex order. Countermonotonic pairs of random variables are minimal with respect to the supermodular order resp. their sums are minimal with respect to the convex order.

For the maximization problem M_Ψ, a unique solution is given by the comonotonic vector. For the minimization problem m_Ψ, for $n \ge 2$ one obtains as a result of Proposition 2.6 a) the following necessary countermonotonicity condition.

Proposition 2.7 *Let $X_* = (f_1(U), \ldots, f_n(U))$ with $f_i \sim_r F_i^{-1}$ be a solution of the minimization problem m_Ψ for $\psi(x) = \varphi(\sum_{i=1}^n x_i)$ with a strictly convex function φ, i.e.,*

$$m_\varphi = \inf\left\{E\varphi\left(\sum_{i=1}^n X_i\right); \ X_i \sim F_i, 1 \le i \le n\right\}$$

$$= E\varphi(f_1(U) + \cdots + f_n(U)),$$

then, necessarily,

$$f_j \perp \sum_{l \neq j} f_l \quad \text{for all } j, \tag{2.15}$$

where $f \perp g$ means that f, g are countermonotonic. More generally: for all $J \subset \{1, \ldots, n\}$, it holds that

$$\sum_{j \in J} f_j \perp \sum_{l \in J^c} f_l. \tag{2.16}$$

Thus to solve the minimization problem m_φ (under the conditions as in Proposition 2.7), we require that all variables $X_j, 1 \leq j \leq n$ are countermonotonic to $\sum_{l \neq j} X_l$ or more generally that for any $J \subset \{1, \ldots, n\}$, $\sum_{j \in J} X_j$ is countermonotonic with $\sum_{l \in J^c} X_l$. The former condition leads to the rearrangement algorithm (RA) and the latter to the block rearrangement algorithm (BRA). Similar countermonotonicity properties can also be formulated for the minimization problem with respect to a function Ψ that allows for all i a representation as $\psi(x) = \psi_i^2(x_i, \varphi_i(x_{(i)}))$, with $x_{(i)} = (x_j)_{j \neq i}$, where ψ_i^2 is supermodular and φ_i are increasing functions of $n - 1$ arguments. Examples are functions of the form $\Psi(x) = \varphi(\sum_{i=1}^n x_i)$, $\varphi(\prod_{i=1}^n x_i)$, $\varphi(\min\{x_i\})$ with φ convex.

The countermonotonicity conditions are in particular satisfied if $X = (X_1, \ldots, X_n)$ is mixing, i.e., for some constant μ it holds that

$$\sum_{i=1}^n X_i = \sum_{i=1}^n f_i(U) = \mu. \tag{2.17}$$

However, in general the countermonotonicity condition in (2.14) or in (2.15) is not sufficient for optimality.

Example 2.8

a) **$U(0, 1)$ distributed random variables**

Consider the case $n = 3$, and let $U \sim U(0, 1)$. The random vector $X = (U, U, 1 - U)$ satisfies the countermonotonicity condition in (2.14) as well as in (2.15). For the variance function $\Psi(x) = (\sum_{i=1}^3 x_i)^2$, we obtain

$$E\Psi(X) = E(1 + U)^2 = \frac{7}{3}. \tag{2.18}$$

Gaffke and Rüschendorf (1981) construct for this case a triplet of mixing uniformly distributed random variables X_1, X_2, X_3 given by

$$X_1 = U, \quad X_2 = \left(U + \frac{1}{2}\right)1_{[0, \frac{1}{2}]} + \left(U - \frac{1}{2}\right)1_{[\frac{1}{2}, 1]}, \tag{2.19}$$

$$X_3 = (1 - 2U)1_{[0, \frac{1}{2}]} + (2 - 2U)1_{[\frac{1}{2}, 1]}.$$

Then by construction $X_1 + X_2 + X_3 = \frac{3}{2}$, and thus $X = (X_1, X_2, X_3)$ is a minimizing solution with $E(X_1 + X_2 + X_3)^2 = \frac{9}{4} = 2\frac{1}{4} < 2\frac{1}{3} = \frac{7}{3}$, showing that the countermonotonicity conditions in (2.15), (2.16) are not sufficient for optimality.

For the product $\Psi(x) = x_1 x_2 x_3$ in holds that $EU \cdot U \cdot (1 - U) = \frac{1}{12} = 0.083 \ldots$, and for the second construction it holds that $EX_1 X_2 X_3 = \frac{1}{16} = 0.0625$, while the minimizing value in this case is $0.0548 \ldots$ (see Example 2.9).

b) **Discrete uniform distribution, sum case**

Let X, Y, Z be $U_{\{1,\dots,8\}}$, i.e., uniformly distributed on $\{1, \dots, 8\}$; then X, Y, Z can be formally written as rearrangements f_1, f_2, f_3 of a uniform random variable U on $[0, 1]$. Restricting to rearrangements that are given by permutations π_2, π_3 of $\{a_1, \dots, a_8\} = \{1, \dots, 8\}$, i.e., $f_1^{-1}(i) = f_2^{-1}(\pi_2(i)) = f_3^{-1}(\pi_3(i))$, we do not split the probabilities and consider the following rearrangements (in fact assignments):

$$
\begin{array}{c|cccccccc}
X & 1 & 4 & 2 & 3 & 8 & 7 & 6 & 5 \\
\hline
Y & 3 & 5 & 7 & 8 & 4 & 1 & 2 & 6 \\
\hline
Z & 8 & 5 & 4 & 2 & 1 & 7 & 6 & 3 \\
\hline
\text{sum } S_3 & 12 & 14 & 13 & 13 & 13 & 15 & 14 & 14
\end{array}
\tag{2.20}
$$

Each combination is taken with probability $\frac{1}{8}$. We obtain

$$
ES_3^2 = E(X + Y + Z)^2 = \frac{1}{8} \cdot 1464.
\tag{2.21}
$$

It is easy to check that the countermonotonicity conditions in (2.15) resp. in (2.16) hold for these rearrangements. An alternative rearrangement (assignment) is given by

$$
\begin{array}{c|cccccccc}
U & 1 & 3 & 2 & 4 & 8 & 7 & 6 & 5 \\
\hline
V & 4 & 5 & 7 & 8 & 3 & 2 & 1 & 6 \\
\hline
W & 8 & 6 & 5 & 1 & 2 & 4 & 7 & 3 \\
\hline
\text{sum } T_3 & 13 & 14 & 13 & 13 & 13 & 13 & 14 & 14
\end{array}
\tag{2.22}
$$

Again the countermonotonicity condition in (2.15) resp. in (2.16) holds for these rearrangements. We get the improved result

$$
ET_3^2 = \frac{1}{8} \cdot 1460 < ES_3^2.
\tag{2.23}
$$

In Rüschendorf (1983a) it is shown that the construction in (2.22) is optimal for convex functions of the sum, and moreover an optimal construction is given there in the case of general N and $n \geq 2$, $F_i = U_{\{1,\dots,N\}}$ and $\Psi(x) = \varphi(\sum_{i=1}^{n} x_i)$. For $n = 3$ and $N = 2k$ (even), there exist and are constructed rearrangements X, Y, Z with $X + Y + Z \in \{N + k + 2, N + k + 1\}$ that are optimal. For $N = 2k + 1$, there exist rearrangements X, Y, Z with $X + Y + Z = N + k + 1$ that are optimal solutions for m_Ψ.

c) **Discrete uniform distributions, product case**

Let $X, Y, Z \sim U_{\{1,\dots,8\}}$ be uniformly distributed on $1, \dots, 8$ and consider the problem of minimizing $EXYZ$, i.e., $\Psi(x) = \prod_{i=1}^{n} x_i$, $n = 3$. In Rüschendorf (1983a) one finds the following construction for $n = 3$:

$$
\begin{array}{c|cccccccc}
X & 1 & 2 & 3 & 4 & 5 & 6 & 7 & 8 \\
\hline
Y & 8 & 5 & 3 & 7 & 4 & 2 & 1 & 6 \\
\hline
Z & 7 & 5 & 6 & 2 & 3 & 4 & 8 & 1 \\
\hline
\text{product } P_1 & 56 & 50 & 54 & 56 & 60 & 48 & 56 & 48
\end{array}
$$

This example fulfills the necessary countermonotonicity conditions $X \perp YZ, Y \perp XZ$, and $Z \perp XY$, and one obtains

$$EXYZ = \frac{1}{8} \cdot 428. \tag{2.24}$$

Also, the rearrangements U, V, W given by

U	1	2	3	4	5	6	7	8
V	7	4	5	3	6	8	1	2
W	7	6	4	5	2	1	8	2
product P_2	49	48	60	60	60	48	56	48

satisfy the countermonotonicity condition, and one obtains

$$EUVW = \frac{1}{8} \cdot 429 > EXYZ. \tag{2.25}$$

In particular we see that the countermonotonicity condition is not a sufficient condition for the minimization problem.

The attained values for different countermonotonic constructions in these examples are astonishingly close and indicate that it might be useful to determine rearrangements that satisfy the necessary countermonotonicity conditions in order to obtain an approximation of the optimization problems considered in this section. This proposal, together with a method for how to determine the bounds, will lead in the following section to the rearrangement algorithm (RA) and similarly in Section 2.3 to an extension, the block rearrangement algorithm (BRA).

2.2   The Rearrangement Algorithm

As described in Section 2.1, problems of extremal dependence, like risk bounds, can be formulated as rearrangement problems. To turn the results into an algorithm, the first step is to approximate the optimization problem M_Ψ or m_Ψ on Fréchet classes as in (2.2), (2.3), or (2.4) by a discretized problem, i.e., by discretizing the marginals F_i by some discrete distribution functions, say $F_{i,N}$ with finite support, resulting in a problem $M_{\Psi,N}$ resp. $m_{\Psi,N}$. Since the max-tail risk problem $M_\Psi(s)$ can be represented by Theorem 2.1 by a minimal rearrangement problem (see (2.5)), we deal in detail in the following with the min-optimization problem m_Ψ and its discretized version $m_{\Psi,N}$. By general results on mass transportation problems, this approximation can be achieved under weak regularity conditions on Ψ, leading to optimization problems over the class of discrete distributions with support in the discrete product of the marginal supports $\mathcal{X}_1 \times \cdots \times \mathcal{X}_n$ and with fixed marginals, say $p_1, \ldots, p_n$. This is still an involved problem, also called the continuous assignment problem in the combinatorial optimization literature.

An essential step to simplify this problem is to assume that all marginal support sets $\mathcal{X}_j$ have the same size N, i.e., $\mathcal{X}_j = \{x_{1j}, \ldots, x_{Nj}\}$ and further that the probability

p_{ij} of x_{ij} with respect to P_{F_j} is identical to $\frac{1}{N}$ for all i, j. The main reduction then is to restrict the scope to the class of discrete distributions determined by permutations $\pi_2, \ldots, \pi_n$, i.e., when

$$p_{i_1,\ldots,i_n} = p_{i_1,\ldots,i_n}^{\pi_2,\ldots,\pi_n} = P(X_1 = x_{i_1,1}, \ldots, X_n = x_{i_n,n}) \qquad (2.26)$$

$$= \begin{cases} \frac{1}{N}, & \text{if } i_j = \pi_j(i_1), 2 \le j \le n, \\ 0, & \text{else.} \end{cases}$$

The argument for a reduction to the assignment problem in (2.26) is the fact that any discrete distribution $(p_{i_1,\ldots,i_n})$ can be uniformly approximated by a permutation induced distribution $(p_{i_1,\ldots,i_n}^{\pi_2,\ldots,\pi_n})$, when formally splitting all marginal masses into small pieces of size ε, i.e., formally enlarging the size of the marginal sets to sets of size N of sufficient size. This argument can be made formal. As a result of these approximation steps, one can approximate the original optimization problem m_Ψ by an optimization problem m_ψ^N over classes of permutations, i.e., by a multidimensional assignment problem. In other words, we reduce the class of all rearrangements $f_1, \ldots, f_n$ to a very particular class of rearrangements $f_1^N, \ldots, f_n^N$, which are induced by permutations $\pi_2, \ldots, \pi_n$. Formally choosing $x_{ij} = F_j^{-1}(\frac{i-0.5}{N})$, $1 \le i \le N$, $1 \le j \le n$, and the corresponding matrix

$$X = [x_1, \ldots, x_n] = \begin{bmatrix} x_{11} & \cdots & x_{1n} \\ \vdots & & \vdots \\ x_{N1} & \cdots & x_{Nn} \end{bmatrix}, \qquad (2.27)$$

we consider the problem of rearranging the elements x_{ij} of the matrix X such that after rearrangement the sums of the rearranged matrix $X^{\pi_2,\ldots,\pi_d}$, obtained by applying the permutations π_j to the columns x_j, are minimal with respect to a given convex function φ (compare with Proposition 2.7), i.e., the optimization problem is reduced to

$$m_\varphi^N = \inf \left\{ \frac{1}{N} \sum_{i=1}^N \varphi\left(\sum_{j=1}^n x_{\pi_j(i),j} \right); \pi_j \in S_N \right\}. \qquad (2.28)$$

Denoting by $X_j^{\pi_j}$ the discretized random variables associated to the permutations π_j (resp. to rearrangements $f_j^{\pi_j}$), this can be formulated as

$$m_\varphi^N = \inf \left\{ E\varphi(X_1 + X_2^{\pi_2} + \cdots + X_n^{\pi_n}); \pi_j \in S_N \right\}$$
$$= \inf \{ E\varphi(f_1(U) + f_2^{\pi_2}(U) + \cdots + f_n^{\pi_n}(U)); f_i^\pi \sim \pi \}. \qquad (2.29)$$

For example, to deal with the variance, we may choose $\varphi(x) = \left(\sum_{j=1}^n x_j \right)^2$, in which case we aim to determine the $X_j^{\pi_j}$ with given marginals such that their sum has minimum variance. The main idea of the rearrangement algorithm (RA) is, motivated by Proposition 2.7, to iteratively choose π_j such that the jth permuted column of the matrix X is countermonotonic to the sum of the other columns, i.e., such that

$$x_j^{\pi_j} \perp \sum_{\ell \ne j} x_\ell. \qquad (2.30)$$

This rearrangement algorithm was suggested first for pure assignment problems in Rüschendorf (1983a). It was then introduced and formulated in various forms for the problem of risk bounds in Puccetti and Rüschendorf (2012a,b) and Embrechts et al. (2013, 2014). After a finite number of steps, one reaches a permutation matrix $X = X^{\pi_2, \ldots, \pi_n}$ such that for any j, the jth column x_j of X is countermonotonic to $s_{(j)} = \sum_{\ell \neq j} x_\ell$, which by Proposition 2.7 is a necessary condition for an optimal solution of the minimization problem in (2.29). As shown in previous examples, this algorithm produces only local minima of $m_\varphi = \inf\{E\varphi(X_1 + X_2^{\pi_2} + \cdots + X_n^{\pi_n}); \pi_i \in S_N\}$; the necessary condition is not sufficient for optimality in general. The experience of a great number of examples shows, however, that the local minima are typically close to the value of a global solution.

Formally, the rearrangement algorithm is defined as follows:

Rearrangement algorithm

Let $X = [x_1, \ldots, x_n]$ be an assignment matrix, and for $j \leq n$ let $s_{(j)} = \sum_{\ell \neq j} x_\ell = (s_{1(j)}, \ldots, s_{N(j)})^\top$ denote the vector of row sums over $\ell \neq j$.

1. For $j = 1, \ldots, n$ choose $\pi_j \in S_N$ such that $x_j^{\pi_j} = (x_{\pi_j(i), j})$ is countermonotonic to $s_{(j)}$, i.e., rearrange the elements of the jth column such that the rearranged column is oppositely ordered to the vector of row sums of the other columns.

2. Iterate step 1 until there is no improvement of $v_\varphi^N = \sum_{i=1}^N \varphi(\sum_{j=1}^n x_{ij})$, i.e., $x_j \perp s_{(j)}$ for all j; all columns x_j are countermonotonic to the sums $s_{(j)}$ of the remaining columns. Then output the current assignment matrix X.

At each step of the algorithm, the value v_φ is reduced. Since the reduction of the value is defined on a finite set – the set of all assignment matrices – the algorithm converges in finite time in a local minimum. As shown in Haus (2015) (for the case $\varphi(x) = x^2$), the assignment problem is NP-complete, which implies that most likely there is no deterministic algorithm with polynomial time complexity that solves the problem exactly.

We end this section with a discussion of various applications of the RA to different risk functionals and demonstrate the very good quality of the RA approximations to the optimal solution through examples.

A) Numerical ranges of the RA-approximation

One can get an impression of the quality of the RA approximation to the optimal risk bounds by constructing numerical ranges for them. For discretized marginals F_j, $1 \leq j \leq n$, one can get estimates for the range of the optimal values m_Ψ, M_Ψ of the best/worst case risk problem in (2.2), as described in Puccetti and Rüschendorf (2012b). Let $\underline{F}_j, \overline{F}_j$ be n distribution functions uniformly distributed on their supports $\underline{F}_j \leq F_j \leq \overline{F}_j$ with associated assignment matrices $\underline{X}, \overline{X}$. Then we get for $\underline{m}_\Psi = m_\Psi(\underline{X})$, $\overline{m}_\Psi = m_\Psi(\overline{X})$:

$$\underline{m}_\Psi \leq m_\Psi \leq \overline{m}_\Psi \quad \text{and} \quad \underline{M}_\Psi \leq M_\Psi \leq \overline{M}_\Psi. \tag{2.31}$$

The corresponding numerical ranges implied by the RA, $[\underline{m}_\Psi^n, \overline{m}_\Psi^n]$ and $[\underline{M}_\Psi^n, \overline{M}_\Psi^n]$, then define the numerical ranges for the RA approximations as m_Ψ, M_Ψ.

Example 2.9

a) **Product of uniform $U[0, 1]$ random variables**

For $F_j \sim U[0, 1]$, $1 \le j \le n$, and $\Psi(x) = \prod_{i=1}^n x_i$, it holds that $M_\psi = \frac{1}{n+1}$, while the lower bound m_Ψ is given in analytical form in Wang and Wang (2011, Table 4.1). Numerical ranges for m_Ψ, M_Ψ are determined by the RA with $N = 10^5$ and show a remarkable precision. We report the results in Table 2.1.

n	m_Ψ (RA range)	m_Ψ (analytical)	M_Ψ (RA range)	M_Ψ (analytical)
3	$5.4800\times10^{-2} - 5.4807\times10^{-2}$	5.4803×10^{-2}	$0.2500 - 0.2500$	0.2500
4	$1.9096\times10^{-2} - 1.9100\times10^{-2}$	1.9098×10^{-2}	$0.2000 - 0.2000$	0.2000
5	$6.8594\times10^{-3} - 6.8615\times10^{-3}$	6.8605×10^{-3}	$0.1667 - 0.1667$	0.1667
10	$5.6486\times10^{-5} - 4.5435\times10^{-5}$	4.5410×10^{-5}	$0.0909 - 0.0909$	0.0909
20	$2.0553\times10^{-9} - 2.0639\times10^{-9}$	2.0612×10^{-9}	$0.0476 - 0.0476$	0.0476
50	$1.8865\times10^{-22} - 1.9352\times10^{-22}$	1.9286×10^{-22}	$0.0196 - 0.0196$	0.0196
100	$3.3861\times10^{-44} - 3.7452\times10^{-44}$	3.7201×10^{-55}	$0.0099 - 0.0099$	0.0099

Table 2.1 Numerical ranges of m_Ψ, M_Ψ for the product of uniforms.

Also, for inhomogeneous uniform distributions where no analytical results are available, Table 2.2 indicates estimates of a similar precision.

n	m_Ψ (RA range)	M_Ψ (RA range)
10	$1.5470\times10^{-1} - 1.5473\times10^{-1}$	$4.2191\times10^{0} - 4.2194\times10^{0}$
20	$5.0315\times10^{-2} - 5.0333\times10^{-2}$	$1.0764\times10^{2} - 1.0766\times10^{2}$
50	$1.6784\times10^{-3} - 1.6794\times10^{-3}$	$4.8464\times10^{6} - 4.8482\times10^{6}$
100	$5.7255\times10^{-6} - 5.7362\times10^{-6}$	$6.0091\times10^{14} - 6.0133\times10^{14}$

Table 2.2 Numerical ranges for the product of $U_j \sim U[a_j, a_{j+1}]$, $a_j = \frac{j-1}{n}$, $1 \le j \le n$, $N = 10^5$.

b) **Min-max stop loss of exponential random variables:**

For $F_j \sim \text{Exp}(1)$, $1 \le j \le n$ and $\psi(x) = \left(\sum_{j=1}^n x_j - k \right)_+$, it holds that $M_\Psi = \int_{k/n}^\infty (nx - k)e^{-x}\,dx$, while m_ψ can be calculated analytically using Theorem 3.5 in Wang and Wang (2011). Since $\text{Exp}(1)$ is unbounded from above, only numerical lower approximations are determined and show good accuracy of the RA approximation. We report the results in Table 2.3.

k	m_Ψ (RA lb)	M_Ψ (analytical)	M_Ψ (RA lb)	M_Ψ (analytical)
0	2.9998	3.0000	2.9998	3.0000
1	1.9998	2.0000	2.1494	2.1496
2	0.9998	1.0000	1.5401	1.5403
3	0.1693	0.1695	1.1035	1.1036
4	0.0570	0.0571	0.7906	0.7907
5	0.0203	0.0204	0.5664	0.5666

Table 2.3 Numerical lower bounds (lbs) for stop loss function of $n = 3$ $\text{Exp}(1)$ variables.

c) **Min-max stop loss of inhomogeneous Pareto random variables**
   For $n = 3$ random variables $X_j \sim \text{Pareto}(j+1)$, numerical estimates of m_Ψ, M_Ψ for
   stop loss functions Ψ as in example b) are calculated. An analytical value is only
   available for $k = 0$ and indicates good quality of the RA estimates; see Table 2.4.

k	m_Ψ (numerical lb)	M_Ψ (numerical lb)
0	1.8281	1.8281 (exact = 1.8333)
1	0.8281	1.1499
2	0.4027	0.8114
3	0.2829	0.6144
4	0.2181	0.4877
5	0.1772	0.4004

Table 2.4 Numerical lower bounds (lbs) for stop loss function of inhomogeneous Pareto random variables.

B) Application to the max resp. min tail risk problem
For increasing functions Ψ and for any real thresholds $s \in \mathbb{R}^1$, a rearrangement representation of the tail risk problems $M_\Psi(s)$, $m_\Psi(s)$, i.e., the maximal/minimal probability that $\Psi(X) \geq s$ defined in (2.4), is given in Theorem 2.1 and in Puccetti and Rüschendorf (2012a). Approximating the (discretized) marginal distribution functions F_j from below and above by $\underline{F}_j \leq F_j \leq \overline{F}_j$, one gets as in Section A) approximations from below and above thus:

$$\underline{M}_\Psi(s) \leq M_\Psi(s) \leq \overline{M}_\Psi(s), \quad \underline{m}_\Psi(s) \leq m_\Psi(s) \leq \overline{m}_\Psi(s) \tag{2.32}$$

and the corresponding numerical ranges $[\underline{m}_\Psi^N(s), \overline{m}_\Psi^N(s)]$, $[\underline{M}_\Psi^N(s), \overline{M}_\Psi^N(s)]$. In Figure 2.1, the numerical RA-based approximation is compared with the dual bound $M_\Psi^n(s)$ as discussed in Section 3.2.

The dual bound $M_\Psi^n(s)$ is an upper bound for $M_\Psi(s)$ and can be computed for homogeneous marginals; in the following example, $X_i \sim \text{Pareto}(2)$ and $\Psi(x) = \sum_{i=1}^n x_i$, i.e., we consider the tail risk of the joint portfolio; denote $M_+(s) = M_\Psi(s)$, $m_+(s) = m_\Psi(s)$, see Figure 2.1.

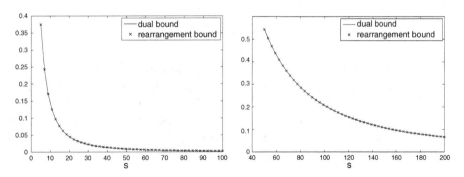

Figure 2.1 Upper dual bound and RA approximation for the joint portfolio, left $n = 3$, right $n = 30$.

	$\underline{M}^N_+(s)$	$\overline{M}^N_+(s)$		$\underline{m}^N_+(s)$	$\overline{m}^N_+(s)$	dual bound
$s = 0.5$	0.5101	0.5102	$s = 10$	0.1419	0.1420	0.1420
$s = 1.0$	0.2500	0.2500	$s = 15$	0.0740	0.0740	0.0740
$s = 1.5$	0.1599	0.1600	$s = 20$	0.0453	0.0453	0.0454
$s = 2.0$	0.1110	0.1111	$s = 25$	0.0305	0.0306	0.0306
$s = 2.5$	0.0816	0.0816	$s = 30$	0.0220	0.0220	0.0220

Table 2.5 Numerical ranges $[\underline{M}^N_+(s), \overline{M}^N_+(s)]$, $[\underline{m}^N_+(s), \overline{m}^N_+(s)]$ and upper dual bound.

One sees an extremely good coincidence. For $\Psi(x) = \sum_{i=1}^n x_i$ one obtains the numerical ranges in Table 2.5, which also show a good coincidence.

For $\Psi(x) = \max x_i$, a sharp analytical result for $M_{\max}(s)$ is available from Lai and Robbins (1978). Again for $X_i \sim \text{Pareto}(2)$, $n = 3$, one finds good coincidence and precision of the RA approximations, as presented in Table 2.6.

s	$\underline{m}^N_{\max}(s)$	analyt. result	$\overline{m}^N_{\max}(s)$	$\underline{M}^N_{\max}(s)$	analyt. result	$\overline{M}^N_{\max}(s)$
1	0.2500	0.2500	0.2500	0.7500	0.7500	0.7500
2	0.1103	0.1111	0.1113	0.3330	0.3333	0.3339
3	0.0625	0.0625	0.0625	0.1875	0.1875	0.1875
4	0.0390	0.0400	0.0400	0.1191	0.1200	0.1201
5	0.0273	0.0277	0.0283	0.0830	0.0833	0.0839

Table 2.6 Numerical ranges and analytic result for the max aggregation.

In the example shown in Table 2.7, the product case $\Psi(x) = \prod_{i=1}^n x_i$ is considered for an inhomogeneous Pareto(ϑ_j) portfolio. Let $n = 5$ and $\vartheta = (1.5, 1.8, 2.0, 2.2, 2.5)$, and denote $m_\Pi(s) = m_\Psi(s)$, etc.

s	$\underline{m}^N_\Pi(s)$	$\overline{m}^N_\Pi(s)$	s	$\underline{M}^N_\Pi(s)$	$\overline{M}^N_\Pi(s)$
$s = 0.001$	0.1611	0.1621	$s = 100$	0.2158	0.2167
$s = 0.002$	0.0985	0.0986	$s = 200$	0.1787	0.1796
$s = 0.003$	0.0634	0.0644	$s = 300$	0.1591	0.1601
$s = 0.004$	0.0410	0.0419	$s = 400$	0.1464	0.1474
$s = 0.005$	0.0244	0.0253	$s = 500$	0.1376	0.1386

Table 2.7 Numerical ranges for $m_\Pi(s)$, $M_\Pi(s)$.

All examples indicate that the RA-based approximations are precise and fast. The RA allows one to deal with general inhomogeneous portfolios and, due to the introduction of a stopping rule in Embrechts et al. (2013), allows one to deal with high-dimensional problems with n in the thousands. More details on the implementation and time complexity can be found in the abovementioned papers. Hofert (2020) gives an overview on computational techniques to implement the RA.

C) Approximation of upper and lower VaR bounds

The application of the RA to the approximation of upper and lower VaR bounds for a portfolio $\sum_{i=1}^{n} X_i$ has been described in Embrechts et al. (2013) and, together with applications to (VaR-)aggregation of risks, in Embrechts et al. (2014).

For a given confidence level α and marginal distribution functions $F_j, 1 \le j \le d$, the modified RA algorithm to determine the upper risk bound $\overline{\text{VaR}}_\alpha$ is based on Theorem 2.1 as well as on the discretization steps described in the introduction of this section.

RA to compute $\overline{\text{VaR}}_\alpha \left(\sum_{i=1}^{n} X_i \right)$ and $\underline{\text{VaR}}_\alpha \left(\sum_{i=1}^{n} X_i \right)$

The computation of the upper bounds uses the following steps:

U_1. Define the assignment matrices $X^\alpha = (x_{ij}^\alpha)$, $Y^\alpha = (y_{ij}^\alpha)$ as

$$x_{ij}^\alpha = F_j^{-1}\left(\alpha + \frac{(1-\alpha)(i-1)}{N}\right), \qquad y_{ij}^\alpha = F_j^{-1}\left(\alpha + \frac{(1-\alpha)i}{N}\right),$$
$$1 \le i \le N, 1 \le j \le n. \tag{2.33}$$

X^α, Y^α correspond to lower and upper approximations $\underline{F}_j^\alpha \le F_j^\alpha \le \overline{F}_j^n$ in the upper α-quantile region F_j^α of F_j.

U_2. Iteratively apply the rearrangement step to successively arrange all columns in countermonotonic order to the sum of the other columns to X^α and to Y^α until countermonotonic rearranged matrices $X^* = (x_{ij}^*)$, $Y^* = (y_{ij}^*)$ are obtained, such that

$$x_i^* \perp \sum_{j \ne i} x_j^* \quad \text{and} \quad y_i^* \perp \sum_{j \ne i} y_j^* \quad \text{for all } i \le n.$$

U_3. Define

$$\underline{s}_N = \min_{1 \le i \le N} \sum_{j=1}^{n} x_{ij}^*, \qquad \overline{s}_N = \min_{1 \le i \le N} \sum_{j=1}^{n} y_{ij}^*, \tag{2.34}$$

then

$$\underline{s}_N \le \overline{s}_N. \tag{2.35}$$

$[\underline{s}_N, \overline{s}_N]$ is the numerical range induced by the RA for $\overline{\text{VaR}}_\alpha$ for the sum $\sum_{i=1}^{n} X_i$, corresponding to the numerical range $[\underline{M}_\psi^N, \overline{M}_\psi^N]$ in (2.32).

For the approximation of the lower VaR bound $\underline{\text{VaR}}_\alpha$ for the sum $\sum_{i=1}^{n} X_i$, a similar procedure is used.

L_1. The lower and upper approximations of the upper quantile area of F_j^α are replaced by those of $F_{j,\alpha}$, i.e., F_j restricted to the lower quantile area.

Define: $X^{(\alpha)} = (x_{ij}^{(\alpha)})$, $Y^{(\alpha)} = (y_{ij}^{(\alpha)})$, where

$$x_{ij}^{(\alpha)} = F_j^{-1}\left(\frac{\alpha(i-1)}{N}\right), \quad y_{ij}^{(\alpha)} = F_j^{-1}\left(\frac{\alpha i}{N}\right), \qquad 1 \le i \le N, 1 \le j \le n. \tag{2.36}$$

L₂. Iteratively apply the rearrangement step as in L_1 to the assignment matrices $X^{(\alpha)}$, $Y^{(\alpha)}$ in the lower α quantile region $F_{j,\alpha}$ of F_j, leading to countermonotonic rearranged matrices X^*, Y^*.

L₃. Define

$$\underline{t}_N = \max_{i \leq N} \sum_{j=1}^{n} x_{ij}^*, \quad \bar{t}_N = \max_{i \leq N} \sum_{j=1}^{n} y_{ij}^*; \tag{2.37}$$

then $\underline{t}_N \leq \bar{t}_N$, and $[\underline{t}_N, \bar{t}_N]$ is the numerical range implied by the RA for $\underline{\mathrm{VaR}}_\alpha$ for the sum $\sum_{i=1}^{n} X_i$.

As described before, earlier the approximations converge for $N \to \infty$ a random permutation step of the columns x_j of X^α to the lower and upper VaR bounds. Embrechts et al. (2013) suggest adding before the rearrangement step U_2 resp. L_2 in order to avoid artificial obstructions. The following examples from Embrechts et al. (2013) show the good approximation properties of the RA based algorithm.

Example 2.10

a) **VaR bounds for a Pareto portfolio**

We consider a homogeneous Pareto portfolio with $\vartheta = 2.5$, i.e., $X_i \sim \text{Pareto}(2.5)$ for $n = 3$, which allows us to compare the numerical result for $\overline{\mathrm{VaR}}_\alpha = \overline{\mathrm{VaR}}_\alpha(\sum_{i=1}^{3} X_i)$ with the known analytical result in this case. Discretizing the upper 1 % quantile area as in (2.33) with $N = 50$ and choosing as starting assignment matrix the matrix with comonotone columns, the row sums vary from 15.9287 to 87.5126 (for the whole table, cf. Embrechts et al. (2013)); after the rearrangement step they vary from 24.4653 to 39.8412.

Thus the $\mathrm{VaR}_{0.99}$ of the comonotone risk vector given by $15.92\ldots$ is much smaller than $\underline{s}_{50}$ given by $24.47\ldots$, the lower estimate of $\overline{\mathrm{VaR}}_{0.99}$. The rearrangement of Y^α yields $\bar{s}_{50} = 25.1200$, and as a result we obtain as empirical range for $\overline{\mathrm{VaR}}_{0.99}$ the interval $[24.47, 25.12]$. The exact analytic value in this case is given by $\overline{\mathrm{VaR}}_{0.99} = 24.93$, which lies within the numerical interval. It is obtained as a consequence of the sharpness of the dual bounds (see Theorem 3.15).

b) **An application from extreme value theory**

As a model for analyzing operational risk, Moscadelli (2004) used generalized Pareto distributions (GPD) F_i given by $F_i(x) = 1 - (1 + \xi_i \frac{x}{\beta_i})^{-\frac{1}{\xi_i}}$, $x \geq 0$, for the main $d = 8$ risk categories. Under these marginal assumptions, the risk portfolio $X = (X_1, \ldots, X_8)$ shows a heavy tail behavior for six out of eight losses X_i, which exhibit infinite means, while two losses have finite means but do not have a finite variances.

The following Figure 2.2 (see Puccetti and Rüschendorf, 2012a) shows the RA-based numerical approximation of the range $[\underline{\mathrm{VaR}}_\alpha, \overline{\mathrm{VaR}}_\alpha]$ and also the co-monotonic VaR for $\alpha \in [0.95, 1)$. It shows the wide range of the VaR values and the somewhat astonishing fact that the comonotonic VaR in this example is closer to the lower VaR bound than to the upper bound. The large difference between the worst and the best case VaR under the same marginal assumptions implies a significant model uncertainty issue underlying VaR calculations for confidence levels close to 1.

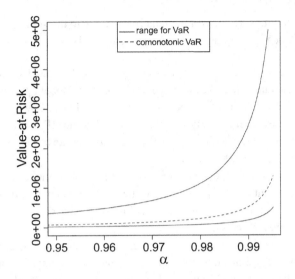

Figure 2.2 VaR range and comonotonic VaR for $n = 8$ GPD risks with parameters following Moscadelli (2004) based on RA for $N = 10^5$.

Consequently, the ample experience with the RA shows that it can be used with confidence for general inhomogeneous and also high-dimensional portfolios.

2.3   The Block Rearrangement Algorithm

Bernard et al. (2017c) proposed generalizing the RA by considering the simultaneous rearrangement of multiple columns that are stacked together (called blocks). This idea led to the block rearrangement algorithm (BRA) and has been developed in detail in a number of papers (Bernard and McLeish, 2016; Boudt et al., 2018; Bernard et al., 2018b). Specifically, the BRA builds on the insight that in order to obtain solutions for problem m_Ψ as in Proposition (2.7), it must hold that for any partition of $\{1, 2, \ldots, n\}$ into subsets J and J^c, the sums $s_J := \sum_{j \in J} X_j$ and $s_{J^c} := \sum_{l \in J^c} X_l$ are anti-monotonic (and not only for sets of the form $J = \{j\}$, as is exploited in the RA). This gives rise to the block rearrangement algorithm.

Block rearrangement algorithm

Let $X = [x_1, \ldots, x_n]$ be an assignment matrix, and for $J \subset \{1, 2, \ldots, n\}$ denote the column vectors of row sums $s_J = \sum_{j \in J} x_j$ resp. $s_{J^c} = \sum_{l \in J^c} x_l$.

1. For $J \subset \{1, 2, \ldots, n\}$ choose $\pi \in S_N$ such that $s_J^\pi = \sum_{j \in J} x_j^\pi$, $x_j^\pi = (x_{\pi(i),j})$ is countermonotonic to $s_{J^c} = \sum_{l \in J^c} x_l$, i.e., rearrange the elements of all selected columns (with index in J) such that after rearrangement the row sums over the rearranged columns are oppositely ordered to the vector of row sums of all remaining columns.

2. Iterate step 1 until there is no improvement in $v_\varphi^N = \sum_{i=1}^N \varphi(\sum_{j=1}^n x_{ij})$, i.e., $s_J \perp s_{J^c}$ for all $J \subset \{1, 2, \ldots, n\}$; all vectors of row sums obtained by taking the sum s_J over

columns x_j, $j \in J$ are countermonotonic to the sum s_{J^c} of the remaining columns. Then output the current assignment matrix X.

The matrices $X_J = (x_j)_{j \in J}$ and X_{J^c} are called blocks. Similarly to the RA, there is no guarantee that the BRA will converge to a global optimum and thus that a solution to the problem m_φ^N is obtained. The experience, however, is that the local optima are of excellent quality and that – provided the BRA is run until convergence – they improve on those obtained by the RA. To illustrate these points we repeat the following experiment 10,000 times:

- Initialize the $N \times n$ matrix X by randomly simulating N values from a $U(0, 1)$ distribution. We place these values in the first column of X. Next we place values in each of the subsequent $n - 1$ columns by each time randomly permuting the values in the first column.
- If $N \le 10$ rows and $n \le 4$ columns, permute columns $2, 3, \ldots, n - 1$ in all $(N!)^{n-2}$ ways, and arranging column n so that it is countermonotonic with the sum of the other columns. Among all these configurations, find the matrix X^* whose row sums have the global minimum variance V^* (here we use the variance function $\varphi(\sum_{i=1}^n x_i) = (\sum_{i=1}^n x_i)^2$).
- Apply the RA to X to obtain a local minimum X_{ra} in which all columns are countermonotonic to the sum of the others. Compute the variance V_{ra} of the row sums of X_{ra}.
- Apply the BRA to X_{ra} to obtain the matrix X_{bra} and the variance V_{bra} of the row sums of X_{bra}.

We first compute the average value for V_{ra} and for V_{bra} (based on the 10,000 experiments) and report the results in Table 2.8.

	$n = 4$		$n = 7$		$n = 10$	
$N =$	V_{ra}	V_{bra}	V_{ra}	V_{bra}	V_{ra}	V_{bra}
10	0.001	0.0006	0.0004	1.1×10^{-5}	0.00018	1.8×10^{-7}
100	1.2×10^{-5}	5.5×10^{-6}	3.4×10^{-6}	8×10^{-8}	1.6×10^{-6}	1.3×10^{-9}
1,000	1.2×10^{-7}	5.5×10^{-8}	3.2×10^{-8}	7.6×10^{-10}	1.6×10^{-8}	1.2×10^{-11}

Table 2.8 Average value of V_{ra} and V_{bra} for different values of n and N. Both averages are estimated using the 10,000 experiments as described above. All digits reported in the table are significant. The RA and BRA are both run until convergence, and thus the extra accuracy obtained for the BRA comes at a very high computational cost.

We make the following observation in Table 2.8: The larger the number of variables n or the number of discretization steps N, the larger the improvement of the BRA over the RA. We have performed other experiments with other distributions and we obtain similar results.

Next, in Figure 2.3, we plot the percentage of cases in which V_{ra}, resp. V_{bra} is within a given tolerance[1] of V^*. It shows that this percentage decreases quickly to 0 as

---

[1] 10^{-6} in this case.

N increases. Table 2.9 reports the averages of the difference $V_{ra} - V^*$ and $V_{bra} - V^*$. We find that the BRA outperforms the RA by several orders of magnitude.

	$N = 4$	$N = 5$	$N = 6$	$N = 7$
RA	0.0020	0.0015	0.0026	0.0015
BRA	0.0001	0.0002	0.0003	0.0003

Table 2.9 Average distance between V_{ra} and V^* resp. V_{bra} and V^* for $n = 4$ variables.

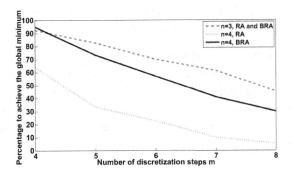

Figure 2.3 Percentage of cases that the minimum variance obtained by the RA (V_{ra}) resp. BRA (V_{bra}) match with the global minimum V^* (within 10^{-6}). Note that when $n = 3$, RA and BRA are equivalent.

Remark 2.11 The comparison in Table 2.9 is necessarily done with very small matrices, as the global minimum V^* is obtained by computing the variance for all possible permutations of the matrix. For larger matrices, such a technique cannot be used. In fact, there are very few examples for which the value of V^* is known. One specific result is given in Cuesta-Albertos et al. (1993), used also by Haus (2015), which gives the minimum variance in the case of an N by n matrix that contains in each column the integers $1, 2, \ldots, N$. In this case (at the minimum variance matrix), the mean of the sum is $\mu := n \frac{1+2+\cdots+N}{N}$, and the sum takes two values, $M = \lfloor \mu \rfloor$ with probability $q := \mu - \lfloor \mu \rfloor$ and $M + 1$ with probability $(1 - q)$, so that the minimum global variance can be computed explicitly. We are then able to check the percentage of the time the RA or the BRA achieves the global minimum by starting from a randomized matrix (where each column has been randomly permuted). We obtain similar conclusions as in Figure 2.3 and Table 2.9, namely the percentage of the cases in which one achieves the global minimum decreases with n and N, and the BRA is closer to the global minimum by several orders of magnitude. ◇

The above discussion confirms that both the RA and BRA are accurate tools for obtaining approximate solutions to assignment problems. While the BRA appears to lead to better results, one can, however, not conclude that the BRA is outperforming the RA. In the experiments we run both the RA and BRA until convergence, but for the BRA to converge it must hold that that for all $2^n - 1$ partitions of $\{1, 2, \ldots, n\}$ into subsets J and J^c, the row sums of the block X_J are countermonotonic to those of X_{J^c}. By contrast, for the RA this countermonotonicity property only needs to hold for sets of the form $J = \{j\}, j = 1, 2, \ldots, n$. In other words, the BRA converges much slower than the RA, and it remains to be seen whether for a given time budget, approximate solutions for an assignment problem that are obtained by applying the BRA still dominate those that are obtained by the RA. For instance, if for a particular application the RA turns out to be 100 times faster than the BRA, then one could run the RA 100 times, each time randomly permuting the initial assignment matrix X, and take the best of the 100 candidate solutions. In fact, we merely observe that the RA can in principle be improved by permuting entire blocks rather than only columns; a clever choice for the blocks could result in an algorithm that outperforms the BRA for a given time budget. This will be discussed next.

Block selection and comparison between RA and BRA

When the assignment matrix X contains only three columns, the RA and the BRA are equivalent. In what follows, we concentrate on the rearrangement of $N \times n$ matrices in which $n \geq 4$. We run experiments in which each *step* of the RA and of the BRA requires exactly the same computational time. What we call a "step" consists of the following:

- For RA: Choose one column randomly between 1 and n, i.e., select a subset $J = \{j\}$, $j \in \{1, 2, \ldots, n\}$. As a result this yields two blocks: block X_J has 1 column, and X_{J^c} contains the remaining $n - 1$ columns.

- For BRA: Choose *randomly* a subset $J \subset \{1, 2, \ldots, n\}$. As a result this yields a block X_J with #J columns and another one containing the $n - $#$J$ remaining ones.

Then compute the row sums of each block, and rearrange block X_J (permuting entire rows) such that its row sums are countermonotonic to the row sums of the other one.

At each step of the BRA, one needs to pick "randomly" a subset J of $\{1, 2, \ldots, n\}$. Originally in Bernard and McLeish (2016) and Bernard et al. (2018b), the BRA was implemented such that each $J \subset \{1, 2, \ldots, n\}$ is equally likely.[2] Out of the $2^{n-1} - 1$ possible subsets of $\{1, 2, \ldots, n\}$, there are many more that have cardinality close to $\left\lceil \frac{n}{2} \right\rceil$ than there are with cardinality close to one or two. In this implementation of the BRA, a lot of the steps are thus performed using blocks of similar sizes rather than using one small block and one big block. To contrast this point, we study a version of the BRA in which we first pick the block size k uniformly in $\{1, 2, \ldots, \left\lceil \frac{n}{2} \right\rceil\}$ and next select randomly k columns out of n (where each column is equally likely to be chosen) to form a block X_J. We refer to this version of the BRA as "BRA unif."

We then apply RA, BRA, and BRA unif on an $N \times n$ assignment matrix X that is obtained by simulating random numbers from a $U(0, 1)$ distribution or from a Pareto distribution with tail index 2. We compute the log variance of the row sums after each step and display how this quantity decreases over time; we take the log variance for

---

[2] This can be achieved by choosing random numbers between 0 and $2^{n-1} - 1$ and using their binary decomposition to decide which elements $j \in \{1, 2, \ldots, n\}$ are contained in J.

a better visualization, as the variance decreases exponentially fast at the beginning of the RA or BRA. We report the average across K experiments,

$$\frac{1}{K} \sum_{i=1}^{K} \log \left(\mathrm{var} \left(\sum_{j=1}^{n} X_j \right) \right). \tag{2.38}$$

Numerical results are displayed in Figures 2.4 and 2.5.

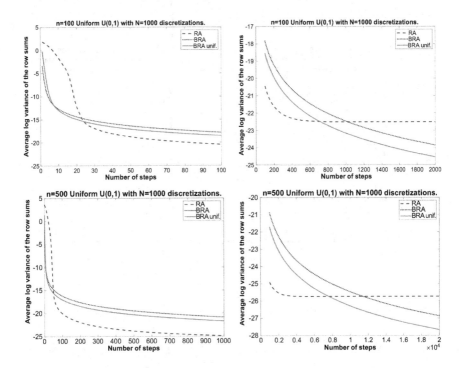

Figure 2.4 Uniform risks: We display log variances as in (2.38) over $K = 100$ experiments when the initial $N \times n$ matrix X contains random draws from a $U(0, 1)$ distribution. Here, $N = 1000$, and $n = 100$ for the top two graphs whereas $n = 500$ for the bottom two graphs. On the left, we display the evolution of the average log variance during the first 100 (top graph) resp. 1000 steps (bottom graph). The right panel displays this average for the later steps up to 2000 (top graph) resp. 20,000 steps (bottom graph).

Comparing the top graphs from Figure 2.4 with Figure 2.5, both obtained in the case of $n = 100$ columns for X, we observe that the RA and BRA converge faster when the values x_{ij} of X are independent random draws from a uniform distribution rather than from a Pareto one, which is a heavy tailed distribution. Furthermore, when comparing the bottom graphs from Figure 2.5 with Figure 2.4 corresponding to $n = 500$ columns, it is clear that the impact of heavy tailedness on convergence speed is more pronounced when the number of columns is larger.

For instance, when $n = 500$ and draws are taken from the Pareto distribution, it would be necessary to run more than 2000 steps to ensure that the RA is close enough to its state of convergence.

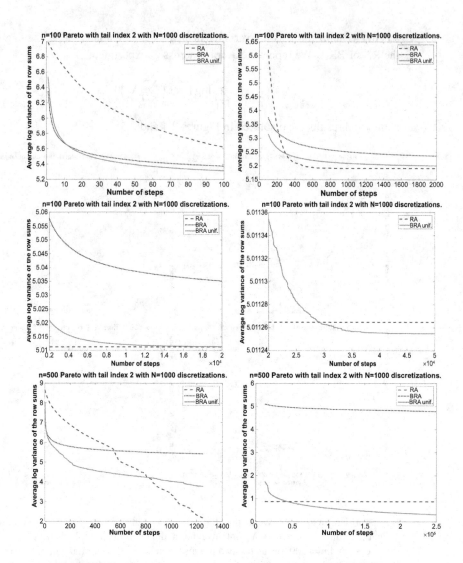

Figure 2.5 Pareto risks: We display log variances as in (2.38) over $K = 100$ experiments when the initial $N \times n$ matrix X contains random draws from a Pareto distribution with tail index 2. Here, $N = 1000$, and $n = 100$ for the top four graphs whereas $n = 500$ for the bottom two graphs. Each graph displays the evolution of the average log variance during a number of steps that correspond to a certain stage of the algorithm (e.g., first graph shows this during the first 100 steps).

Furthermore, from Figure 2.5 and Figure 2.4 we observe that the BRA leads to a higher variance reduction during the very first steps, but the RA catches up quickly and leads to a much higher variance reduction during the intermediate range of steps. When sampling from the Pareto distribution, both in the case of $n = 100$ and $n = 500$, the RA performs better than the BRAs for the first 2000 steps.

In addition, we observe from these figures that in all our experiments the BRA unif ultimately leads to a slightly better variance reduction. The convergence of the (standard) BRA, in which all partitions of the assignment matrix into two blocks are equally likely to be chosen, is the slowest out of the three considered approaches.

These observations motivate us to design a BRA such that the size of the blocks is not uniformly distributed over $\{1, 2, \ldots, \frac{n}{2}\}$ but is such that the first steps of the BRA are performed with bigger block sizes, whereas the last steps of the BRA are performed with small block sizes. That is, we aim to design a BRA that behaves like the standard BRA during the first series of steps and like the RA during the later steps performed.

To do so, at each step t we draw the block size from a Beta distribution with parameters $\alpha(t)$ and $\beta(t)$ that are adequately chosen. The Beta distribution is scaled to have support in $[1, \frac{n}{2} + 1]$. At each step $t \in \{1, 2, \ldots, T\}$, the integer part of a simulated value from the Beta distribution with parameters $\alpha(t)$ and $\beta(t)$ then gives the block size chosen in the BRA at step t. For the tth step, we use the following parametrization of the Beta distribution with parameters $\alpha(t)$ and $\beta(t)$:

$$\alpha(t) = A - \left(\frac{t-1}{T-1}\right)^{\frac{1}{B}}(A-1), \qquad \beta(t) = 1 + \left(\frac{t-1}{T-1}\right)^{\frac{1}{B}}(A-1), \qquad (2.39)$$

in which A and B are tuning parameters. When $A = B = 1$, the Beta distribution coincides with the uniform distribution, and the block size at each step t is thus uniformly drawn from the set $\{1, 2, \ldots, \frac{n}{2}\}$. Recall that the mean of a Beta distribution with parameters $\alpha(t)$ and $\beta(t)$ is given by $\frac{\alpha(t)}{\alpha(t)+\beta(t)}$. When $B = 1$, we have a linear decrease in the average size of the blocks, as displayed in Figure 2.6. Finally, when $B > 1$, we observe a sharper decrease in the average size of the blocks.

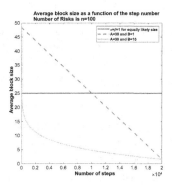

Figure 2.6 **Average block size:** We show the average block size at each tth step of the BRA for various choices of the Beta distribution (see equation (2.39) for its parametrization).

In the next two figures, we reproduce our previous experiments with $n = 100$ and $n = 500$, in which draws are taken from a $U(0, 1)$ distribution (Figure 2.7) or from a Pareto distribution with tail index 2 (Figure 2.8) using various choices for A and B in the parametrization (2.39) of the Beta distribution. In order to facilitate the comparison, note that $A = B = 1$ corresponds to the uniform distribution of the block sizes and was already reported in our first experiment as "BRA unif."

Figures 2.7 and 2.8 provide evidence that the choice of the blocks in the BRA plays an important role, especially when one wants to achieve a quick convergence. Here

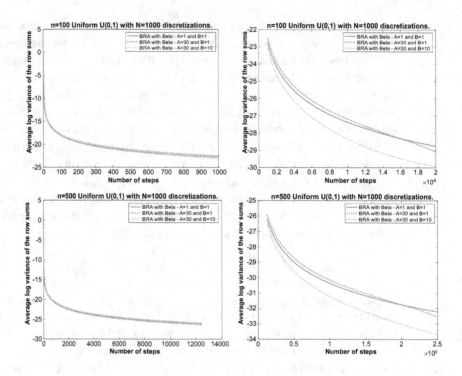

Figure 2.7 Uniform risks: We display log variances as in (2.38) over $K = 100$ experiments when the initial $N \times n$ matrix X contains random draws from a $U(0, 1)$ distribution. Here, $N = 1000$, and $n = 100$ for the top two graphs whereas $n = 500$ for the bottom two graphs. Each graph displays the evolution of the average log variance during a number of steps that correspond to a certain stage of the algorithm (e.g., the first graph shows this during the first 1000 steps).

we run 20,000 steps for each BRA. Such an experiment requires a few minutes on a standard laptop. However, if one has no constraints on their time and has access to more computational power, then all these variants of the BRA appear to lead to similar results. Using a Beta distribution in (2.39), which involves a choice for $B > 1$, leads to more variance reduction during the first series of steps, but in the long run, the linear decrease in the average block size under the BRA unif appears to be outperforming the RA and all the other BRA versions.

To conclude, the RA provides a fast, good quality result. However, to achieve more accuracy one needs to implement a BRA. It is advisable to make sure that not all partitions of $\{1, 2, \ldots, n\}$ into sets J and J^c have equal probability, especially in the case in which the assignment matrix X contains draws coming from a heavy tailed distribution. Instead, it is recommended to require that the cardinalities of the sets J are roughly equal to $\frac{n}{2}$ during the first series of steps, to obtain a sharp decrease in variance at the beginning of the algorithm, and then to gradually move to partitions in which the cardinalities of the sets J become smaller and smaller.

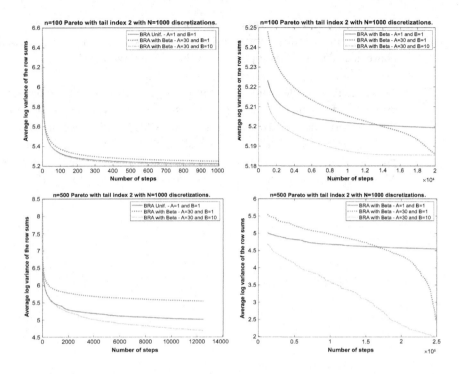

Figure 2.8 Pareto risks: We display averages of log variance over $K = 100$ experiments for an $N \times n$ assignment matrix X with $N = 1000$ and $n = 100$ (for the top two graphs) or $n = 500$ (for the bottom two graphs). Draws are randomly made from a Pareto distribution with tail index 2. We then compute (2.38) with $K = 100$. The left panel displays average of log variance of the row sums during the first 100 steps. The right panel displays the same quantity during the later steps.

Further applications

There are a number of optimization problems that the (B)RA can help solving. We have already discussed in Section 2.2 how these algorithms can be used to obtain approximate solutions for the tail risk problem as well as approximate bounds on value-at-risk. In operations research, Boudt et al. (2018) show how the BRA can be used to deal in an efficient manner with the k-partitioning problem. Cornilly et al. (2022) propose an algorithm that is inspired by the RA for solving the "fair allocation of goods problems" under various fairness criteria.

Another application of the (B)RA consists of inferring the joint distribution among variables $(X_1, \ldots, X_n)$ for which the marginals F_i and the distribution G of the sum S are known (the consistency problem). Note that the consistency problem is equivalent to the problem of whether $F_1, \ldots, F_n, G_-$, where G_- is the distribution of $-S$, are mixable (see Chapter 1.4). The RA or BRA is then a tool to produce (approximatively) a corresponding mixing dependence structure. In the ideal case it produces random variables $X_i \sim F_i$ and $S \sim G$ such that $\sum_{i=1}^{n} X_i = S$.

For instance, when prices of basket options on an index S are available as well as prices of regular options written on individual stocks $X_i, i = 1, \ldots, n$, then as described above, the (B)RA makes it possible to obtain a multivariate distribution for the assets $(X_1, \ldots, X_n)$ that is consistent with this information.

While in general there are many multivariate distributions that are consistent with the available information, Bernard et al. (2018b) provide evidence that the BRA singles out one particular multivariate distribution, which in the case of normal marginal distributions for the $X_i, i = 1, \ldots, n$ and $-S$ has maximum entropy among all other possible multivariate distributions. For more details we refer the reader to Bernard and McLeish (2016), Bernard et al. (2017c), and Bernard et al. (2018b, 2021).

3   Dual Bounds

Hoeffding–Fréchet functionals aim to determine the largest and smallest value of a risk functional over all possible dependence structures, when the marginal distributions are held fixed. In the late 1970s and early '80s, some basic duality results for these kinds of problems were developed, leading to dual representations of these functionals. It was soon realized that these duality results are far reaching extensions of the dual representation in the Monge–Kantorovich theory, which originated in early work in the 1940s and '50s. Even if the dual representations typically pose significant challenges to being evaluated in explicit form, it was found in some important work on risk bounds from around 2006 that relaxations of the dual representations lead to solvable problems and to meaningful risk bounds. The range and quality of these "dual bounds" is the content of this chapter.

3.1   Hoeffding–Fréchet Functionals

The aim of this chapter is to describe the method of dual bounds for developing bounds for various risk functionals. Our aim is to describe the possible influence of dependence between the components of a risk vector $X = (X_1, \ldots, X_n)$ on the expectation of a real function $\varphi(X)$, i.e., to determine the range of $E\varphi(X)$ under all possible dependence structures.

When the distribution of X_i is given by P_i, $1 \le i \le n$, then the set of all possible dependence structures with margins P_i is given by the Fréchet class $M(P_1, \ldots, P_n)$. We assume that P_i are probability measures on measure spaces $(E_i, \mathcal{A}_i)$, $1 \le i \le n$, typically $E_i = \mathbb{R}$ or $E_i = \mathbb{R}^d$ or more generally E_i, some Polish space. In the following we generally make this assumption. Then the **Fréchet class** $M(P_1, \ldots, P_n)$ is defined as

$$M(P_1, \ldots, P_n) = \{P \in M^1(E, \mathcal{A}); \ P^{\pi_i} = P_i, 1 \le i \le n\}, \tag{3.1}$$

where $(E, \mathcal{A}) = \bigotimes_{i=1}^{n}(E_i, \mathcal{A}_i)$ is the product space, and π_i are the projections of the product space E on the ith component.

For a measurable, real function $\varphi\colon E \to \mathbb{R}$, $\varphi(x) = \varphi(x_1, \ldots, x_n)$ of n variables, we consider the **Hoeffding–Fréchet functionals**

$$M(\varphi) = \sup\left\{\int \varphi\,\mathrm{d}P;\ P \in \mathcal{M}(P_1, \ldots, P_n)\right\}$$
$$\text{and} \quad m(\varphi) = \inf\left\{\int \varphi\,\mathrm{d}P;\ P \in \mathcal{M}(P_1, \ldots, P_n)\right\}. \tag{3.2}$$

The open or closed interval $[(m(\varphi), M(\varphi))]$ describes exactly the possible influence of dependence on the expectation $E\varphi(X)$. The classical results of Fréchet (1951) and Hoeffding (1940) concern the real case $E_i = \mathbb{R}$ and $n = 2$ and the following sharp bounds.

Fréchet bounds

$$F^-(x_1, x_2) := (F_1(x_1) + F_2(x_2) - 1)_+ \le F(x_1, x_2)$$
$$\le F^+(x_1, x_2) := \min\{F_1(x_1), F_2(x_2)\} \tag{3.3}$$

and **Hoeffding bounds**

$$\iint (F^-(x_1, x_2) - F_1(x_1)F_2(x_2))\,\mathrm{d}x_1\,\mathrm{d}x_2$$
$$\le \mathrm{Cov}(X_1, X_2) \le \iint (F^+(x_1, x_2) - F_1(x_1)F_2(x_2))\,\mathrm{d}x_1\,\mathrm{d}x_2, \tag{3.4}$$

where F_1, F_2 are the distribution functions of X_1, X_2, F is the distribution function of X, and where F^-, F^+ denote the Fréchet bounds. The inequality in (3.4) is a consequence of (3.3) and the **covariance representation**

$$\mathrm{Cov}(X_1, X_2) = \iint (F(x, y) - F_1(x)F_2(y))\,\mathrm{d}x\,\mathrm{d}y. \tag{3.5}$$

The upper and lower Fréchet bounds for the distribution function F in (3.3) yield also the joint distribution function having the maximal resp. minimal possible covariance and correlation in (3.4). From the covariance representation in (3.5) it is obvious that the upper respectively lower bounds in (3.4) are attained only by the upper esp. lower Fréchet bounds, i.e., there are unique extremal solutions in (3.4).

The statements in (3.3) and (3.4) concern the Hoeffding–Fréchet functionals for the case $\varphi_1(y) = 1_{(-\infty, y]}$, resp. for $\varphi_2(y) = y_1 y_2$, $y \in \mathbb{R}^2$. For the risk bounds we will consider various functionals of the form $f\left(\sum_{i=1}^n y_i\right)$, f convex, or $1_{[t, \infty)}\left(\sum_{i=1}^n y_i\right)$. While the first of these yields, for example the stop loss premium for $\left(\sum_{i=1}^n y_i - t\right)_+$, the second yields the tail risk of the joint portfolio and thus by inversion the value-at-risk. The main results of this chapter give various forms of dual representations of the Hoeffding–Fréchet functionals that yield extensions of the classical results of Fréchet and Hoeffding.

3.2   Dual Representation of Hoeffding–Fréchet Functionals

The problem of determining the Hoeffding–Fréchet functionals in (3.2) can be considered as an infinite-dimensional linear optimization problem. The linear functional

$\int \varphi \, \mathrm{d}P$, which is linear in φ and P, has to be maximized resp. minimized with respect to P over the convex Fréchet class $\mathcal{M} = \mathcal{M}(P_1, \ldots, P_n)$. From experience with finite dimensional linear programming it is natural to formulate the following dual functionals:

$$U(\varphi) := \inf \left\{ \sum_{i=1}^{n} \int f_i \, \mathrm{d}P_i; \ \ f_i \in L^1(P_i), \bigoplus_{i=1}^{n} f_i \geq \varphi \right\}, \tag{3.6}$$

where $\bigoplus_{i=1}^{n} f_i(x) := \sum_{i=1}^{n} f_i(x_i)$, and

$$I(\varphi) := \sup \left\{ \sum_{i=1}^{n} \int f_i \, \mathrm{d}P_i; \ \ f_i \in L^1(P_i), \bigoplus_{i=1}^{n} f_i \leq \varphi \right\}. \tag{3.7}$$

Then we obviously have the inequalities

$$M(\varphi) \leq U(\varphi) \quad \text{and} \quad m(\varphi) \geq I(\varphi); \tag{3.8}$$

the dual bounds $U(\varphi)$ and $I(\varphi)$ give generally upper bounds for the "worst case" $M(\varphi)$ resp. lower bounds for the "best case" $m(\varphi)$.

The basic question is whether "**duality holds**," i.e., $M(\varphi) = U(\varphi)$ and $m(\varphi) = I(\varphi)$.

The basic duality theorem states that for general classes of functions and on general spaces, duality holds – for example, for upper-semicontinuous functions or for lower majorized measurable functions. For a detailed formulation of conditions on the validity of the duality theorem and for the existence of solutions for $M, U, m,$ and I we refer the reader to the exposition in Rachev and Rüschendorf (1998a,b). We state the duality theorem in this book as a principle.

Duality Theorem $M(\varphi) = U(\varphi) \quad \text{and} \quad m(\varphi) = I(\varphi). \tag{3.9}$

Remark 3.1   In the case $n = 2$ when $\varphi(x_1, x_2)$ is a cost function, $m(\varphi)$ can be understood as the minimal expected cost of transport from a mass distribution P_1 to a mass distribution P_2. For this case, Kantorovich (1942) formulated this classical mass transportation problem and established duality $m(\varphi) = I(\varphi)$ in the case where φ is a metric. The general duality theorem in (3.9) thus generalizes the classical mass transportation problem.   ◇

The duality theorem simplifies in the homogeneous case where $E_1 = \cdots = E_n$ and $P_1 = \cdots = P_n$, if the risk function φ is symmetric: $\varphi(x) = \varphi(x_{\pi(1)}, \ldots, x_{\pi(n)})$ for all permutations $\pi \in \mathcal{S}_n$; see Rüschendorf (1982) or Rüschendorf (2013, Proposition 2.5). A relevant example of this situation is the case where $\varphi(x) = f\left(\sum_{i=1}^{n} x_i\right)$.

Theorem 3.2 (Duality in the homogeneous case)   *If $E_1 = \cdots = E_n$, $P_1 = \cdots = P_n =: Q$ and if the risk function $\varphi(x)$ is symmetric and satisfies the assumption of the duality theorem, then:*

$$M(\varphi) = \inf \left\{ n \int h \, \mathrm{d}Q; \ \ h \in L^1(Q), \sum_{i=1}^{n} h(x_i) \geq f(x) \right\}, \tag{3.10}$$

and an optimal dual solution h exists.

Proof For admissible $f_i \in L^1(Q)$ in the definition of $U(f)$, i.e., $\oplus f_i \geq \varphi$, we define

$$h(y) := \frac{1}{n} \sum_{i=1}^{n} f_i(y),$$

then $h \in L^1(Q)$, and by symmetry of φ we obtain that $\sum_{i=1}^{n} h(x_i) \geq \varphi(x)$.

In consequence we obtain

$$n \int h \, dQ = \int \left(\sum_{i=1}^{n} f_i \right) dQ = \sum_{i=1}^{n} \int f_i \, dQ,$$

which implies by the duality theorem in (3.9) the statement in (3.10). The existence of an optimal dual solution in (3.10) is a consequence of the existence result in the general duality theorem (3.9).   □

Theorem 3.2 implies that in the homogeneous case we can expect simplified formulas. In general the dual representation is not easy to calculate in explicit form. For some classes of examples, however, explicit solutions have been found. An important consequence of the duality theorem in (3.9) and (3.10) is that it yields the possibility to consider various classes of relaxations of the dual problems. These relaxations can in several cases be numerically or exactly solved and give relevant bounds for the Hoeffding–Fréchet functionals.

An interesting consequence of the duality theorem is the following result on sharpness of Fréchet bounds (see Rüschendorf, 1981b).

Theorem 3.3 (Sharpness of Fréchet bounds) *Let $P_i \in M^1(E_i, \mathcal{A}_i)$ and $A_i \in \mathcal{A}_i$, $1 \leq i \leq n$. Then for any $P \in M(P_1, \ldots, P_n)$, it holds that*

$$\left(\sum_{i=1}^{n} P_i(A_i) - (n-1) \right)_+ \leq P(A_1 \times \cdots \times A_n) \tag{3.11}$$

$$\leq \min\{P_i(A_i); \ 1 \leq i \leq n\},$$

and the upper and lower bounds in (3.11) are sharp.

In particular, Theorem 3.3 implies the sharpness of the classical Fréchet bounds in the general case $n \geq 2$:

For any distribution function $F \in \mathcal{F}(F_1, \ldots, F_n)$, i.e., F has marginal distribution functions F_i, $1 \leq i \leq n$, it holds that

$$F^-(x) \leq F(x) \leq F^+(x), \tag{3.12}$$

where $F^-(x) := \left(\sum_{i=1}^{n} F_i(x_i) - (n-1) \right)_+$ and

$$F^+(x) := \min\{F_i(x_i); \ 1 \leq i \leq n\} \tag{3.13}$$

are the lower and the upper Fréchet bounds. The bounds in (3.12) are sharp.

Note that this result also holds true for multivariate marginals $F_i \in \mathcal{F}_{k_i}$.

This duality proof was the first proof of sharpness of the lower Fréchet bounds in the literature. In the case $n = 2$, the duality theorem also implies a classical result of Strassen (1965) on the maximal resp. minimal probability of general sets A.

Let $n = 2$ and $A \in \mathcal{A}_1 \otimes \mathcal{A}_2$, and define

$$M(A) := M(\varphi_A), \quad m(A) := m(\varphi_A), \tag{3.14}$$

where $\varphi_A(x) = 1_A(x)$.

Theorem 3.4 (Duality theorem for sets)   *For $A \in \mathcal{A}_1 \otimes \mathcal{A}_2$ it holds that*

$$M(A) = \inf\{P_1(A_1) + P_2(A_2); \quad A_i \in \mathcal{A}_i, A \subset A_1 \times E_2 \cup E_1 \times A_2\} \tag{3.15}$$

and

$$m(A) = \sup\{(P_1(A_1) + P_2(A_2) - 1)_+; \quad A_i \in \mathcal{A}_i, A \supset A_1 \times A_2\}. \tag{3.16}$$

If A is closed then the A_i can be restricted to the class of closed sets.

Proof   For the proof of (3.15) one shows that one can restrict in the dual problem to two valued functions $f_i = 1_{A_i}$, $i = 1, 2$. Equation (3.16) can be reduced to (3.15) by taking the complements. □

Theorem 3.4 gives fairly explicit results for the problem of determining the maximum resp. minimum probability of sets. In the case where $E_1 = E_2 = E$ and $\mathcal{A}_1 = \mathcal{A}_2$, where $(E, \leq)$ is a partially ordered set with a closed order $\leq$, it implies the famous Strassen representation theorem. Define the stochastic order:

$$P_1 \leq_{st} P_2 \quad \text{if} \quad \int f \, dP_1 \leq \int f \, dP_2 \text{ for all integrable increasing functions } f.$$

Theorem 3.5 (Strassen a.s. representation theorem)   *For $P_i \in \mathcal{M}^1(E, \mathcal{A}_1)$, it holds that: $P_1 \leq_{st} P_2$ if and only if there exists a probability space $(\Omega, \mathcal{B}, P)$ and random variables X_1, X_2 on Ω with values in E such that $X_i \sim P_i$, $i = 1, 2$ and*

$$X_1 \leq X_2 \, [P].$$

Proof   The proof is a consequence of Theorem 3.4, choosing $A = \{(x_1, x_2) \in E \times E; \ x_1 \leq x_2\}$. Then it is easy to see that in the dual problem in (3.15) one can restrict to the case where A_2 is an isotonic set and $A_1 = A_2^c$. In consequence the condition $P_1 \leq_{st} P_2$ implies

$$P_1(A_1) + P_2(A_2) = 1 - P_1(A_2) + P_2(A_2) \geq 1,$$

and the result follows, choosing for example $\Omega = E \times E$, $\mathcal{B} = \mathcal{A}_1 \otimes \mathcal{A}_1$, P the measure with $P(A) = M(A)$, and X_i the projections π_i, $i = 1, 2$. □

An interesting consequence of Theorem 3.3 is the following result on maximal concentration.

Proposition 3.6   *Let $F_i \in \mathcal{F}$ be one-dimensional distribution functions, $1 \leq i \leq n$, and let X_i be random variables with $X_i \sim F_i$. Then*

a) **Maximal/minimal concentration** For all $a = (a_1, \ldots, a_n) \le b = (b_1, \ldots, b_n)$ it holds that

$$\Big(\sum_{i=1}^{n} (F_i(b_i) - F_i(a_i)) - (n-1) \Big)_+ \qquad (3.17)$$

$$\le P(X_i \in (a_i, b_i]; 1 \le i \le n) \le \min\{F_i(b_i) - F_i(a_i); \quad 1 \le i \le n\}.$$

b) **Maximal risk**

$$H^-(t) := \Big(\sum_{i=1}^{n} F_i(t) - (n-1) \Big)_+ \le P\Big(\max_{1 \le i \le n} X_i \le t \Big) \qquad (3.18)$$

$$\le H^+(t) := \min\{F_i(t); \quad 1 \le i \le n\},$$

and the bounds in a) and b) are sharp.

Let for $U \sim U(0,1)$, $X^c := (F_1^{-1}(U), \ldots, F_n^{-1}(U))$ be a **comonotonic risk vector** with marginals F_i, then the distribution function of $M_n(X^c)$, the maximum of the components of X^c, is given by

$$P(M_n(X^c) \le t) = P(X_1^c \le t, \ldots, X_n^c \le t)$$
$$= F_+(t, \ldots, t) = \min\{F_i(t); \quad 1 \le i \le n\} \qquad (3.19)$$
$$= H^+(t).$$

Thus for the comonotonic vector X^c, the distribution function for the maximum is the largest possible. Lai and Robbins (1976, 1978) have recursively constructed for continuous F_i a random vector $\widetilde{X}$ such that the distribution function of $M_n(\widetilde{X})$ is given by H_- (for general F_i, see Rüschendorf, 1980). This random vector is called a **maximally dependent** random vector. It can be constructed by iteratively determining $\widetilde{X}_j$ as countermonotonic to $M_{j-1}(\widetilde{X})$, the maximum of the first $j-1$ random variables $\widetilde{X}_j$. The comonotonic random vector X^c has the largest distribution functions of the maximum, so it has the smallest maximum in the stochastic order $\le_{\mathrm{st}}$. The maximally dependent random vector $\widetilde{X}$ has the smallest distribution function of the maximum, so it has the largest maximum in the stochastic order $\le_{\mathrm{st}}$.

Corollary 3.7 (Maximal risks) *For any random vector X with marginal distribution function F_i, i.e., $X_i \sim F_i$, it holds that*

$$M_n(X^c) \le_{\mathrm{st}} M_n(X) \le_{\mathrm{st}} M_n(\widetilde{X}). \qquad (3.20)$$

Remark 3.8 There is a simple proof of the upper bound in (3.20) by a typical duality argument without resort to the duality theorem. This is based on the particular form of dual admissible functions given by the inequality

$$M_n(X) \le \sum_{n=1}^{n} v_i + \sum_{i=1}^{n} (X_i - v_i)_+, \qquad (3.21)$$

choosing for $v \in \mathbb{R}^n$ the dual functionals $f_i(x_i) = v_i + (x_i - v_i)_+$. This inequality leads to

$$M_n(X) \le \inf_{v \in \mathbb{R}^n} \Big\{ \sum_{i=1}^{n} v_i + \sum_{i=1}^{n} (X_i - v_i)_+ \Big\}. \qquad (3.22)$$

For exactness of the upper bound in (3.22), we need pairwise disjointness of $\{X_i \geq v_i\}$ and, further, that $\Omega = \bigcup_{i=1}^{n}\{X_i \geq v_i\}$. These two conditions are fulfilled for the maximally dependent risk vector $\underset{\sim}{X}$.   ◊

For the tail risk $P\left(\sum_{i=1}^{n} X_i \geq s\right)$, we obtain simplified upper resp. lower bounds from the duality theorem in (3.9); the exact dual problems are not easy to solve in this case. Define

$$M_n^+(s) := \sup\left\{P\left(\sum_{i=1}^{n} X_i \geq s\right); \ X_i \sim F_i, 1 \leq i \leq n\right\}$$

$$\text{and } \ m_n^+(s) := \inf\left\{P\left(\sum_{i=1}^{n} X_i \geq s\right); \ X_i \sim F_i, 1 \leq i \leq n\right\}; \tag{3.23}$$

then the following relaxed dual bounds were established in Embrechts and Puccetti (2006b).

Theorem 3.9 (Dual bounds for the tail risk)   *Let $X_i \sim F_i$, and let $\overline{F}_i = 1 - F_i$ be the survival function of F_i, $1 \leq i \leq n$. Then for any $s \in \mathbb{R}$ we have*

$$M_n^+(s) \leq D(s) := \inf_{u \in \overline{\mathcal{U}}(s)} \ \min\left\{\frac{\sum_{i=1}^{n} \int_{u_i}^{s - \sum_{j \neq i} u_j} \overline{F}_i(t)\, dt}{s - \sum_{i=1}^{n} u_i}\right\} \tag{3.24}$$

$$\text{and } \ m_n^+(s) \geq d(s) := \sup_{u \in \underline{\mathcal{U}}(s)} \ \max\left\{\frac{\sum_{i=1}^{n} \int_{u_i}^{s - \sum_{j \neq i} u_j} \overline{F}_i(t)\, dt}{\sum_{i=1}^{n} u_i - s}\right\}, \tag{3.25}$$

where

$$\overline{\mathcal{U}}(s) := \left\{u \in \mathbb{R}^n; \ \sum_{i=1}^{n} u_i < s\right\} \ \text{and} \ \underline{\mathcal{U}}(s) := \left\{u \in \mathbb{R}^n; \ \sum_{i=1}^{n} u_i > s\right\}.$$

Proof The bound in (3.24) is obtained from the duality theorem in (3.9) by proving that the piecewise linear functions $g_i \colon \mathbb{R} \to \mathbb{R}$, $1 \leq i \leq n$, defined by

$$g_i(x_i) = \begin{cases} 0, & \text{if } x_i \leq u_i, \\ \dfrac{x_i - u_i}{s - \sum_{i=1}^{n} u_i}, & \text{if } u_i < x_i \leq s - \sum_{j \neq i} u_j, \\ 1, & \text{otherwise,} \end{cases}$$

are admissible for the dual problem. This can be seen by direct calculation and considering case-by-case analysis. Substituting these functions into the dual problem in (3.9) and taking the infimum over all $u \in \overline{\mathcal{U}}(s)$ yields the result.

The proof of (3.25) is similar. Here it is sufficient to prove that the functions f_i, $1 \leq i \leq n$, defined by

$$f_i(x_i) = \begin{cases} 1, & \text{if } x_i \leq s - \sum_{j \neq i} u_i, \\ \dfrac{u_i - x_i}{\sum_{i=1}^{n} u_i - s}, & \text{if } s - \sum_{j \neq i} u_j < x_i \leq u_i, \\ 0, & \text{otherwise,} \end{cases}$$

are an admissible choice in the duality theorem (3.9) for the lower bound. The result follows from taking the suprema over $u \in \underline{\mathcal{U}}(s)$.   □

In the above given proof if we choose a vector $u \in \mathcal{U}(s)$, that is, when $\sum_{i=1}^n u_i = s$, the piecewise linear dual admissible choices become piecewise constant. Consequently we obtain the standard bounds (see Section 1.2). Therefore, the dual bounds improve the corresponding standard bounds.

The dual bounds turn out in examples to be strong improvements over the standard bounds (see Section 3.3). However, calculation of the dual bounds requires one to solve an n-dimensional optimization problem, which typically will only be possible for small values of n, say $n \le 5$. The situation changes in the homogeneous case where $F_i = F$, $1 \le i \le n$. Then we obtain based on Theorem 3.2 a considerably simplified expression and a one-dimensional problem to solve, independent of the dimension n. For details see Embrechts and Puccetti (2006a).

Theorem 3.10 (Dual bounds for the tail risk; homogeneous case)
Let $F_1 = \cdots = F_n =: F$ be a distribution function on $\mathbb{R}_+$. Then for any $s \ge 0$ it holds that

$$M_n^+(s) \le D(s) = \min\left\{ \inf_{t \in [0, \frac{s}{n}]} \frac{n \int_t^{s-(n-1)t} \overline{F}(x)\, dx}{s - nt}, 1 \right\}. \tag{3.26}$$

Proof As in the proof of Theorem 3.9, we introduce for $t < \frac{s}{n}$ piecewise linear functions g_t by

$$g_t(x) := \begin{cases} 0, & \text{if } x < t, \\ \frac{x-t}{s-nt}, & \text{if } t \le x \le x - (n-1)t, \\ 1, & \text{otherwise.} \end{cases} \tag{3.27}$$

Then $g_t \in \mathcal{D}(s)$: $\{g : \mathbb{R} \to \mathbb{R} \text{ bounded}, \sum_{i=1}^n g(x_i) \ge 1_{[s,\infty)}(\sum x_i), x_i \in \mathbb{R}\}$, i.e., g_t is admissible for the duality theorem, Theorem 3.2. Minimizing over t yields the dual bounds in (3.26), which can also be derived from the dual bounds in (3.24).   □

The dual bounds in the homogeneous case in (3.26) are easy to calculate and, as illustrated in Figure 3.1, they essentially improve the standard bounds (see Embrechts and Puccetti, 2006b).

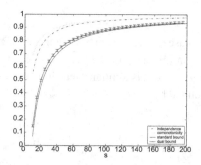

Figure 3.1 Dual and standard lower bounds for $P(\sum_{i=1}^n X_i < s)$ for $n = 3$, $X_i \sim \text{Pareto}(1.5, 1)$.

Remark 3.11 (value-at-risk bounds) The tail risk bounds for $M_n^+(t)$ resp. $m_n^+(t)$ imply by inversion value-at-risk (VaR) bounds for the joint portfolio. Define for a random variable Y with distribution function H the **value-at-risk** $\text{VaR}_\alpha(Y)$ at level α as the (lower) α-quantile of the distribution function, i.e.,

$$\text{VaR}_\alpha(Y) = H^{-1}(\alpha) = \inf\{t \in \mathbb{R}; \ H(t) \geq \alpha\}$$
$$= \inf\{t \in \mathbb{R}; \ \overline{H}(t) = 1 - H(t) \leq 1 - \alpha\} \qquad (3.28)$$
$$=: \overline{H}^{-1}(1 - \alpha).$$

Therefore, we obtain for the distribution function H of the portfolio sum $\sum_{i=1}^{n} X_i$ the bound $\overline{H}(t) \leq M_n^+(t)$, and thus as a result of (3.28),

$$\text{VaR}_\alpha \left(\sum_{i=1}^{n} X_i \right) \leq (M_n^+)^{-1}(1 - \alpha); \qquad (3.29)$$

the upper VaR_α bound is given by the inversion of M_n^+ at level $1 - \alpha$. ◊

3.3 Sharpness of the Dual Bounds for the Tail Risk

The dual bounds in (3.26) in the homogeneous case have been proved in Puccetti and Rüschendorf (2013) to be sharp under some additional assumptions. The sharpness result is based on the simplified representation of the duality theorem in the homogeneous case in Theorem 3.2, which states for the portfolio tail risk case with $F_1 = \cdots = F_n = F$ that

$$M(s) := M_n^+(s) = \inf \left\{ n \int g \, dF; \ g \in \mathcal{D}(s) \right\}, \qquad (3.30)$$

where

$$\mathcal{D}(s) = \left\{ g : \mathbb{R} \to \mathbb{R}; \ g \text{ bounded}, \sum_{i=1}^{n} g(x_i) \geq 1_{[s,\infty)} \left(\sum_{i=1}^{n} x_i \right) \text{ for } x_i \in \mathbb{R} \right\}.$$

An optimal dual solution $g \in \mathcal{D}(s)$ exists.

An extension of this duality result characterizes optimal solutions.

Proposition 3.12

a) *A random vector X^* with distribution function $F_{X^*} \in \mathcal{F}(F, \ldots, F)$, i.e., with marginal distribution functions F, is a solution of*

$$M_n^+(s) = P\left(\sum_{i=1}^{n} X_i^* \geq s \right) \qquad (3.31)$$

if and only if there exists an admissible function $g^* \in \mathcal{D}(s)$ such that

$$P\left(\sum_{i=1}^{n} g^*(X_i^*) = 1_{[s,\infty)} \left(\sum_{i=1}^{n} X_i^* \right) \right) = 1. \qquad (3.32)$$

b) *A solution X^* of (3.31) exists.*

So a worst case solution X^* of (3.31), a coupling with maximal tail risk, exists, and it is characterized by equation (3.32). We call such couplings "**optimal couplings**."

The proof of sharpness of the dual bounds in (3.26) uses the concept of n-mixability as in Chapter 1.

A distribution (function) F on $\mathbb{R}$ is called **n-mixable** if there exist n random variables $X_i \sim F$, $1 \le i \le n$, such that

$$\sum_{i=1}^{n} X_i = n\mu \ [P] \tag{3.33}$$

for some $\mu \in \mathbb{R}$. Any such μ is called a center of F.

The structure of optimal couplings (X_i^*) is described in the following proposition (see Proposition 3 of Rüschendorf, 1982).

Proposition 3.13 *For any optimal coupling X^* with $F_{X^*} \in \mathcal{F}(F, \ldots, F)$ of (3.31), it holds for any $i \le n$ that*

$$\{X_i^* > F^{-1}(1 - M(s))\} \subset \Big\{ \sum_{j=1}^{n} X_j^* \ge s \Big\} \subset \{X_i^* \ge F^{-1}(1 - M(s))\} \ a.s. \tag{3.34}$$

If F is continuous, then for any $i \le n$,

$$\Big\{ \sum_{j=1}^{n} X_j^* \ge s \Big\} = \{X_i \ge F^{-1}(1 - M(s))\} \ a.s.$$

Formula (3.34) implies for an optimal dual solution $g^* \in \mathcal{D}(s)$ that $P(g^*(X_i^*) = 0 \mid X_i^* < a^*) = 1$, $1 \le i \le n$, where $a^* = F^{-1}(1 - M(s))$. This implies that it is sufficient to determine the behavior of an optimal function g^* above the threshold a^*. This behavior is described in the following proposition.

Proposition 3.14 *Let $a^* = F^{-1}(1 - M(s))$. A random vector X^* with distribution function $F_{X^*} \in \mathcal{F}(F, \ldots, F)$ is a solution of $M(s) = P(\sum_{i=n}^{n} X_i^* \ge s)$ if and only if there exists a function $g^* \in \mathcal{D}(s)$ such that the conditional distribution of $g^*(X_1) \mid X_1 \ge a^*$ is n-mixable with center $\mu = \frac{1}{n}$.*

By Proposition 3.14, the construction of optimal couplings (X_i^*) is connected to a mixing property. This leads to the following assumptions.

(A1) There exists $a < \frac{s}{n}$ such that

$$D(s) = \inf_{t < \frac{s}{n}} \frac{n \int_{t}^{s - (n-1)t} \overline{F}(x)\, dx}{s - nt} = \frac{n \int_{a}^{b} \overline{F}(x)\, dx}{b - a}, \tag{3.35}$$

where $b = s - (d - 1)a$.

(A2) For $a^* = F^{-1}(1 - D(s))$, suppose that F is n-mixable on (a, b).

(A3) $(n - 1)(F(y) - F(b)) \le F(a) - F\left(\dfrac{s - y}{n - 1}\right)$ for all $y \ge b$.

Then we get as a consequence of (A1) and (A2) that the conditional distribution H of $g_a(X_1) \mid X_1 \geq a^*$ is n-mixable with center $\mu = \frac{1}{n}$. Since $g_a \in \mathcal{D}(s)$ (see (3.27)), we obtain

$$M(s) \leq n \int g_a \, dF = \frac{n \int_a^b \overline{F}(x) \, dx}{b - a} = \overline{F}(a) + (d - 1)\overline{F}(b) \qquad (3.36)$$

and, in particular, $a^* \leq a$.

Under the ordering condition in (A3), we can construct a random vector $Y \in \mathcal{F}(F_{a^*}, \ldots, F_{a^*})$, where $F_{a^*} \sim (X_1 \mid X_1 \geq a^*)$, such that:

a) When one of the Y_i lies in the interval (a, b), then all Y_j are in (a, b) and

$$P\left(\sum_{j=1}^{n} Y_j = s \mid Y_i \in (a, b) \right) = 1. \qquad (3.37)$$

b) For $1 \leq i \leq n$ it holds that

$$P(Y_j = F_{a^*}^{-1}((n - 1)\overline{F}_{a^*}(Y_i)) \mid Y_i \geq b) = 1, \quad \text{for all } j \neq i.$$

If Y_i has a value $\geq b$, then it is antimonotonically coupled to any Y_j, $j \neq i$, and all Y_j, $j \neq i$ have values in $[a^*, a]$. The ordering condition (A3) implies that in this case $\sum_{i=1}^{n} Y_i^* \geq s$, too.

Consequently we obtain that

$$M(s) = \overline{F}(a^*) = D(s) = \overline{F}(a) + (n - 1)\overline{F}(b), \qquad (3.38)$$

and we get the following result.

Theorem 3.15 (Sharpness of the dual bounds)   *Under the attainment condition (A1), the mixing condition (A2), and the ordering condition (A3), the dual bounds are sharp, i.e.,*

$$M(s) = D(s) = \frac{n \int_a^b \overline{F}(x) \, dx}{b - a}.$$

Remark 3.16

a) **Monotone densities.** If F is continuous and has a positive decreasing density on (a^*, ∞), then conditions (A2) and (A3) are satisfied. The corresponding mixing result is due to Wang and Wang (2011). If condition (A1) also holds, then we obtain optimality of the dual bounds and thus the optimality result of Wang and Wang (2011). The couplings used in the above sketched proof are similar to those in Wang et al. (2013).

More generally, sharpness of the dual bound $D(s)$ can be stated for large enough s for all continuous unbounded distribution functions that have an ultimately decreasing density. In particular, assumptions (A1)–(A3) hold for the Pareto distribution $F(x) = 1 - (1 + x)^{-\vartheta}$, $x > 0$ with tail parameter $\vartheta > 0$ for all $s \in \mathbb{R}$ with $D(s) < 1$. Therefore, the bounds in Section 5.2 of Embrechts and Puccetti (2006b) are sharp. See also Figure 3.2.

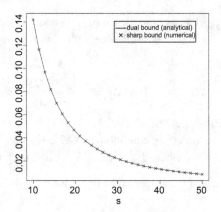

Figure 3.2 Dual bounds $D(s)$ for the sum of three Pareto(2)-distributed risks.

b) Concave densities. If F has a concave density f on (a, b) then it satisfies the mixing assumption (A2) (see Theorem 4.3 in Puccetti et al., 2012). Conditions (A1) and (A3) have to be checked analytically or numerically, see the following example with $n = 3$ and risks distributed as LN(2, 1) resp. Gamma(3, 1) in Figure 3.3.   ◊

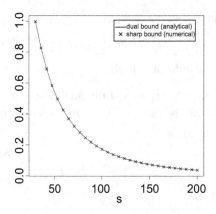

Figure 3.3 Dual bounds $D(s)$ for the sum of three lognormal(2,1)-distributed risks.

For further examples of sharp dual bounds, see Embrechts and Puccetti (2006a,b).

4   Asymptotic Equivalence Results

The main result in this chapter states the asymptotic equivalence of the sharp upper VaR bounds with the sharp upper TVaR bounds. This result is of interest since the upper TVaR bounds are easy to calculate. In examples it is found that the asymptotic equivalence already holds true for relatively small numbers of risks, n, depending however on the tail behavior. For heavy tails the approximation gets worse, and in the infinite mean limit, the equivalence no longer remains valid. As a consequence of the equivalence results, the limiting behavior of the worst case superadditivity and the worst case diversification are also obtained.

4.1   Introduction and Motivation

The basic upper VaR bound in Section 1.1 relates the worst case $\overline{\mathrm{VaR}}_\alpha(S_n)$, $S_n = \sum_{i=1}^n X_i$ to the worst case TVaR bound:

$$\overline{\mathrm{VaR}}_\alpha(S_n) \le \overline{\mathrm{TVaR}}_\alpha(S_n) = \sum_{i=1}^n \mathrm{TVaR}_\alpha(X_i) := B, \tag{4.1}$$

where we use the notation $\overline{\mathrm{VaR}}_\alpha(S_n)$ and similarly $\overline{\mathrm{TVaR}}_\alpha(S_n)$ to make clear the dependence of the upper bound $\overline{\mathrm{VaR}}_\alpha$ resp. $\overline{\mathrm{TVaR}}_\alpha$ on the portfolio size n.

The TVaR bound B is easy to calculate, and as demonstrated by several examples it is a reasonably good bound for $\overline{\mathrm{VaR}}_\alpha$. As seen in Section 1.1, the worst case dependence structure for the TVaR is given by the comonotonic dependence, i.e.,

$$\overline{\mathrm{TVaR}}_\alpha(S_n) = \mathrm{TVaR}_\alpha(S_n^c), \tag{4.2}$$

where $S^c = \sum_{i=1}^n X_i^c$ is the comonotonic sum. By contrast, for the VaR the comonotonic sum is seen in many examples not to be the worst case dependence, i.e., for certain dependence structures, it holds that

$$\mathrm{VaR}_\alpha(S_n) > \mathrm{VaR}_\alpha(S_n^c) = \sum_{i=1}^n \mathrm{VaR}_\alpha(X_i). \tag{4.3}$$

This superadditivity of the VaR was considered in the beginning of a more detailed study as counterintuitive and also contradicting common industrial practice, namely determining the "worst case risk capital" as

$$(1 - \beta) \operatorname{VaR}_\alpha(S_n^c) = (1 - \beta) \sum_{i=1}^n \operatorname{VaR}_\alpha(X_i) \tag{4.4}$$

with diversification factor $1 - \beta \approx 0.7$. Motivated by this idea of diversification, the so-called **diversification benefit** for an aggregate loss $S_n = \sum_{i=1}^n X_i$ was introduced in Cope et al. (2009) as

$$D_\alpha := 1 - \frac{\operatorname{VaR}_\alpha(S_n)}{\operatorname{VaR}_\alpha(S_n^c)} = D_\alpha(S_n), \tag{4.5}$$

and equivalently,

$$\operatorname{VaR}_\alpha(S_n) = 1 - D_\alpha(S_n) \operatorname{VaR}_\alpha(S_n^c). \tag{4.6}$$

The corresponding **worst case diversification benefit** $\overline{D}_\alpha$ is given by

$$\overline{D}_\alpha := 1 - \frac{\overline{\operatorname{VaR}}_\alpha(S_n)}{\operatorname{VaR}_\alpha(S_n^c)}. \tag{4.7}$$

Closely related is the **worst case superadditivity ratio**

$$\delta_\alpha = \frac{\overline{\operatorname{VaR}}_\alpha(S_n)}{\operatorname{VaR}_\alpha(S_n^c)} = \frac{\overline{\operatorname{VaR}}_\alpha(S_n)}{\sum_{i=1}^n \operatorname{VaR}_\alpha(X_i)} = \delta_\alpha(n). \tag{4.8}$$

The expectation is that "typically" $0 \le D_\alpha \le 1$ and $1 \le \delta_\alpha(n)_0$. This is the typical situation occurring in elliptical distributions (see Theorem 6.8 in McNeil et al., 2005), where $D_\alpha \in [0, 1]$. Further, by definition $D_\alpha = 0$ if $S_n = S_n^c$ is the strongest positive dependence structure, the comonotonic dependence structure, which is generally considered as the worst case dependence structure. However, this view soon turned out not to be correct, and several examples were found with $D_\alpha(S_n) < 0$, i.e., diversification may have a negative impact on the risk of a joint portfolio $X = (X_1, \ldots, X_n)$. One can see this effect in an analytical way in the following simple example (see Mainik and Rüschendorf, 2010). Let $X_1, \ldots, X_n$ be independent such that $X_i \sim S_\vartheta$, $1 \le i \le n$, i.e., all X_i are symmetric ϑ-stable distributed with $\vartheta < 1$. Then $S_n \sim n^{1/\vartheta} X_1$ and thus

$$\operatorname{VaR}_\alpha(S_n) = n^{1/\vartheta} \operatorname{VaR}_\alpha(X_1) > n \operatorname{VaR}_\alpha(X_1)$$
$$= \operatorname{VaR}_\alpha(S_n^c) = \sum_{i=1}^n \operatorname{VaR}_\alpha(X_i). \tag{4.9}$$

The total risk for a financial institution may in general exceed the sum of the risks across the individual marginals. Negative diversification benefits are obtained in the case of heavy-tailed and/or skew marginals and/or marginals coupled by a non-elliptical copula (see Nešlehová et al., 2006). This property is not a defect of the VaR; by contrast, as (4.9) shows, subadditive risk measures do not correctly describe the diversification property in its full range.

The main aim of this chapter is to establish that under general conditions in the finite mean case, there is an asymptotic equivalence result between the worst case VaR and the TVaR, i.e.,

$$\lim_{n \to \infty} \frac{\overline{\text{TVaR}}_\alpha(S_n)}{\overline{\text{VaR}}_\alpha(S_n)} = 1. \tag{4.10}$$

This surprising property was found first in a paper by Puccetti and Rüschendorf (2013) in the homogeneous case; examples indicating this phenomenon in the inhomogeneous case were also given in that paper. It was then established in a series of papers that this kind of equivalence holds true quite generally.

4.2 Asymptotics for Conservative Capital Charges: The Homogeneous Finite Mean Case

In this section we assume that $X_i \geq 0$ are a non-negative, homogeneous portfolio, $X_i \sim F$, $1 \leq i \leq n$, with finite mean μ. Our aim is to establish that

$$\frac{\overline{\text{TVaR}}_\alpha(S_n)}{\overline{\text{VaR}}_\alpha(S_n)} \xrightarrow[n \to \infty]{} 1. \tag{4.11}$$

The asymptotic equivalence result is based on the dual bound theorem (Theorem 3.10 in Chapter 3). More precisely, let $\mathcal{F}_\alpha$ denote the set of unbounded continuous distribution functions on $\mathbb{R}_+$ having a positive density f that is decreasing on $(F^{-1}(\alpha), \infty)$. Distributions in $\mathcal{F}_\alpha$ cover a large domain in quantitative risk management, where VaR_α or TVaR_α are typically calculated for high quantiles α. Theorem 3.10 states in this case:

Theorem 4.1 *If $F \in \mathcal{F}_\alpha$ and if there exists $t^* < \frac{s}{n}$ such that*

$$D(s) = \inf_{0 \leq t < \frac{s}{n}} \frac{n \int_t^{s-(n-1)t} \overline{F}(x)\,dx}{s - nt} = \frac{n \int_{t^*}^{s-(n-1)t^*} \overline{F}(x)\,dx}{s - nt^*} \tag{4.12}$$

and $\quad t^* \leq F^{-1}(1 - D(s))$,

then

$$M_n^+(s) = \sup \left\{ P\left(\sum_{i=1}^n X_i \geq s \right); \ X_i \sim F, 1 \leq i \leq n \right\} = D(s). \tag{4.13}$$

The asymptotic equivalence result is based on the dual bounds in (4.12). Define for $X \sim F$ the function $h: [\mu, \infty) \to [0, \infty)$ as the unique solution of the equation

$$E(X \mid X \geq h(s)) = h(s) + \frac{\int_{h(s)}^\infty \overline{F}(x)\,dx}{\overline{F}(h(s))} = s. \tag{4.14}$$

Note that in the case $h(s) = F^{-1}(\alpha) = \text{VaR}_\alpha(X)$,

$$E(X \mid X \geq \text{VaR}_\alpha(X)) = \text{TVaR}_\alpha(X) \tag{4.15}$$

and thus

$$h^{-1}(F^{-1}(\alpha)) = s = E(X \mid X \geq h(s)) = \text{TVaR}_\alpha(X). \tag{4.16}$$

For the asymptotic equivalence result, we need a bound for the maximum tail probability of S_n at the level ns.

Lemma 4.2 *Let (X_i) be a sequence of random variables $X_i \sim F \in \mathcal{F}_\alpha$ with mean μ. For any $s \geq \mu$, it holds that $\lim_{n \to \infty} M_n^+(ns) \leq \overline{F}(h(s))$.*

Proof For any $s \geq \mu$, it holds that $0 \leq h(s) \leq s$. From the dual bound $M_n^+(s) \leq D_n(s)$, we obtain

$$M_n^*(ns) \leq \inf_{0 \leq t < s} \frac{\int_t^{n(s-t)+t} \overline{F}(s)\,dx}{s-t} \leq \frac{\int_{h(s)}^{n(s-h(s))+h(s)} \overline{F}(x)\,dx}{s-h(s)}.$$

From the definition of $h(s)$ in (4.14), this implies

$$\overline{\lim_{n \to \infty}} M_n^+(ns) \leq \frac{\int_{h(s)}^\infty \overline{F}(x)\,dx}{s-h(s)} = \overline{F}(h(s)). \qquad \square$$

Theorem 4.3 (**Asymptotic equivalence result in the homogeneous, finite mean case**) *For $\alpha \in (0, 1)$, let (X_i) be a sequence of random variables with finite mean μ and $X_i \sim F \in \mathcal{F}_\alpha$. Suppose that for a fixed threshold $s \geq \mu$, the assumption (4.16) in Theorem 4.1 is satisfied for all $n \geq n_0$. Then*

$$\lim_{n \to \infty} \frac{\overline{\text{VaR}}_\alpha(S_n)}{n} = \text{TVaR}_\alpha(X_1) \tag{4.17}$$

and

$$\lim_{n \to \infty} \frac{\overline{\text{VaR}}_\alpha(S_n)}{\text{TVaR}_\alpha(S_n)} = 1. \tag{4.18}$$

Proof For $n \geq n_0$, let $t_n = \arg\inf_{0 \leq t < s} f(t, n)$, where $f(t, n) := \frac{\int_t^{n(s-t)+t} \overline{F}(x)\,dx}{s-t}$. From the dual bound theorem, Theorem 4.1, we get

$$M_n^+(ns) = \inf_{0 \leq t \leq s} f(t, n) = f(t_n, n). \tag{4.19}$$

From the boundedness of $M_n(ns)$ and (4.19), we obtain by discussing some subcases: There exists $\varepsilon > 0$ such that $\lim_{n \to \infty} t_n = s - \varepsilon$. This implies that for $n \leq n_0$,

$$M_n^*(ns) = \inf_{0 \leq t \leq s-\varepsilon} f(t, n).$$

Then from the finiteness of $u = \int_0^\infty \overline{F}(x)\,dx$, we conclude that

$$\lim_{n \to \infty} \sup_{0 \leq t \leq s-\varepsilon} |f(t, n) - f(t)| = 0,$$

where $f(t) = \frac{\int_t^\infty \overline{F}(x)\,dx}{s-t}$, i.e., $f(t, n)$ converges uniformly to $f(t)$, and as a result we obtain

$$\lim_{n \to \infty} M_n^*(ns) = \inf_{0 \leq t \leq s-\varepsilon} f(t).$$

n	$\overline{\text{VaR}}_{0.995}$	$\overline{\text{TVaR}}_{0.995}$	ratio
3	66.3830	81.8528	1.2349
10	258.4381	272.8420	1.0562
50	1250.0000	1365.2140	1.0105
100	2714.2490	2727.4270	1.0052

n	$\overline{\text{VaR}}_{0.999}$	$\overline{\text{TVaR}}_{0.999}$	ratio
3	151.9194	186.7367	1.2292
10	589.9999	633.4555	1.0550
50	3080.4950	3112.2780	1.0103
100	6192.8530	6224.5550	1.0051

Table 4.1 Ratio between the worst TVaR and the worst VaR for the sum of n Pareto(2) marginals.

n	$\overline{\text{VaR}}_{0.995}$	$\overline{\text{TVaR}}_{0.995}$	ratio
3	399.0155	420.5341	1.0539
10	1398.8790	1401.7800	1.0021
50	7008.8790	7008.9020	1.0000
100	14017.8000	14017.8000	1.0000

n	$\overline{\text{VaR}}_{0.999}$	$\overline{\text{TVaR}}_{0.999}$	ratio
3	640.0679	668.7629	1.0448
10	2225.8490	2229.2100	1.0015
50	11146.0300	11146.0500	1.0000
100	22292.1000	22292.1000	1.0000

Table 4.2 Ratio between the worst TVaR and the worst VaR for the sum of n LN(2, 1) marginals.

Recalling the definition of $h(s)$ in (4.14) and checking the first- and second-order conditions for $f(t)$, we obtain

$$\lim_{n \to \infty} M_n^+(ns) = \frac{\int_{h(s)}^{\infty} \overline{F}(x)\,dx}{s - h(s)} = \overline{F}(h(s)),$$

which implies

$$\lim_{n \to \infty} M_n^+(nh^{-1}(F^{-1}(\alpha))) = 1 - \alpha. \tag{4.20}$$

From (4.20), we obtain using (4.16) that

$$\lim_{n \to \infty} \frac{\overline{\text{VaR}}_\alpha(S_n)}{n} = h^{-1}(F^{-1}(\alpha)) = \text{TVaR}_\alpha(X_1).$$

Equation (4.18) is then a consequence of (4.17) and (4.1). □

In examples we find that the asymptotic equivalence result holds approximatively true already for small examples; see Tables 4.1 and Table 4.2 for the case of a Pareto(2) distribution resp. of an LN(2,1) distribution Puccetti and Rüschendorf, 2013.

The asymptotic equivalence result in Theorem 4.3 implies asymptotics for the worst case diversification $\overline{D}_\alpha$ of the aggregate risk $S = S_n$ and also for the worst case superadditivity ratio δ_α in (4.8).

In the homogeneous case, the worst case diversification benefit $\overline{D}_\alpha = \overline{D}_\alpha(n)$ and the superadditivity ratio $\delta_\alpha = \delta_\alpha(n)$ depend on n. The asymptotic equivalence theorem implies that these indices have limits as $n \to \infty$ which are easy to describe.

Theorem 4.4 (**Asymptotics of worst case diversification limit and superadditivity ratio**)

For $\alpha \in (0, 1)$, let (X_i) be a sequence of random variables with finite mean and $X_i \sim F$, for all $i \in \mathbb{N}$ satisfying the assumptions of Theorem 4.3. Then it holds that:

a) $$\lim_{n\to\infty} \overline{D}_\alpha(n) = 1 - \frac{\text{TVaR}_\alpha(X_1)}{\text{VaR}_\alpha(X_1)} =: 1 - d_\alpha; \qquad (4.21)$$

b) $$\lim_{n\to\infty} \delta_\alpha(n) = \frac{\text{TVaR}_\alpha(X_1)}{\text{VaR}_\alpha(X_1)} = d_\alpha. \qquad (4.22)$$

Proof a) From (4.17) it follows that

$$\frac{\overline{\text{VaR}}_\alpha(S_n)}{n\,\text{VaR}_\alpha(X_1)} = \frac{\overline{\text{VaR}}_\alpha(S_n)}{\text{VaR}_\alpha(S_n^c)} \to \frac{\text{TVaR}_\alpha(X_1)}{\text{VaR}_\alpha(X_1)} = d_\alpha.$$

b) follows from a) and the definition of $\delta_\alpha(n)$ in (4.8). □

Recalling (4.6), the VaR of a homogeneous risk portfolio could be in the worst case d_α times larger than the VaR in the comonotonic portfolio. Table 4.3 gives values of the factor d_α in the range $d_\alpha \in [1.4, 11.2]$ for Pareto distributions with tail exponent ϑ varying between 4 and 1.1.

α	$\vartheta = 1.1$	$\vartheta = 1.5$	$\vartheta = 2$	$\vartheta = 3$	$\vartheta = 4$
0.990	11.154337	3.097350	2.111111	1.637303	1.487492
0.995	11.081599	3.060242	2.076091	1.603135	1.343080
0.999	11.018773	3.020202	2.032655	1.555556	1.405266

Table 4.3 Values for the constant d_α for Pareto(ϑ) distributions.

From Table 4.3 it is also evident that d_α settles down to a limit when $\alpha \to 1-$. Indeed in the case that X_1 has a Pareto distribution with tail exponent $\vartheta > 1$, it holds (see formula (5.2) of Mainik and Rüschendorf, 2010) that

$$\lim_{\alpha\to 1-} d_\alpha = \lim_{\alpha\to 1-} \frac{\text{TVaR}_\alpha(X_1)}{\text{VaR}_\alpha(X_1)} = \frac{\vartheta}{\vartheta - 1}. \qquad (4.23)$$

In Figure 4.1, $\overline{D}_\alpha(n)$ is plotted as function of n for a Pareto distribution having different tail exponents. For a Pareto distribution, a smaller exponent ϑ corresponds to a heavier tail. For $\vartheta \le 1$, the Pareto distribution has an infinite first moment. Figure 4.1 clearly shows that $\overline{D}_\alpha(n) \to 1 - d_\alpha$ fairly fast for all Pareto distributions under study. However, the heavier the tail of the Pareto distribution, the slower the convergence.

In Figure 4.2 the behavior of $\overline{D}_\alpha(n)$ is compared for Pareto, Gamma, and log normal distributions. Again the heaviness of the tail appears to play a crucial role for the speed of convergence of $\overline{D}_\alpha(n)$.

4.3   Asymptotics for VaR in the Infinite Mean Case

In the case of an infinite mean, we will see in this section that the quotient $\overline{\text{VaR}}_\alpha(S_n)/\text{VaR}_\alpha(S_n^c)$ converges to infinity, i.e., the worst case VaR capital can be arbitrarily large compared to $\text{VaR}_\alpha(S_n^c) = \sum_{i=1}^n \text{VaR}_\alpha(X_i)$, the VaR for comonotonic risks. A similar difference between the finite mean and the infinite mean case is known,

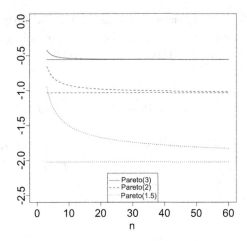

Figure 4.1 Dual bounds $D(s)$, for the sum of three Pareto(2)-distributed risks.

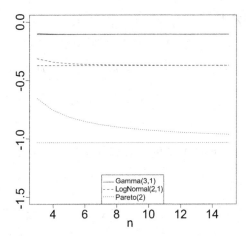

Figure 4.2 Dual bounds $D(s)$, for the sum of three LN(2, 1)-distributed risks.

when considering the limit $\alpha \to 1$ of this quotient in the framework of regular variation (extreme value theory) (see Nešlehová et al., 2006; Mainik and Rüschendorf, 2010).

Theorem 4.5   *Let $\alpha \in (0, 1)$, and let (X_i) be a finite sequence of random variables with $X_i \sim F_i$ such that*

$$\liminf_{x \to \infty} \inf_{i \geq 1} x \overline{F}_i(x) = \infty, \tag{4.24}$$

and for some constant c, $F_i^{-1}(\alpha) \leq c$, for all $i \geq 1$.
    Then

$$\lim_{n \to \infty} \frac{\overline{\mathrm{VaR}}_\alpha(S_n)}{\mathrm{VaR}_\alpha(S_n^c)} = \infty.$$

Proof We give the proof in the case of continuous distribution functions. By Corollary 3.7 in Chapter 3 (see (4.19) and (4.18)) on maximally dependent random variables, we obtain for the maximal risk $M_n(X) = \max\{X_1, \ldots, X_n\}$,

$$\overline{M}_n(x) := \sup\{P(M_n(X) \geq x; \quad X - i \sim F_i, 1 \leq i \leq n\}$$
$$= P(M_n(\widetilde{X}) \geq x) = \min\{1, \overline{G}_n(x)\},\tag{4.25}$$

where $\widetilde{X}$ is the maximally dependent risk vector and $\overline{G}_n(x) := \sum_{i=1}^n \overline{F}_i(x)$.

With $\overline{G}_n^{-1}(\beta) = \inf\{x \in \mathbb{R}^1; \ \overline{G}_n(x) \leq \beta\}$, $\beta \in (0, 1)$, we define $y_n := \overline{G}_n^{-1}(1 - \alpha)$ and obtain $\overline{M}_n(y_n) \geq \min\{1, \overline{G}_n(y_n)\} = 1 - \alpha$.

This implies $\overline{\mathrm{VaR}}_\alpha(S_n) \geq y_n$, and using the assumption $F_i^{-1}(\alpha) \leq c$, for all i, we obtain

$$\frac{\overline{\mathrm{VaR}}_\alpha(S_n)}{\sum_{i=1}^n \mathrm{VaR}_\alpha(X_i)} \geq \frac{y_n}{\sum_{i=1}^n \mathrm{VaR}_\alpha(X_i)} \geq \frac{y_n}{nc}.\tag{4.26}$$

As $y_n = \overline{G}_n^{-1}(1 - \alpha)$, this inequality leads to

$$1 - \alpha = \overline{G}_n(y_n) = \sum_{i=1}^n \overline{F}_i(y_n) = \frac{\sum_{i=1}^n y_n \overline{F}_i(y_n)}{y_n}$$
$$\geq \frac{n \min_{i \leq n} y_n \overline{F}_i(y_n)}{y_n} \quad \text{for all } n.\tag{4.27}$$

Since by assumption $\min_{i \leq n} y_n \overline{F}_i(y_n) \to \infty$, (4.27) implies that $\frac{n}{y_n} \to 0$, and thus from (4.26) we conclude that

$$\lim_{n\to\infty} \frac{\overline{\mathrm{VaR}}_\alpha(S_n)}{\sum_{i=1}^n \mathrm{VaR}(X_i)} \geq \lim_{n\to\infty} \frac{y_n}{nc} = \infty. \qquad \square$$

Remark 4.6 Condition (4.24) in Theorem 4.5 is a bit stronger than the assumption of an infinite mean of all marginal distributions. In the Pareto case where $\overline{F}_i(x) = \frac{1}{1+x}$, $i \geq 1$, the mean is infinite but $\lim_{x\to\infty} x\overline{F}_i(x) = 1$. In the Pareto case $\overline{F}_{i,\vartheta}(x) = \frac{1}{(1+x)^\vartheta}$ with $0 < \vartheta < 1$, condition (4.24) is satisfied; more generally it is satisfied for $F_i = F_{i,\vartheta_i}$ where $0 < \vartheta_i < 1$ and $\sup_i \vartheta_i < 1$. This holds more generally for regular varying distributions with tail index $0 < \vartheta_i < 1$.

Condition (4.25) can be relaxed to $\lim_{x\to\infty} \inf_{i \in J} x\overline{F}_i(x) = \infty$, where for some fixed $\varepsilon > 0$, $J \subset \mathbb{N}$ satisfies

$$\#(J \cap \{1, \ldots, n\}) \geq \varepsilon n, n \geq 1,$$

i.e., a positive part of the distributions satisfies the tail condition.   ◊

Table 4.4 demonstrates the result of Theorem 4.5 for the example of Pareto marginals with $\vartheta_{3k+1} = 0.9$, $\vartheta_{3k+2} = 2$, $\vartheta_{3k+3} = 3$, $k \geq 0$.

n	$\overline{\text{VaR}}_{0.995}(S_n)$	$\text{VaR}_{0.995}(S_n^c)$	ratio		n	$\overline{\text{VaR}}_{0.999}(S_n)$	$\text{VaR}_{0.999}(S_n^c)$	ratio
3	377.32	463.73	1.2290		3	2,193.06	2,458.23	1.1209
9	1,131.96	3,502.01	3.0938		9	6,579.17	20,323.17	3.0890
30	3,773.20	20,788.17	5.5094		30	21,930.57	122,718.50	5.5958
99	12,451.55	99,600.49	7.9990		99	72,370.90	590,636.20	8.1612

Table 4.4 Ratio between the worst VaR and the sum of marginal VaRs for the sum $S_n := \sum_{i=1}^{n} X_i$ of Pareto type marginals with tail coefficients $\vartheta_{3k+1} = 0.9$, $\vartheta_{3k+2} = 2$, $\vartheta_{3k+3} = 3$, $k = 0, \ldots, \frac{n}{3} - 1$.

4.4   Asymptotic Equivalence in the Inhomogeneous Case

After the discovery of the surprising asymptotic equivalence of worst case VaR bounds with the simple-to-calculate TVaR bounds in the homogeneous case in (4.10), it was soon found in several papers that this phenomenon holds quite generally and is connected with some form of asymptotic mixability, which can be seen as a strengthened version of the law of large numbers.

For the discussion of sharpness, it is useful to use the rearrangement representation of the risks X_i, i.e., $X_i = f_i(U)$, $1 \le i \le n$ for some $U \sim U(0, 1)$, where f_i are **rearrangements** of F_i^{-1}, $f_i \sim_r F_i^{-1}$, i.e.,

$$\lambda(\{u \in (0, 1); \; f_i(u) \le x\}) = \lambda(\{u \in (0, 1); \; F_i^{-1}(u) \le x\}) = F_i(x) \qquad (4.28)$$

for all $x \in \mathbb{R}$. $f_i \sim_r F_i^{-1}$ amounts to saying that for any $V \sim U(0, 1)$, $f_i(V) \sim F_i^{-1}(V)$ have the same distribution. By the rearrangement method, in order to determine the worst case VaR, we can restrict to those risks $X_i = f_i(U)$ such that

$$\{S_n \ge \text{VaR}_\alpha(S_n)\} = \{U \in [\alpha, 1]\} \quad \text{a.s.,} \qquad (4.29)$$

i.e., the upper $(1 - \alpha)$-part of the distribution of S_n is given by the upper part $\{U \ge \alpha\}$ of the random variable U. The following characterization of sharpness was noted in Puccetti and Rüschendorf (2012b) (who used the terminology "conditional moment bounds" for TVaR bounds) and in Bernard et al. (2017c).

Theorem 4.7 (Sharpness of TVaR bounds)
Let $X_i = f_i(U) \sim F_i$, $1 \le i \le n$, and let $S_n = \sum_{i=1}^{n} X_i$ be the portfolio sum. Then

a) *The TVaR upper bound $B := \text{TVaR}_\alpha(S_n^c) = \sum_{i=1}^{n} \text{TVaR}_\alpha(X_i)$ (see Theorem 1.6) is sharp, i.e., $\overline{\text{VaR}}_\alpha(S_n) = B$ if and only if*
   1) *the f_i are rearrangements of F_i^{-1} on $[\alpha, 1]$, $1 \le i \le n$, and*
   2) *$X_1, \ldots, X_n$ are mixing on $\{U \ge \alpha\}$, i.e., $\sum_{i=1}^{n} f_i(u) = c$ a.s. on $[\alpha, 1]$ for some constant c.*
b) *The lower TVaR bound $A := \text{LTVaR}(S_n^c) = \sum_{i=1}^{n} \text{LTVaR}_\alpha(X_i)$ is sharp, i.e., $\underline{\text{VaR}}_\alpha(S_n) = A$ if and only if*
   1) *the f_i are rearrangements of F_i^{-1} on $[0, \alpha]$ and*
   2) *$X_i, \ldots, X_n$ are mixing on $\{U < \alpha\}$, i.e., $\sum_{i=1}^{n} f_i(u) = c'$ a.s. on $[0, \alpha)$ for some constant $c' \in \mathbb{R}^1$.*

The mixing properties in Theorem 4.7 may be too strong to achieve, but a weakened form of asymptotic mixability, i.e., for $n \to \infty$, may hold true in great generality. The following result of Bernard et al. (2017c) says that one can expect approximative sharpness of the TVaR upper bound B in the case of large portfolios; a similar result also holds for the lower bound A.

Theorem 4.8 (Asymptotic sharpness of upper TVaR bounds)
Let $X_i^n = f_i^n(U) \sim F_i$, $1 \le i \le n$, such that for some sequences $\alpha_n \downarrow \alpha$, $\beta_n \uparrow 1$, the asymptotic mixability condition (AM) holds

$$\text{(AM)} \qquad \sum_{i=1}^{n} X_i^n = c_n + R_n \text{ on } [\alpha_n, \beta_n], \qquad (4.30)$$

where c_n are constants and $R_n = r_n(U)$ satisfy $Er_n(U_{[\alpha,1]}) \to 0$, $u \to 0$, i.e., $S_n^\alpha = \sum_{i=1}^{n} X_i^n$ is **asymptotically mixing** on $[\alpha, 1]$. Then

$$\frac{\overline{\text{VaR}}_\alpha(S_n^\alpha)}{\text{TVaR}_\alpha(S_n^c)} \to 1, \qquad (4.31)$$

i.e., the upper TVaR bounds are asymptotically sharp.

The proof of Theorem 4.8 is similar to that in the exact mixing case. The asymptotic mixing condition (AM) can be verified in particular in the homogeneous case with decreasing densities (see Wang and Wang, 2011) or in the more general case with finitely many distributions considered in Puccetti et al. (2013). More general discussion of the phenomenon and more general conditions are given in Wang (2014) and in Embrechts et al. (2015). From the last paper we describe the following result with general easy-to-verify conditions (see Theorem 3.3 in Embrechts et al., 2015).

Theorem 4.9 (Sharpness of TVaR bounds, general conditions)
Let $X_i \sim F_i$ be random variables, and $\alpha \in (0, 1)$ and $\vartheta > 1$:

a) Assume
   a_1) $E|X_i - EX_i|^\vartheta < M$ for some $M > 0$,
   a_2) $\varliminf_{n \to \infty} n^{-1/\vartheta} \sum_{i=1}^{n} \text{TVaR}_\alpha(X_i) = \infty$,

   then $\qquad \dfrac{\overline{\text{VaR}}_\alpha(S_n)}{\text{TVaR}_\alpha(S_n)} \to 1.$

b) If a_1) holds and
   a_2^*) $c_0 := \varliminf_{n \to \infty} \frac{1}{n} \sum_{i=1}^{n} \text{TVaR}_\alpha(X_i) > 0$, then for $n \ge n_0$ it holds that:

$$1 \ge \frac{\overline{\text{VaR}}_\alpha(S_n)}{\text{TVaR}_\alpha(S_n)} \ge 1 - Cn^{-1+1/\vartheta}, \qquad (4.32)$$

   where $C = (\frac{1}{1-\alpha} \frac{\vartheta}{\vartheta-1} + 1) \frac{M^{1/\vartheta}}{c_0} > 0$.

Proof For the proof, in a first step the following approximation results for $\overline{\text{VaR}}_\alpha$ and $\underline{\text{VaR}}_\alpha$ are established.

Define for $0 \le q \le 1$, $\mu_q^{(i)} := \frac{1}{q-\alpha} \int_\alpha^q F_i^{-1}(u)\, du$, then from Theorem 3.1 in Embrechts et al. (2015),

$$\sup_{q>\alpha} \left\{ \sum_{i=1}^n \mu_q^{(i)} - \max_{i \le n}(F_i^{-1}(q) - F_i^{-1}(\alpha)) \right\} \le \underline{\text{VaR}}_\alpha(S_n) \le \overline{\text{TVaR}}_\alpha(S_n) \qquad (4.33)$$

and

$$\sum_{i=1}^n \mu_{0,\alpha}^{(i)} \le \underline{\text{VaR}}_\alpha(S_n) \le \inf_{q \in [0,\alpha)} \left\{ \sum_{i=1}^n \mu_{q,\alpha}^{(i)} + \max_{i \le n}(F_i^{-1}(q) - F_i^{-1}(\alpha)) \right\}. \qquad (4.34)$$

In particular, if $F_1, \ldots, F_n$ are supported on $[a, b]$, $a < b$, then

$$\overline{\text{TVaR}}_\alpha(S_n) - (b - a) \le \underline{\text{VaR}}_\alpha(S_n) \le \overline{\text{TVaR}}_\alpha(S_n). \qquad (4.35)$$

Consequently, for bounded random variables, $\underline{\text{VaR}}_\alpha(S_n)$ and $\overline{\text{TVaR}}_\alpha(S_n)$ differ at most by a constant independent of n. The proof of (4.33) and (4.34) is based on the following interesting coupling result (see Lemma A1 in Embrechts et al., 2015), which we formulate as a proposition.

Proposition 4.10 (Coupling of bounded random variables)

a) *If F_i have supports in $[0, 1]$, then there exist random variables $X_i \sim F_i$, $i \in \mathbb{N}$, such that for all $n \in \mathbb{N}$,*

$$\text{ess sup } S_n - \text{ess inf } S_n \le 1.$$

b) *If F_i have bounded supports, $i \in \mathbb{N}$, then there exist random variables $X_i \sim F_i$, $i \in \mathbb{N}$, such that for all $n \in \mathbb{N}$,*

$$|S_n - ES_n| \le L_n, \qquad (4.36)$$

where $L_n = \max\{\text{ess sup } X_i - \text{ess inf } X_i; \; X_i \sim F_i, i \le n\}$.

If $F_{\alpha,q}^{(i)}$ denotes the conditional distribution of $X_i \sim F_i$ on the interval $[F_i^{-1}(\alpha), F_i^{-1}(q)]$, then by (4.36) one can construct a coupling $Y_i \sim F_{\alpha,q}^{(i)}$, $1 \le i \le n$, such that

$$Y_1 + \cdots + Y_n \ge \sum_{i=1}^n \mu_{\alpha,q}^{(i)} - \max_{i \le n}(F_i^{-1}(q) - F_i^{-1}(\alpha)).$$

This construction can then easily be extended to a coupling X_i of F_i, $i \le n$, such that (4.33) (without the sup) holds. $\qquad\qquad\qquad\qquad\qquad\qquad\qquad\qquad\qquad\qquad\qquad$ □

Part II

Additional Dependence Constraints

In comparison to Part I, we assume in Part II that we not only know the marginal distributions of the risk vector $X = (X_1, \ldots, X_n)$, say $X_i \sim F_i$, $1 \leq i \leq n$, but we also have additional information on the dependencies between these n variables.

In Chapter 5 we present the method of improved standard bounds. We review bounds on the distribution function that improve upon the Hoeffding–Fréchet bounds by including, in addition to the marginal information, some positive or negative dependence information on the distribution functions of the form $F \leq G$ or $F \geq G$. These improved Hoeffding–Fréchet bounds lead to improved upper and lower VaR bounds. We build on results of Williamson and Downs (1990), Denuit et al. (1999), Embrechts et al. (2003), Rüschendorf (2005), Embrechts and Puccetti (2006a,b), and Puccetti et al. (2016), among others. We present some examples in which the information significantly reduces the standard bounds. Further examples of this type are discussed in Rüschendorf (2017a,b). The last section of this chapter presents an extension of the method of improved standard bounds by Lux and Rüschendorf (2018) to include two-sided dependence information of the form $G \leq F \leq H$ or related bounds for the copula C or for the survival function $\overline{F}$.

Chapter 6 deals with bounds on the value-at-risk (VaR) under the additional constraint that a bound on the variance of the sum is known (Bernard et al., 2017c) or that information on higher moments of the sum is available (Bernard et al., 2017a). As compared to assuming only knowledge of the mean and variance of the portfolio loss (so-called moment bounds), the additional knowledge of the marginal distributions of the portfolio components may lead to an improvement of the moment bounds. However, when supplementing the information on the marginal distributions with information on the portfolio variance, it turns out that when the variance constraint is "low" enough, the VaR-bounds coincide with moment bounds, as studied in detail in Part IV. We propose the ERA-algorithm, an extension of the RA-algorithm, a numerical approach to approximate sharp VaR bounds given the marginal constraints and additionally bounds on the variance of the sum. A corresponding rearrangement algorithm is no longer available if only moment bounds are assumed. We show that for large portfolios, our analytical bounds are nearly sharp. However, in the context of smaller portfolios (or when the portfolio depicts significant concentration), one cannot expect the bounds to be sharp. The ERA-algorithm can accommodate this situation and makes it possible to approximate sharp bounds. As a by-product, this algorithm also sheds light on the composition of extreme portfolios.

Chapter 7 is devoted to a study of the case in which information on the joint distribution of the risk vector $X = (X_1, \ldots, X_n)$ is partially available, either by knowing the conditional distribution of X on a subset S in R^n, the marginal distributions, and the probability of the subset, or alternatively by knowing bounds for the distribution function F_X resp. the copula C_X on the corresponding subsets. For example, it may be the case that C_X of X is known on a central domain due to the availability of sufficient statistical data to allow for precise modeling within this domain. Alternatively, some positive or negative dependence information may be available in the upper or lower tail area.

We derive VaR bounds in several situations of interest in which the conditional distribution of X is known on a central domain. The corresponding improvements of the Hoeffding–Fréchet bounds were derived in Rachev and Rüschendorf (1994), Nelsen et al. (2004), Tankov (2011), and Bernard and Vanduffel (2014) in the two-dimensional case and in Puccetti et al. (2016) and Lux and Papapantoleon (2019) in the n-dimensional case.

We solve several cases exactly or numerically, whether we know the distribution function on a subset or if we know the conditional distribution on a subset (Bernard and Vanduffel, 2015). While knowing the distribution function F on an open domain A implies knowledge of P^X on A, the converse direction is not valid, even for rectangle domains A. So the two types of information are typically different. Based on a numerical approach using the RA-algorithm, the value of this type of information for improving VaR bounds for the sum is determined.

5 Improved Standard Bounds

The subject of this chapter is the method of improved standard bounds, which is designed as an improvement of the method of standard bounds for the upper and lower VaR bounds. This method allows one to use one-sided bounds on the distribution function, which improves upon the Hoeffding–Fréchet bounds to include some positive or negative dependence information. The degree of reduction of the risk bounds by this method is determined in several examples, and some techniques for evaluation of these bounds are described. As a result it turns out that reasonable improvements are obtained if the dependence restriction is strong enough and the dimensionality of the problem is not too high. An extension of this method to include two-sided dependence bounds yields a considerable performance improvement. Various examples show the wide applicability of this method.

5.1   Standard and Improved Standard Bounds for Monotonic Functions

The standard bounds for the distribution function of the joint portfolio $\sum_{i=1}^{n} X_i$ with $X_i \sim F_i$ state

$$\left(\bigvee_{i=1}^{n} F_i(t) - (n-1) \right)_{+} \leq P\left(\sum_{i=1}^{n} X_i \leq t \right) \leq \min \left(\bigwedge_{i=1}^{n} F_i(t), 1 \right), \qquad (5.1)$$

where $\bigwedge_{i=1}^{n} F_i(t) := \inf \{ \sum_{i=1}^{n} F_i(u_i); \sum_{i=1}^{n} u_i = t \}$ is the **infimal convolution** of the (F_i), and $\bigvee_{i=1}^{n} F_i(t) := \sup \{ \sum_{i=1}^{n} F_i(u_i); \sum_{i=1}^{n} u_i = t \}$ is the **supremal convolution** of the (F_i).

Equivalently, this implies for the tail risk of the sum the bounds

$$\left(\bigvee_{i=1}^{n} \overline{F}_i(t) - (n-1) \right)_{+} \leq P\left(\sum_{i=1}^{n} X_i \geq t \right) \leq \min \left(\bigwedge_{i=1}^{n} \overline{F}_i(t), 1 \right), \qquad (5.2)$$

where $\overline{F}_i(t) = P(X_i \geq t) = 1 - F_i(t-)$ are the survival functions of F_i. (We remark that in some literature $\overline{F}_i(t)$ is defined as $1 - F_i(t)$.)

The standard bounds in (5.1) and (5.2) have been extended in slightly varied form in a series of papers to monotonically non-decreasing functions $\Psi(x_1, \ldots, x_n)$: see Williamson and Downs (1990), Denuit et al. (1999), Embrechts et al. (2003), Rüschendorf (2005), and Embrechts and Puccetti (2006b).

For componentwise non-decreasing aggregation functions, define

$$A_\Psi^+(t) := \{u = (u_1, \ldots, u_n) \in \mathbb{R}^n; \ u \text{ is a maximal point in } \{\Psi \leq t\}\}$$

and the **Ψ-inf-** resp. **Ψ-sup-convolutions**

$$\bigwedge_\Psi^+ F_i(t) := \inf\left\{\sum_{i=1}^n F_i(u_i); \ u \in A_\Psi^+(t)\right\}$$

$$\text{resp.} \quad \bigvee_\Psi^+ F_i(t) := \sup\left\{\sum_{i=1}^n F_i(u_i); \ u \in A_\Psi^+(t)\right\}. \tag{5.3}$$

Similarly, we define

$$A_\Psi^-(t) := \{u \in (u_1, \ldots, u_n) \in \mathbb{R}^n; \ u \text{ is a minimal point of } \{\Psi \geq t\}\}$$

and correspondingly (as in (5.3))

$$\bigwedge_\Psi^- \overline{F}_i(t) := \inf\left\{\sum_{i=1}^n \overline{F}_i(u_i); \ u \in A_\Psi^-(t)\right\};$$

$$\bigvee_\Psi^- \overline{F}_i(t) := \sup\left\{\sum_{i=1}^n \overline{F}_i(u_i); \ u \in A_\Psi^-(t)\right\}. \tag{5.4}$$

Theorem 5.1 (Standard bounds for monotonic functions) *For a risk vector $X = (X_1, \ldots, X_n)$ with $X_i \sim F_i$ and a monotonically non-decreasing function Ψ, it holds that*

a)

$$\left(\bigvee_\Psi^+ F_i(t) - (n-1)\right)_+ \leq P(\Psi(X) \leq t) \leq \bigwedge_\Psi^+ F_i(t); \tag{5.5}$$

b)

$$\left(\bigvee_\Psi^- \overline{F}_i(t) - (n-1)\right)_+ \leq P(\Psi(X) \geq t) \leq \bigwedge_\Psi^- \overline{F}_i(t). \tag{5.6}$$

Proof a) To determine the distribution function of $\Psi(X)$, let $u \in A_\Psi^+(t)$. Then the maximality of u in $\{\Psi \leq t\}$ implies

$$P(\Psi(X) \leq t) \leq P\left(\bigcup_{i=1}^n \{X_i \leq u_i\}\right) \leq \sum_{i=1}^n F_i(u_i).$$

This implies the upper bound in a). Furthermore,

$$P(\Psi(X) \leq t) \geq P(X_1 \leq u_1, \ldots, X_n \leq u_n) \geq \sum_{i=1}^n F_i(u_i) - (n-1)$$

by the lower Fréchet bounds. This implies the lower bound in a).

b) The proof of b) is similar. $\qquad\qquad\qquad\qquad\qquad\qquad\qquad\qquad\qquad\quad\square$

Remark 5.2 Note that the bounds in a) may not be good when Ψ is not lower semicontinuous (lsc), since then $\{\Psi \leq t\}$ is not closed, and $A_\Psi^+(t)$ may therefore even be empty in the worst case. Similarly, the tail risk bounds in b) may not be good when

Ψ is not upper semicontinuous (usc). If Ψ is lsc then one gets from a) the alternative bounds

$$1 - \bigwedge\nolimits_{\Psi}^{+} F_i(t) \le P(\Psi(X) > t) \le \min\left(n - \bigvee\nolimits_{\Psi}^{+} F_i(t), 1\right); \qquad (5.7)$$

resp. if Ψ is usc then one gets from b) the alternative bound

$$1 - \bigwedge\nolimits_{\Psi}^{-} \overline{F}_i(t) \le P(\Psi(X) < t) \le \min\left(n - \bigvee\nolimits_{\Psi}^{-} \overline{F}_i(t), 1\right). \qquad (5.8)$$

Note that several further related forms of the bounds can be given. ◊

It was soon found that the standard bounds may be widespread, i.e., the spread between the upper and lower bounds may be quite wide. Also better upper and lower bounds as the dual bounds were found that give strongly improved bounds, particularly for higher dimensionality. A natural idea to improve the standard bounds was followed from early papers on, namely to include additional one-sided information on the joint distribution function F of the risk vector X. If we know, for example, that X is positive lower orthant dependent (PLOD) resp. positive upper orthant dependent (PUOD), then

$$F_X(x) \ge F_{X^\perp}(x) = \prod_{i=1}^{n} F_i(x_i), \qquad (5.9)$$

where $X^\perp$ is a vector with independent components $X_i^\perp \sim X_i$, or that

$$\overline{F}_X(x) \ge \overline{F}_{X^\perp}(x) = \prod_{i=1}^{n} \overline{F}_i(x_i). \qquad (5.10)$$

In general by the Fréchet bounds for any $F \in \mathcal{F}(F_1, \dots, F_n)$ we know that

$$F^-(x) = \left(\sum_{i=1}^{n} F_i(x_i) - (n-1)\right)_+ \le F(x) \le \min_{i \le n} F_i(x_i) = F^+(x) \qquad (5.11)$$

and

$$\overline{F}^-(x) = \left(\sum_{i=1}^{n} \overline{F}_i(x_i) - (n-1)\right)_+ \le \overline{F}(x) \le \overline{F}^+(x) = \min_{i \le n} \overline{F}_i(x_i). \qquad (5.12)$$

Assume now that $G, \overline{H}$ are increasing resp. decreasing functions on $\mathbb{R}^n$ with

$$F^-(x) \le G(x) \le F^+(x) \quad \text{and} \quad \overline{F}^-(x) \le \overline{H}(x) \le \overline{F}^+(x). \qquad (5.13)$$

In the following we assume that we have additional one-sided dependence information of the form $G \le F_X$ or $\overline{F}_X \ge \overline{H}$, which amounts to improved Fréchet bounds. Then the following improved standard bounds have been given in various forms in the literature: see Williamson and Downs (1990), Denuit et al. (1999), Embrechts et al. (2003), Rüschendorf (2005), Embrechts and Puccetti (2006b).

Theorem 5.3 (Improved standard bounds) *Let* $G, \overline{H}$ *be increasing resp. decreasing functions satisfying* (5.13).

a) *Positive lower orthant dependence condition:* If $G \le F_X$, then

$$P\left(\sum_{i=1}^{n} X_i \le s\right) \ge \bigvee G(s) := \sup\left\{G(u); \sum_{i=1}^{n} u_i = s\right\}. \tag{5.14}$$

b) *Positive upper orthant dependence condition:* If $\overline{F}_X \ge \overline{H}$, then

$$P\left(\sum_{i=1}^{n} X_i \ge s\right) \ge \bigvee \overline{H}(s) = \sup\left\{\overline{H}(u); \sum_{i=1}^{n} u_i = s\right\}. \tag{5.15}$$

Proof

a) If $G \le F_X$, i.e., $G(x) \le F_X(x)$ for all x, then for all $u \in U(s) = \{v \in \mathbb{R}^n; \sum_{i=1}^{n} v_i = s\}$, we have

$$P\left(\sum_{i=1}^{n} X_i \le s\right) \ge P(X_1 \le u_1, \ldots, X_n \le u_n) = F_X(u) \ge G(u),$$

which implies (5.14).

b) If $\overline{F}_X \ge \overline{H}$, then we obtain similarly for $u \in U(s)$,

$$P\left(\sum_{i=1}^{n} X_i \ge s\right) \ge P(X_1 \ge u_1, \ldots, X_n \ge u_n) = \overline{F}_X(u) \ge \overline{H}(u),$$

which implies (5.15).   □

Remark 5.4

a) For the tail risk $P(\sum_{i=1}^{n} X_i \ge s)$, the lower bound in b) typically improves upon the standard bound related to (5.1). The upper tail risk bound resulting form (5.14), namely

$$P\left(\sum_{i=1}^{n} X_i > s\right) \le 1 - \bigvee G(s), \tag{5.16}$$

is typically not a substantial improvement over the standard tail risk bound related to (5.1). This means that positive dependence information typically does not essentially improve the upper tail risk bounds but may improve the lower bounds for the tail risk.

b) If X is **positive orthant dependent** (POD), i.e., X is PUOD and PLOD, then we obtain as a result of Theorem (5.2):

$$\bigvee F^{\perp}(s) \le P\left(\sum_{i=1}^{n} X_i \le s\right), \quad P\left(\sum_{i=1}^{n} X_i \ge s\right) \ge \bigvee \overline{F}^{\perp}(s), \tag{5.17}$$

where $F^{\perp}(x) = \prod_{i=1}^{n} F_i(x_i)$, $\overline{F}^{\perp}(x) = \prod_{i=1}^{n} \overline{F}_i(x_i)$.

c) The bounds can also be formulated with respect to bounds on the copula $C = C_X$. If, for example, $C_L \le C$ for an increasing function C_L with $W \le C_L \le M$, W, M the Fréchet bounds, then this implies that

$$G(x) = C_L(F_1(x_1), \ldots, F_n(x_n)) \le F_X(x), \tag{5.18}$$

and thus we can apply (5.14); similarly for the upper bound we can apply (5.15). ◊

As in Theorem 5.1, the improved standard bounds also extend to monotonically non-decreasing aggregation functions $\Psi(X)$, $\Psi = \Psi(x_1, \ldots, x_n)$.

Theorem 5.5 (**Improved standard bounds for monotonic functionals I**) *Let Ψ be a monotonically non-decreasing function and $X = (X_1, \ldots, X_n)$ a risk vector with $X_i \sim F_i$, and let G, $\overline{H}$ be increasing resp. decreasing functions satisfying (5.13).*

a) If $G \leq F_X$, then

$$P(\Psi(X) \leq x) \geq \bigvee{}_{\Psi}^{+} G(s) = \sup\{G(z); u \in A_{\Psi}^{+}(s)\}. \tag{5.19}$$

b) If $\overline{F}_X \geq \overline{H}$, then

$$P(\Psi(X) \geq x) \geq \bigvee{}_{\Psi}^{-} G(s) = \sup\{\overline{H}(z); u \in A_{\Psi}^{-}(s)\}. \tag{5.20}$$

Proof The proof is similar to that of Theorem 5.3 and is, therefore, omitted. □

We also give an alternative formulation of the bounds that may have advantages for the numerical calculation. For given marginals $F_1, \ldots, F_n$ for a non-decreasing function Ψ and for a copula $C = C_X$ of X, let

$$\sigma_{C,\Psi}(s) := P_C(\Psi(X) < s). \tag{5.21}$$

For a non-decreasing function C_ℓ on $[0,1]^n$, define

$$\tau_{C_\ell,\Psi}(s) := \sup_{x_{(n-1)} \in \mathbb{R}^{n-1}} C_\ell(F_1(x_1), \ldots, F_{n-1}(x_{(n-1)}), F_n^-(\overline{y}_\Psi)), \tag{5.22}$$

where $\overline{y}_\psi = \overline{y}_\Psi(s) := \sup\{y; \Psi(x_{(n-1)}, y) < s\}$, and F_n^- is the left continuous version of F_n. The formulation in (5.22) avoids the problem of the existence of maximal points as in the formulation of (5.19). By considering the probability $P(\Psi(X) < s)$, it also avoids the problem of jumps of the marginal distribution function F_n. A similar alternative formulation can also be given for part b) of Theorem 5.5. As result we get the following variant of Theorem 5.5 a).

Theorem 5.6 (**Improved standard bounds for monotonic functionals II**) *Let Ψ be monotonically non-decreasing, and let C_ℓ be non-decreasing on $[0,1]^n$ with $W \leq C_\ell \leq M$. If $C \geq C_\ell$, then*

$$\sigma_{C,\Psi}(s) \geq \tau_{C_\ell,\Psi}(s), \quad \text{for all } s \in \mathbb{R}. \tag{5.23}$$

As a result we obtain that the standard bounds for risk aggregation in (5.1) and in (5.5) and (5.6) can be improved if further positive dependence information of the form $F_X \geq G$ or $\overline{F}_X \geq \overline{H}$ is available. The improved standard bounds are given in terms of generalized infimal or supremal Ψ-convolutions. In general these convolutions are typically not available in analytic form but have to be approximated numerically.

5.2 Computation of Improved Standard Bounds

The standard bounds and improved standard bounds are in general not easy to determine. There are, however, some results in the literature that are useful for determining these bounds or at least establishing upper resp. lower estimates of them: see Williamson and Downs (1990), Denuit et al. (1999), Embrechts and Puccetti (2006a), Puccetti et al. (2016).

Denoting the standard bounds in (5.1) by

$$F_{\min}(t) := \left(\bigvee_{i=1}^{n} F_i(t) - (n-1) \right)_+, \quad F_{\max}(t) := \min \left\{ \bigwedge_{i=1}^{n} F_i(t), 1 \right\}, \qquad (5.24)$$

the inequalities in (5.1) yield the VaR bounds

$$F_{\min}^{-1}(\alpha) \leq \mathrm{VaR}_\alpha \left(\sum_{i=1}^{n} X_i \right) \leq F_{\max}^{-1}(\alpha). \qquad (5.25)$$

With

$$\tau(s) := \tau_+(s) := \sup_{x_{(n-1)} \in \mathbb{R}^{n-1}} \left(\sum_{i=1}^{n-1} F_i(x_i) + F_n^- \left(s - \sum_{i=1}^{n-1} x_i \right) - (n-1) \right)_+$$

$$=: \sup_{x_{(n-1)} \in \mathbb{R}^{n-1}} \Phi(x_{(n-1)}), \qquad (5.26)$$

the standard bounds can be formulated (analogously to (5.23)) as

$$P \left(\sum_{i=1}^{n} X_i < s \right) \geq \tau(s) \qquad (5.27)$$

$$\text{resp.} \quad \mathrm{VaR}_\alpha(S_n) \leq \tau^{-1}(\alpha), \quad \alpha \in (0,1).$$

For the calculation of the standard bounds, the following result of Embrechts and Puccetti (2006a) is useful.

Let F_i have the support $[a_i, \infty)$, $a_i \in \mathbb{R}$, and assume that F_i has a continuous, strictly decreasing density $f_i = F_i'$ on $(\overline{x}_{F_i}, \infty)$ for some $\overline{x}_{F_i} \geq a_i$. Under these assumptions we have the following method to calculate the standard bounds.

Theorem 5.7 (Calculation of standard bounds) *Let s be a threshold such that $\tau(s) > p := \max_{i \leq n} F_i(\overline{x}_{F_i})$; then $\tau(s) = \Phi(x_{(n-1)}^*)$, where $x_{(n-1)}^*$ is the unique vector in $\prod_{i=1}^{n-1} (\overline{x}_{F_i}, \infty)$ such that*

$$f_1(x_1^*) = \cdots = f_{n-1}(x_{n-1}^*) = f_n \left(s - \sum_{i=1}^{n-1} x_i^* \right). \qquad (5.28)$$

Proof In the proof one first sees that for finding the sup of $\Phi(x_{(n-1)})$ one can restrict to $\prod_{i=1}^{n-1} (\overline{x}_{F_i}, \infty)$. This is due to the condition $\tau(s) > p$.

Then (5.28) describes the necessary first order condition. Strict decreasingness of the f_i then gives uniqueness of the solution in (5.28). □

The condition $\tau(s) > p$ in Theorem 5.7 is fulfilled for large enough thresholds.

Proposition 5.8   *If $s > \sum_{i=1}^{n} F_i^{-1}(\frac{p+n-1}{n})$, then the condition $\tau(s) > p$ holds true.*

Proof   For $\widehat{x}$ with components $\widehat{x}_i = F_i^{-1}(\frac{p+n-1}{n})$, it holds, using the fact that $\frac{p+n-1}{n} > p$, that

$$F_n^-\left(s - \sum_{i=1}^{n-1} \widehat{x}_i\right) > F_n^-\left(F_n^{-1}\left(\frac{p+n-1}{n}\right)\right) = \frac{p+n-1}{n},$$

and, therefore,

$$\tau(s) = \sup_{x_{(n-1)} \in \mathbb{R}^{n-1}} \left(\Phi(x_{(n-1)})\right)_+ \geq \left(\Phi(\widehat{x}_{(n-1)})\right)_+$$

$$> n\,\frac{p+n-1}{n} - (n-1) = p. \qquad \square$$

In the homogeneous case $F_1 = \cdots = F_n = F$, the standard bound $\tau(s)$ simplifies strongly, and as a result one gets the following simplified version of Theorem 5.7 (see Theorem 2.8 in Puccetti, 2005).

Theorem 5.9 (Standard bounds in the homogeneous case)   *Let $F_1 = \cdots = F_n = F$ have an unbounded support $[a, \infty)$ and a positive density, strictly decreasing on $(\overline{x}_F, \infty)$ with $\overline{x}_F \geq a$. Then*

$$\tau(s) = \left(nF\left(\frac{s}{n}\right) - (n-1)\right)_+, \quad \text{for all } s \geq nF^{-1}\left(\frac{F(\overline{x}_F)+n-1}{n}\right), \tag{5.29}$$

and

$$\tau^{-1}(\alpha) = nF^{-1}\left(\frac{\alpha+n-1}{n}\right), \quad \text{for all } \alpha \in [F(\overline{x}_F), 1). \tag{5.30}$$

Proof   The proof makes use of the following duality principle of Frank and Schweizer (1979).

Theorem 5.10 (Duality principle)   *Let $-\infty \leq a < b \leq \infty$, and let $\Psi : [a, b]^n \to [a, b]$ be an increasing continuous function with range in $[a, b]$. Then for the improved standard bound*

$$\tau_{C_\ell, \Psi}(s) = \sup_{x_{(n-1)} \in \mathbb{R}^{n-1}} C_\ell(F_1(x_1), \ldots, F_{n-1}(x_{n-1}), F_n(\overline{y}_\Psi(s))),$$

it holds that

$$\tau_{C_\ell, \Psi}^{-1}(\alpha) = \inf_{C_\ell(u_1, \ldots, u_n) = \alpha} \Psi(F_1^{-1}(u_1), \ldots, F_n^{-1}(u_n)). \tag{5.31}$$

In consequence, for the case $\Psi(x) = \sum_{i=1}^{n} x_i$ and $C_\ell = W$ we obtain

$$\tau_{W,+}^{-1}(\alpha) = \inf_{\substack{0 \leq u_i \leq 1 \\ \sum_{i=1}^{n} u_i = \alpha+n-1}} \sum_{i=1}^{n} F^{-1}(u_i). \tag{5.32}$$

Using $\Phi(u_{(n-1)}) := \sum_{i=1}^{n-1} F^{-1}(u_i) + F^{-1}(\alpha + n - 1 - \sum_{i=1}^{n-1} u_i)$, one finds that the reduced problem of optimization of $\Phi(u_{(n-1)})$ has an inner point solution $u^*_{(n-1)}$ that satisfies $\frac{\partial \Phi}{\partial u_i}(u_i^*) = 0, 1 \leq i \leq n - 1$ and which by the assumed monotonicity is unique. Then it is easy to check that $u^*_{(n-1)} = (\frac{\alpha+n-1}{n}, \ldots, \frac{\alpha+n-1}{n})$ is this solution. □

Mesfioui and Quessy (2005, Proposition 3.2) give an easily computable expression for $\tau_{C_\ell,+}^{-1}(\alpha)$ assuming that the marginal distribution F has an ultimately decreasing density and the copula bound C_ℓ is symmetric and satisfies an extra second order condition. The following result from Puccetti et al. (2016) is based on the notion of Schur concavity and uses easy to check conditions. For the notion of Schur concavity we refer the reader to Chapter 3 in Marshall et al. (2011).

The diagonal section of a function C_ℓ is the function $\delta_{C_\ell} : [0, 1] \to [0, 1]$ defined by $\delta_{C_\ell}(u) = C_\ell(u, \ldots, u)$.

Theorem 5.11 (Improved standard bound, homogeneous case)
In the homogeneous case $F_1 = \cdots = F_n = F$ under the lower bound condition $C_\ell \leq C$, assume that additionally

1) C_ℓ is Schur concave and
2) F is concave on $[F^{-1}(\alpha) - \xi, \infty)$ for some $\xi > 0$.

Then it holds that

$$\tau_{C_\ell,+}^{-1}(\alpha) = nF^{-1}(\alpha^*),$$

where $\alpha^* = \delta_{C_\ell}^{-1}(\alpha)$.

Proof By definition

$$\tau_{C_\ell,+}^{-1}(\alpha) = \inf\{t \in \mathbb{R}; \ \tau_{C_{\ell^+}}(t) \geq \alpha\}$$
$$= \inf\{t \in \mathbb{R}; \ \sup_{\sum_{i=1}^n x_i = t} C_\ell(F(x_1), \ldots, F(x_n)) \geq \alpha\}.$$

Note that when replacing x_i by $F^{-1}(\alpha) - \xi$, we have $C_\ell(\ldots) < \alpha$. Thus, we can restrict ourselves to considering $x_i \geq F^{-1}(\alpha) - \xi$ and rewrite

$$\tau_{C_\ell,+}^{-1}(\alpha) = \inf\left\{T \geq n(F^{-1}(\alpha) - \xi): \sup_{\substack{\sum_{i=1}^n x_i = t \\ x_i \geq F^{-1}(\alpha) - \xi}} C_\ell(F(x_1,), \ldots, F(x_n)) \geq \alpha\right\}. \quad (5.33)$$

Since $F(x_i)$ is concave for $x_i \geq F^{-1}(\alpha) - \xi$ and C_ℓ is Schur concave, this implies that $C_\ell(F(x_1), \ldots, F(x_n))$ is Schur concave, and the sup in (5.33) is attained at the point $(\frac{t}{n}, \ldots, \frac{t}{n})$ and is thus given by $\delta_{C_\ell}(F(\frac{t}{n}))$.

Since $W \leq C_\ell \leq M$ and $\delta_{C_\ell}(u) \leq u$, we have that $\alpha^* = \delta_{C_\ell,+}^{-1}(\alpha) \geq \alpha$, and we obtain

$$\tau_{C_\ell,+}^{-1}(\alpha) = \inf\left\{t \geq n(F^{-1}(\alpha) - \xi); \ \delta_{C_\ell}\left(F\left(\frac{t}{n}\right)\right) \geq \alpha\right\}$$
$$= \inf\left\{t \geq n(F^{-1}(\alpha) - \xi); \ F\left(\frac{t}{n}\right) \geq \alpha^*\right\}$$
$$= nF^{-1}(\alpha^*). \qquad \square$$

Note that in the inhomogeneous case, an easily computable VaR bound follows from the improved standard bound $\tau_{C_\ell,+}$ by choosing a vector $(\alpha^*, \ldots, \alpha^*)$ such that $\delta_{C_\ell}(\alpha^*) = C_\ell(\alpha^*, \ldots, \alpha^*) = \alpha$. Since the diagonal section of C_ℓ is continuous such α^* exists. Hence, we get as a consequence

Theorem 5.12 (VaR bounds in the inhomogeneous case) *For marginal distribution functions $F_1, \ldots, F_n$ assume the lower bound condition $C \geq C_\ell$. Then*

$$\text{VaR}_\alpha(S_n) \leq \text{VaR}_{\alpha^*}(S_n^c) = \sum_{i=1}^{n} F_i^{-1}(\alpha^*), \qquad (5.34)$$

where $\delta_{C_\ell}(\alpha^*) = \alpha$.

Remark 5.13

a) The VaR bound $\text{VaR}_{\alpha^*}(S_n^c)$ is the VaR for the comonotonic portfolio S_n^c computed at a distorted confidence level α^*. Recall from the proof of Theorem 5.11 that $\alpha^* \geq \alpha$ and, therefore, that

$$\text{VaR}_{\alpha^*}(S_n^c) \geq \text{VaR}_\alpha(S_n^c). \qquad (5.35)$$

By this bound one can reach at best the comonotonic VaR as upper bound, which is reached in case $C_\ell = M$.

b) An extension of Theorem 5.12 is given in Puccetti et al. (2016) to the case where the lower bound condition $C \geq C_\ell$ is only assumed on the upper part $S = [\beta, 1]^n$ with $\beta \leq \alpha$ of the distribution. Also under this assumption it is shown that

$$\text{VaR}_\alpha(S_n) \leq \text{VaR}_{\alpha^*}(S_n^c) = \sum_{i=1}^{n} F_i^{-1}(\alpha^*)$$

with $\alpha^* = \delta_{C_\ell}^{-1}(\alpha)$. $\diamond$

Example 5.14

a) **Gumbel copula, lower bound**: Consider a risk portfolio X with identical marginal distributions $F_1 = \cdots = F_n = F$ and a portfolio copula $C = C_X$ bounded below by a Gumbel copula,

$$C_X \geq C_\ell = C_\vartheta^{\text{Gu}} \quad \text{on } S = [\alpha, 1]^n, \qquad (5.36)$$

where $C_\vartheta^{\text{Gu}}(u_1, \ldots, u_n) = \exp\left(-[(-\ln u_1)^\vartheta + \cdots + (-\ln u_n)^\vartheta]\right)$ for $\vartheta \geq 1$.

Table 5.1 compares the bounds $\text{VaR}_{\alpha^*}(S_n^c)$ from Theorem 5.12 resp. Remark 5.4 b) for increasing ϑ values corresponding to increasing positive dependence in the tails. It also gives the sharp upper VaR bounds $\overline{\text{VaR}}_\alpha(S)$ without dependence information and the comonotonic VaR, $\text{VaR}_\alpha(S_n^c)$.

One obtains improvements compared to the bounds $\overline{\text{VaR}}_\alpha(S)$ for sufficiently large ϑ. In coincidence with the theoretical results, one finds that for $\vartheta \to \infty$,

$$C_\vartheta^{\text{Gu}} \to M \quad \text{and} \quad \text{VaR}_{\alpha^*}(S_n^c) \to \text{VaR}_\alpha(S_n^c). \qquad (5.37)$$

We also notice that, according to Theorem 5.11, in this homogeneous case one gets

$$\tau_{C_\ell,+}^{-1}(\alpha) = \text{VaR}_{\alpha^*}(S_n^c) = nF^{-1}(\alpha^*). \qquad (5.38)$$

		VaR$_{\alpha^*}(S_n^c)$				
$n = 5$	$\overline{\text{VaR}}_\alpha$	$\vartheta = 1$	$\vartheta = 3$	$\vartheta = 5$	$\vartheta = 10$	VaR$_\alpha(S_n^c)$
$\alpha = 0.990$	84.44	107.38	60.27	53.68	49.17	45.00
$\alpha = 0.995$	121.49	152.33	87.38	78.02	71.62	65.71
$\alpha = 0.999$	277.84	348.20	201.72	180.71	166.36	153.11

Table 5.1 Values for VaR$_{\alpha^*}(S_n^c)$ and VaR$_\alpha(S_n^c)$ for a risk vector of $n = 5$ risks identically distributed as a Pareto(2), i.e., $F_i = F(x) = 1 - (1 + x)^{-2}$, $x > 0$, $1 \leq i \leq n$ and a Gumbel copula lower bound $C_X \geq C_\vartheta^{\text{Gu}}$ on $S = [\alpha, 1]^n$.

b) Gaussian copula, lower bound: Analogous conclusions can be drawn when a Gaussian copula is assumed as a lower bound, i.e.,

$$C_X \geq C_\ell = C_\varrho^{\text{Ga}} \quad \text{on } S = [\alpha, 1]^n, \tag{5.39}$$

where $C_\varrho^{\text{Ga}}(u_1, \ldots, u_n) = \Phi_\varrho(\Phi^{-1}(u_1), \ldots, \Phi^{-1}(u_n))$, where Φ_ϱ denotes the Gaussian distribution with mean vector 0 and equicorrelations, $\text{Cor}(X_i, X_j) = \varrho$. Table 5.2 compares the resulting bounds for various increas-

		VaR$_{\alpha^*}(S_n^c)$				
$n = 5$	$\overline{\text{VaR}}_\alpha(S)$	$\varrho = 0$	$\varrho = 0.8660$	$\varrho = 0.9511$	$\varrho = 0.9877$	VaR$_\alpha(S_n^c)$
$\alpha = 0.990$	84.44	106.58	75.97	63.37	53.96	45.00
$\alpha = 0.995$	121.49	152.96	112.41	93.89	79.15	65.71
$\alpha = 0.999$	277.84	348.48	269.65	222.57	187.32	153.11

Table 5.2 The same as Table 5.1 with a Gaussian copula lower bound $C_X \geq C_\varrho^{\text{Ga}}$ on $S = [\alpha, 1]^n$.

ing values of the correlation parameter ϱ. The chosen correlation parameters match the corresponding values of the pairwise Kendall's rank correlation of the Gumbel copula in Example 5.14 a). Given the same strength of dependence, the Gaussian copula yields a smaller relative improvement compared to the Gumbel copula.

c) **Inhomogeneous portfolio**: For an inhomogeneous portfolio we choose $n = 9$ with 3 Pareto(2), 3 LogN(0.2, 1), and 3 Gamma(3, 2) risks and a Gumbel lower bound $C_X \geq C_\vartheta^{\text{Gu}}$ on $S = [\alpha, 1]^n$. Again we obtain similar conclusions as in a) and b) but at a higher level, as the marginal risks have higher tails; see Table 5.3.   ◊

5.3   VaR Bounds with Two-Sided Dependence Information

The improved standard bounds dealt with in Sections 5.1 and 5.2 include, in addition to the information on the marginal distributions, one-sided information on the distribution functions of the form $F_X \leq G$ of $\overline{F}_X \geq \overline{H}$ or similar. As a result one obtains improved VaR bounds compared to the marginals-only case if the dependence condition is strong enough and the dimensionality is not too high. The examples in Section 5.2 give a

| | | \multicolumn{5}{c}{$\mathrm{VaR}_{\alpha^*}(S_n^c)$} | |
$n = 9$	$\overline{\mathrm{VaR}}_\alpha(S_n)$	$\vartheta = 1$	$\vartheta = 3$	$\vartheta = 5$	$\vartheta = 10$	$\overline{\mathrm{VaR}}_\alpha(S_n^c)$
$\alpha = 0.990$	165.39	232.76	144.77	132.06	123.24	114.06
$\alpha = 0.995$	206.04	289.42	180.52	164.53	153.51	143.23
$\alpha = 0.999$	349.62	510.77	305.86	277.23	257.74	239.79

Table 5.3 The same as Table 5.1 for an inhomogeneous risk vector of $n = 9$ risks with $F_1 = F_2 = F_3 = \mathrm{Pareto}(2)$, $F_4 = F_5 = F_6 = \mathrm{LogN}(0.2, 1)$ and $F_7 = F_8 = F_9 = \mathrm{Gamma}(3, 2)$, and Gumbel copula lower bound $C_X \geq C_\vartheta^{\mathrm{Gu}}$ on $S = [\alpha, 1]^n$.

first impression of the quality of these improvements. Further examples are discussed in Embrechts and Puccetti (2006a), Puccetti et al. (2016), and the survey papers Rüschendorf (2017b,a).

The method of improved standard bounds is extended in Lux and Rüschendorf (2018) to include two-sided dependence information of the form $G \leq F \leq H$ or related bounds for the copula C or for the survival functions $\overline{F}$. Thus the basis of this method is the availability of improved Fréchet bounds for the distribution functions resp. the copula functions.

In the following, let $X = (X_1, \ldots, X_n)$ be a risk vector with fixed marginals $F_1, \ldots, F_n$. If $C = C_X$ is the copula of X then by Sklar's Theorem the distribution function $F(x) = C(F_1(x_1), \ldots, F_n(x_n))$ is uniquely described by C, and we denote the expectation in dependence of C for a measurable function $\varphi : \mathbb{R}^n \to \mathbb{R}^1$ by

$$
E_C \varphi := E\varphi(X) = \int \varphi(x) dF(x)
$$
$$
= \int \varphi(x) dC(F_1(x_1), \ldots, F_n(x_n)). \tag{5.40}
$$

We consider dependence restrictions of the form

$$
\underline{Q} \leq C \leq \overline{Q}, \tag{5.41}
$$

where $\underline{Q}$ and $\overline{Q}$ are quasi-copulas, i.e., are increasing, satisfy the boundary conditions $Q(u_1, \ldots, u_i = 0, \ldots, u_n) = 0$ and $Q(1, \ldots, 1, u_i, 1, \ldots, 1) = u_i$, and are Lipschitz continuous, i.e., $|Q(u) - Q(v)| \leq \sum_{i=1}^n |u_i - v_i|$. Typical examples of quasi-copulas are infima or suprema over sets of copulas.

We generally assume a copula C as above exists and that $W_n \leq \underline{Q} \leq \overline{Q} \leq M_n$, i.e., $\underline{Q}$ and $\overline{Q}$ are improvements over the Fréchet bounds.

The modified Hoeffding–Fréchet functionals are then given by

$$
\underline{P}_\varphi := \inf\{E_C \varphi; \ C \in C^n, \underline{Q} \leq C \leq \overline{Q}\} \tag{5.42}
$$

and

$$
\overline{P}_\varphi := \sup\{E_C \varphi; \ C \in C^n, \underline{Q} \leq C \leq \overline{Q}\}, \tag{5.43}
$$

where C^n is the class of all copulas. As in the case without additional dependence information, as treated in Chapter 3, there is a dual representation of the Hoeffding–Fréchet functionals. Denote

$$\mathfrak{R} = \left\{ h = \sum_{i=1}^{k} \alpha_i \Lambda_{u^i}; \ k \in \mathbb{N}, \alpha_i \geq 0, u^i \in \overline{\mathbb{R}}^n \right\},$$

with $\Lambda_u(x) = 1_{(-\infty, u]}(x), x \in \mathbb{R}^n$. We denote the lower-semicontinuous version of Λ_u by $\Lambda_u^-(x)$ and define for $h \in \mathfrak{R}$, $h^- := \sum_{i=1}^{k} \alpha_i \Lambda_{u^i}^-$. Then for a copula C we obtain

$$E_C \Lambda_u = \int \Lambda_u(x) dC(F_1(x_1), \dots, F_n(x_n)) = C(F_1(u_1), \dots, F_n(u_n))$$

and $E_C \Lambda_u^- = C(F_1^-(u_1), \dots, F_n^-(u_n))$.

For a quasi-copula Q and $h = \sum_{i=1}^{k} \alpha_i \Lambda_{u^i} \in \mathfrak{R}$, we define

$$Q(h) := \sum_{i=1}^{k} \alpha_i Q(F_1(u_1^i), \dots, F_n(u_n^i))$$

$$\text{and} \quad Q(h^-) := \sum_{i=1}^{k} \alpha_i Q(F_1^-(u_1^i), \dots, F_n^-(u_n^i)).$$

Now we can introduce the dual forms of the Hoeffding–Fréchet functionals:

$$\underline{D}_\varphi := \sup \left\{ \underline{Q}(h) - \overline{Q}(g^-) + \sum_{i=1}^{n} E_i f_i; \right. \tag{5.44}$$
$$\left. f_i \in L^1(F_i), h, g \in \mathfrak{R} \text{ such that } h - g^- + \sum_{i=1}^{n} f_i \leq \varphi \right\}$$

and

$$\overline{D}_\varphi := \inf \left\{ \overline{Q}(h) - \underline{Q}(g^-) + \sum_{i=1}^{n} E_i f_i; \right. \tag{5.45}$$
$$\left. f_i \in L^1(F_i), h, g \in \mathfrak{R} \text{ such that } h^- - g + \sum_{i=1}^{n} f_i \geq \varphi \right\}.$$

Then the following basic result is established in Lux and Rüschendorf (2018).

Theorem 5.15 (**Dual bounds with two-sided dependence information**) *Let φ be upper majorized, i.e., $|\varphi| \leq \sum_{i=1}^{n} g_i$ for some $g_i \in L^1(F_i)$. If φ is lower semicontinuous, then*

$$\underline{P}_\varphi = \underline{D}_\varphi. \tag{5.46}$$

If φ is upper semicontinuous, then

$$\overline{P}_\varphi = \overline{D}_\varphi. \tag{5.47}$$

Moreover, there exist copulas $\underline{C}$ and $\overline{C}$ satisfying (5.38) such that

$$E_{\underline{C}} \varphi = \underline{P}_\varphi \quad \text{and} \quad E_{\overline{C}} \varphi = \overline{P}_\varphi.$$

Proof The proof is first given for the case that φ is bounded continuous. In this case it can be reduced by the minimax theorem of Ky-Fan to the duality theorem for Hoeffding–Fréchet functionals in (3.9) applied to a modified function. The general case then follows by an approximation argument. □

In particular, for any admissible $h, g \in \mathfrak{R}$, $f_i \in L^1(F_i)$ for the dual functional $\underline{D}_\varphi$, i.e., $h - g^- + \sum_{i=1}^m f_i \leq \varphi$, we obtain

$$\underline{P}_\varphi \geq \underline{Q}(h) - \overline{Q}(g^-) + \sum_{i=1}^m E_i f_i. \tag{5.48}$$

The dual representation is not easy to determine in explicit form, but it gives rise to a reduction scheme that is a relaxed version of the dual functional.

We consider the case $\varphi(x) = 1_{(-\infty,s)}(\Psi(x))$ for some non-decreasing aggregation function Ψ. Let $U_\Psi(s) = \{x \in \mathbb{R}^n; \Psi(x) < s\}$, and consider the set of admissible dual functions

$$\underline{\mathcal{A}}^r = \{(h, g); h, g \in \mathcal{R}^r, h - g^- \leq \varphi\}, \tag{5.49}$$

where $\mathcal{R}^r = \{\sum_{i=1}^k \Lambda_{u^i}; k \in \mathbb{N}, u_i \in U_\Psi(s)\}$, i.e., we restrict the dual functions in (5.44) to those with $f_i = 0$ and restrict to $h \in \mathfrak{R}$ with coefficients $\alpha_i = 1$. Optimization over $\underline{A}^r$ requires a truncation of the variable k, and we denote by $\mathcal{R}^r(k) \subset \mathcal{R}^r$ those sums with k summands. For an efficient description of the reduction, the notion of multisets is useful.

Definition 5.16 A multiset over a set $\mathcal{B}$ is a pair $\langle \mathcal{B}, f \rangle$, where $f: \mathcal{B} \to \mathbb{N}$ is called a multiplicity function.

The multiplicity function f counts the number of occurrences of each element of $\mathcal{B}$ in the multiset. The following lemma gives a multiset version of the inclusion-exclusion principle.

Lemma 5.17 (Multiset inclusion-exclusion principle) *Let $B_1, \ldots, B_k \subset \mathbb{R}^n$, and define the multisets $\langle \mathcal{B}^0, f^0 \rangle$, $\langle \mathcal{B}^e, f^e \rangle$ by*

$$\begin{aligned}
\mathcal{B}^0 &:= \{B_{i_1} \cap \cdots \cap B_{i_m}; \ 1 \leq i_1 < \cdots < i_m \leq k, m \text{ odd}\}, \\
\mathcal{B}^e &:= \{B_{i_1} \cap \cdots \cap B_{i_m}; \ 1 \leq i_1 < \cdots < i_m \leq k, m \text{ even}\},
\end{aligned} \tag{5.50}$$

where

$$\begin{aligned}
f^0(B) := |\{(i_1, \ldots, i_m); \\
1 \leq i_1 < \cdots < i_m \leq k, m \text{ odd}, B = B_{i_1} \cap \cdots \cap B_{i_m}\}|
\end{aligned}$$

for $B \in \mathcal{B}^0$ and f^e is defined analogously. Then

$$1_{B_1 \cup \cdots \cup B_k} = \sum_{B \in \mathcal{B}^0} (f^0(B) - f^e(B))_+ 1_B - \sum_{B \in \mathcal{B}^e} (f^e(B) - f^0(B))_+ 1_B. \tag{5.51}$$

Proof Equation (5.51) is based on the classical inclusion-exclusion principle stating that

$$1_{B_1 \cup \cdots \cup B_k} = \sum_{B \in \mathcal{B}^0} f^0(B) 1_B - \sum_{B \in \mathcal{B}^e} f^e(B) 1_B.$$

Now rearranging terms and using that $f^e(B) = 0$ for $B \in \mathcal{B}^0 \setminus \mathcal{B}^e$ and $f^0(B) = 0$ for $B \in \mathcal{B}^e \setminus \mathcal{B}^0$, (5.51) follows.   □

Remark 5.18 (Motivation of two-sided bounds)   Lemma 5.17 establishes a non-redundant version of the classical inclusion-exclusion principle. To illustrate this, consider the case $k = 3$ and $B_1, B_2, B_3 \subset \mathbb{R}^n$ such that $B_1 \cap B_2 \cap B_3 = B_1 \cap B_2$, $B_1 \neq B_2$. Then the classical inclusion-exclusion principle yields

$$
\begin{aligned}
{}^1 B_1 \cup B_2 \cup B_3 \\
= {}^1 B_1 + {}^1 B_2 + {}^1 B_3 - {}^1 B_1 \cap B_2 - {}^1 B_2 \cap B_3 - {}^1 B_1 \cap B_3 + {}^1 B_1 \cap B_2 \cap B_3,
\end{aligned}
$$

where the terms $-{}^1 B_1 \cap B_2$ and $+{}^1 B_1 \cap B_2 \cap B_3$ cancel. This superfluous subtraction is avoided by the multiset formulation, which yields

$$
{}^1 B_1 \cup B_2 \cup B_3 = {}^1 B_1 + {}^1 B_2 + {}^1 B_3 - {}^1 B_1 \cap B_3 - {}^1 B_2 \cap B_3.
$$

In order to estimate the probability of $\sum_{i=1}^n B_i$ from above (or from below), one needs upper and lower bounds for the probability of the B_i. In the case that B_i are intervals, say $B_i = (-\infty, u^i]$, we obtain upper and lower bounds by the assumed two-sided bounds $\underline{Q} \leq F \leq \overline{Q}$ on the distribution function and thus the possibility to improve the bounds for the probability of unions.   ◊

In the case where $B_i = (-\infty, u^i]$, the intersections $B_{i_1} \cap \cdots \cap B_{i_m}$ can be identified with $(-\infty, u]$, where $u = \min(u^{i_1}, \ldots, u^{i_m})$ is the componentwise minimum of the u^{i_j}. Thus we define

$$
\mathcal{M}^0(u^1, \ldots, u^k) := \{\min(u^{i_1}, \ldots, u^{i_m}); \ 1 \leq i_1 < \cdots < i_m \leq k, m \text{ odd}\} \tag{5.52}
$$

and

$$
\mathcal{M}^e(u^1, \ldots, u^k) := \{\min(u^{i_1}, \ldots, u^{i_m}); \ 1 \leq i_1 < \cdots < i_m \leq k, m \text{ even}\}, \tag{5.53}
$$

and the corresponding multiplicity function

$$
\ell^0(u) := |\{(i_1, \ldots, i_m);
$$
$$
1 \leq i_1 < \cdots < i_m \leq k, m \text{ odd}, u = \min(u^{i_1}, \ldots, u^{i_m})\}|
$$

resp. $\ell^e(u)$.

For $u^1, \ldots, u^k \in U_\Psi(s)$, monotonicity of Ψ implies that $\bigcup_{i=1}^k B_i \subset U_\Psi(s)$, and the following reduced lower bounds with two-sided dependence information are obtained. The theorem also states monotonicity in k for the reduced bounds.

Theorem 5.19   (Reduced lower bounds with two-sided dependence information)
Let Ψ be non-decreasing, $\varphi(x) = 1_{(-\infty, s)}(\Psi(x))$, and define

$$
\underline{D}_\varphi(k) := \sup \bigg\{ \sum_{u \in \mathcal{M}^0(u^1, \ldots, u^k)} (\ell^0(u) - \ell^e(u))_+ \underline{Q}(F(u))
$$
$$
- \sum_{u \in \mathcal{M}^e(u^1, \ldots, u^k)} (\ell^e(u) - \ell^0(u))_+ \overline{Q}(F(u)); \ u^1, \ldots, u^k \in U_\Psi(s) \bigg\},
$$

then

$$\underline{D}_\varphi(k) \leq \underline{D}_\varphi(k+1) \leq \cdots \leq \underline{D}_\varphi.$$

Remark 5.20

a) A similar reduction scheme can also be given for the upper bound $\overline{D}_\varphi$ (see Theorem 4.10 in Lux and Rüschendorf, 2018).

b) Theorem 5.19 establishes a tractable optimization problem to compute a lower bound on $\underline{D}_\varphi$ and thus also on $\underline{P}_\varphi$. The optimization takes place over vectors in the sublevel set $U_\Psi(s)$, and the tradeoff between the computational effort and the quality of the bound is moderated by the variable k. For fixed k, $\underline{D}_\varphi(k)$ amounts to a kn-dimensional optimization problem that can be solved with standard optimization packages.

c) It is shown in Lux and Rüschendorf (2018) that in the limit when $\underline{Q}^j \to C$ and $\overline{Q}^j \to C$, the corresponding reduction scheme converges to the true value; i.e., $\underline{D}_\varphi(k) \to P_C(\Psi(X) < s)$ if Ψ is lsc or usc. As a result this implies consistency of the reduction scheme for sequentially consistent upper and lower bounds.

d) It is interesting to note that in general the addition of a splitting variable $u^{k+1} \in U_\Psi(s)$ does not necessarily improve the estimate; i.e., with $B_i = (-\infty, u^i)$, the bound for $\bigcup_{i=1}^{k+1} B_i$ is in general not better than that for $\bigcup_{i=1}^{k} B_i$. On the other hand, if $u^{k+1} = u^k$, then the bound for $B_1 \cup \cdots \cup B_{k+1}$, $B_i = (-\infty, u^i]$, is the same as the bound obtained from $B_1 \cup \cdots \cup B_k$. This is the reason for the monotonicity property. The monotonicity still holds for the best choice of the u^i, as Theorem 5.19 asserts.

◊

Next we consider some examples and compare the two-sided improved bounds with the (one-sided) improved bounds. We typically find a strong improvement.

Example 5.21 (Robust neighborhood bounds) Consider a risk vector $X = (X_1, \ldots, X_5)$ with Pareto(5) marginals, and assume that the copula $C = C_X$ lies in the vicinity of a reference copula C^* as measured by the Kolmogorov distance, i.e.,

$$\|C - C^*\|_\infty \leq \delta \quad \text{for some } \delta > 0. \tag{5.54}$$

C^* might be a reference model like a Gaussian model or an estimated (empirical) copula, where $\delta > 0$ amounts to the estimation error and (5.54) corresponds to a robust neighborhood model. Lux and Papapantoleon (2019) establish improved Fréchet bounds for robust neighborhood models for a large class of distances, which yield under the assumption in (5.54):

$$\underline{Q}(u) := \max\{C^*(u) - \delta, W_5(u)\} \leq C(u)$$
$$\leq \min\{C^*(u) + \delta, M_5(u)\} =: \overline{Q}(u), \tag{5.55}$$

with $\underline{Q}(u) = \underline{Q}^{\|\cdot\|_\infty, \delta}(u)$, $\overline{Q}(u) = \overline{Q}^{\|\cdot\|_\infty, \delta}(u)$.

Similarly, for comparison with the (one-sided) improved standard bounds on VaR, we also assume that the survival copula $\overline{C}$ satisfies

$$\|C - \overline{C}^*\|_\infty \leq \delta,$$

	$\varrho = -0.1$			$\varrho = 0.4$			$\varrho = 0.8$		
	i. standard	scheme	impr.	i. standard	scheme	impr.	i. standard	scheme	impr.
α	(low : up)	(low : up)	%	(low : up)	(low : up)	%	(low : up)	(low : up)	%
				$\delta = 0.0001$					
0.950	3.5 : 44.8	8.8 : 24.0	63 %	3.7 : 41.2	8.0 : 26.7	50 %	7.8 : 31.3	9.6 : 24.8	35 %
0.990	9.0 : 106.3	19.9 : 44.8	74 %	9.1 : 102.7	19.0 : 63.5	54 %	17.8 : 82.0	21.5 : 64.5	33 %
0.995	13.3 : 152.1	27.0 : 60.8	76 %	13.3 : 149.4	25.8 : 90.5	52 %	24.3 : 119.0	28.5 : 91.6	33 %
				$\delta = 0.0005$					
0.950	3.4 : 45.0	8.2 : 24.8	60 %	3.6 : 41.2	7.2 : 28.1	44 %	7.8 : 31.4	9.2 : 26.2	28 %
0.990	9.0 : 106.2	15.9 : 56.7	58 %	9.0 : 105.3	14.9 : 80.8	32 %	17.4 : 84.9	18.6 : 82.2	6 %
0.995	13.3 : 153.0	19.0 : 90.0	49 %	13.3 : 153.0	18.0 : 153.0	3 %	23.3 : 126.0	22.8 : 125.0	0 %

Table 5.4 Improved standard bounds on VaR of $X_1 + \cdots + X_5$ and VaR estimates via reduction schemes for different correlation parameters ϱ and distance thresholds δ.

and thus denoting the corresponding lower and upper bounds in (5.55) by $\underline{\widehat{Q}}$ resp. $\overline{\widehat{Q}}$, we obtain

$$\underline{\widehat{Q}} \leq \overline{C} \leq \overline{\widehat{Q}}.$$

With these improved two-sided Fréchet bounds, we calculate by the reduction scheme in Theorem 5.19 the reduced lower and upper bounds and compare them to the (one-sided) improved standard bounds as in Theorem 5.3 with $G = \underline{\widehat{Q}}$ and $H = \overline{\widehat{Q}}$.

For the calculation we assume that the reference copula C^* is an equicorrelated Gaussian with correlation ϱ. The bounds consistently improve for increasing $k = 1, \ldots, 6$, while for $k \geq 7$ no further improvement is obtained. Table 5.4 shows the VaR estimates for different levels of ϱ and for $\delta = 0.0001$ and $\delta = 0.0005$. Note that in order to obtain informative bounds on the tail of the distribution of the sum, the seemingly small choice of the level δ is appropriate. For $\delta = 0.0001$, the improvement by including two-sided information via the reduction scheme ranges from 33 % in the case of high correlation to 76 % in the case of weak negative correlation. In all cases the bounds improve the sharp unconstrained bounds. In the case $\delta = 0.0005$, the improvement is – as expected – considerably weaker.

	$\varrho = -0.1, \varrho = 0.2$			$\varrho = 0.3, \varrho = 0.5$		
	i. standard	scheme	impr.	i. standard	scheme	impr.
α	(low : up)	(low : up)	%	(low : up)	(low : up)	%
0.950	3 : 32	8 : 26	38 %	1 : 30	7 : 29	24 %
0.990	9 : 74	20 : 52	51 %	2 : 74	18 : 63	37 %
0.995	13 : 104	26 : 70	52 %	3 : 104	25 : 86	40 %

Table 5.5 Improved standard bounds on VaR of $X_1 + \cdots + X_4$ and VaR estimates computed via reduction schemes using $C^{\underline{\Sigma}}$ and $C^{\overline{\Sigma}}$.

Example 5.22 (Parameter robustness) Let $X = (X_1, \ldots, X_4)$ be a risk vector with copula C and Pareto(2) marginals. Assume that

$$C^{\underline{\Sigma}} \leq C \leq C^{\overline{\Sigma}}, \tag{5.56}$$

where $C^{\underline{\Sigma}}$ and $C^{\overline{\Sigma}}$ are four-dimensional Gauss copulas with correlation matrices $\underline{\Sigma} = (\underline{\varrho}_{ij})$, $\overline{\Sigma} = (\overline{\varrho}_{ij})$ such that $\underline{\varrho}_{ij} \leq \overline{\varrho}_{ij}$. By Slepian's lemma, it then holds that

$$C^{\underline{\Sigma}} \leq C^{\overline{\Sigma}}.$$

This model corresponds to correlation uncertainty, which is typically the case with estimated correlations. We similarly also assume corresponding bounds for the survival copulas:

$$\overline{C}^{\underline{\Sigma}} \leq \overline{C} \leq \overline{C}^{\overline{\Sigma}}. \tag{5.57}$$

In Table 5.5 we consider the case of equicorrelated parameters $\underline{\varrho}_{ij} = \underline{\varrho}$, $\overline{\varrho}_{ij} = \overline{\varrho}$ for the parameters $k = 5$. The bounds consistently improve for increasing $k = 1, \ldots, 5$, while for $k \geq 6$ no further improvement is obtained. All bounds improve upon the unconstrained VaR bounds.

	$m = 8$			$m = 4$			$m = 2$		
	i. standard	scheme	impr.	i. standard	scheme	impr.	i. standard	scheme	impr.
α	(low : up)	(low : up)	%	(low : up)	(low : up)	%	(low : up)	(low : up)	%
0.950	42 : 113	59 : 86	62 %	22 : 150	39 : 112	43 %	12 : 193	28 : 150	33 %
0.990	82 : 210	108 : 147	70 %	42 : 264	67 : 175	51 %	21 : 329	42 : 218	43 %
0.995	105 : 266	135 : 180	72 %	53 : 329	83 : 206	55 %	43 : 403	51 : 252	44 %

Table 5.6 Improved standard bounds and VaR estimates via reduction schemes for $X_1 + \cdots + X_{16}$ given distributions of subgroups.

Further applications to subgroup models where the risk vector is partitioned into m strongly positive dependent subgroups where the copula C_m of the subgroup sums lies between two Frank or two Gumbel copulas are discussed in Lux and Rüschendorf (2018). Again the results yield strong improvements compared to the (one-sided) improved standard bounds, increasing with the number m of components (see Table 5.6). The case of general dependence within the subgroups is also discussed there (see also Chapter 9 in this book).

6 VaR Bounds with Variance Constraints

In this chapter, we study VaR bounds for sums of risks with known marginal distributions (describing the stand-alone risks) under the additional constraint that a bound on the variance or on higher moments of the sum is known (see Bernard et al., 2017c). Upper bounds on the variance restrict in particular the degree of positive dependence in the upper tail of the components. In consequence this information is expected to yield a reduction of the upper VaR bounds of a portfolio in comparison to the unconstrained case. The main content of this chapter is to confirm this intuition and to investigate the degree of reduction of the VaR bounds due to additionally known bounds on the variance.

This setting is of significant interest as in many practical situations it corresponds closely to the maximum information at hand when assessing the VaR of a portfolio. For example, in the context of credit risk portfolio models, one typically has knowledge regarding the marginal risks (through so-called probability of default (PD), exposure-at-default (EAD), and loss-given-default (LGD) models), and the variance of the aggregate risk (sum of the individual losses) is also often available, as obtained from default correlation models or through statistical analysis of observed credit losses. The same setting also appears in the context of risk aggregation and solvency calculations (Basel III and Solvency II). Indeed, banks and insurers typically have models to estimate risk distributions and VaR per risk type (credit risk, market risk, operational risk, etc.) and per business line, after which they rely on a correlation matrix to obtain the VaR of the aggregated portfolio. Taking this information into account, we assess the extent of the model risk that remains in the computation of VaR.

The basic assumptions in this chapter are related to the setting of moment bounds for VaR, which is quite common in actuarial science and in finance where (only) some of the moments of the portfolio sum are assumed to be known. As we describe in Part IV of this book, this method yields many practical and useful results, often in explicit analytical form. As we confirm in this chapter, the VaR bounds based on marginal and moment information coincide with the moment bounds for VaR when the variance constraint is low enough. We also show that both types of bounds become equivalent for large portfolios but may differ substantially in the case of smaller portfolios or if the portfolio depicts significant concentration.

To obtain sharp VaR bounds for the constrained case, where in addition to marginal information a bound for the variance is also given, we show that it is necessary to make the distribution of the aggregate risk as flat as possible in the upper part as well as in the lower part while at the same time considering the variance constraint. This aim is achieved by an extension of the RA resp. BRA algorithms.

The marginal structure is essential to this algorithm, and thus it is not available when only moment bounds are given. A second important consequence of the marginal assumptions in combination with variance (or moment) bounds is that one obtains by means of the algorithm bounds for more general risk functions $f(X)$ of the risk vector X, as compared to the moment method based on some moments of the portfolio sum S only.

In Section 6.1, we recollect some basic notions, the TVaR bounds, and the connection with convex order from Chapter 1, and we derive improved (constrained) VaR bounds based on the additional assumption on the bounds for the variance. We discuss the connection with the pure moment bounds assumption, the sharpness of the bounds, and the connection with convex order.

In Section 6.2, algorithms based on the RA and BRA algorithms introduced and applied in examples to give approximate sharp bounds and to compare the different setups. In Section 6.3, we discuss further constraints, such as homogeneity, default probabilities, and higher-order moments. Finally, in Section 6.4, we discuss various consequences and applications to credit risk modeling.

6.1 VaR Bounds with an Upper Bound on the Variance

For a portfolio $X = (X_1, \ldots, X_n)$ with marginal distributions $X_i \sim F_i$, $1 \le i \le n$, we derived in Chapter 1 simple to calculate TVaR bounds for the value-at-risk of the sum $S = \sum_{i=1}^n X_i$; more precisely, the following result is given in Theorem 1.6.

Theorem 6.1 (Unconstrained bounds)   *Let* $X_i \sim F_i$, $1 \le i \le n$, $S = \sum_{i=1}^n X_i$, *and let* $S^c = \sum_{i=1}^n X_i^c$ *denote the comonotonic sum. Then for* $\alpha \in [0, 1]$,

$$A_\alpha := \sum_{i=1}^n \mathrm{LTVaR}_\alpha(X_i) = \mathrm{LTVaR}_\alpha(S^c) \le \mathrm{VaR}_\alpha(S)$$

$$\le \mathrm{VaR}_\alpha^+(S) \le B_\alpha := \sum_{i=1}^n \mathrm{TVaR}_\alpha(X_i) = \mathrm{TVaR}_\alpha(S^c). \tag{6.1}$$

Remark 6.2

1. **Worst (best) case dependence:** The upper bound B_α is sharp and attained (worst case risk) if and only if the quantile function of S takes the constant value B_α on $[\alpha, 1]$, i.e., the restricted distributions F_i^α on the upper part are mixing in the upper α-part (see Theorem 4.7). Similarly, the lower bound A_α is attained (best case risk) if and only if the distributions $F_{i,\alpha}$ restricted to the lower part are mixing. The worst case property of a portfolio thus implies a negative dependence property in the upper α-part of the distributions; in particular, the worst case VaR_α portfolio depends on the level α. In comparison, a TVaR_α worst case distribution is attained by the comonotonic sum for all levels α. On the other hand, any coupling of the upper α-parts of the F_i combined with any coupling of the lower α-parts is a worst case TVaR_α-coupling. Similarly, in the case of mixable lower parts $F_{i,\alpha}$, the best case VaR_α distribution is unique and dependent on α on the lower parts but arbitrary on the upper parts.

2. **Connection to stop loss premiums:** Similar conclusions hold for a stop loss premium that becomes maximized when the risks are comonotonic, but this solution is not unique for a given level of retention. Papers that have studied maximization of stop loss premia under partial information of the marginal distributions without assuming knowledge of the dependence structure among the risks include Lo (1987) and Genest et al. (2002).

   The techniques employed here allow one to improve these stop loss bounds, when bounds on the variance of the portfolio sum are given.

3. **Bounds for the tail risk:** By inverting them, the VaR bounds in Theorem 6.1 also imply tail risk bounds, i.e., upper and lower bounds for $P(S \geq t)$ for $t \in \mathbb{R}$.   ◇

As described in Chapter 1, the problem of obtaining sharp upper bounds for the value-at-risk of a sum is closely connected to determining minimal sums in convex order for the distributions F_i^α of restrictions of F_i to the upper α-part of the distribution. For a precise statement of this result, see Theorem 1.15. In particular, if the F_i^α are mixable, i.e., there exist $Y_i^\alpha \sim F_i^\alpha$ such that $\sum_{i=1}^{n} Y_i^\alpha = c$, then for $X_i \sim F_i$ it holds that

$$\mathrm{VaR}_\alpha \left(\sum_{i=1}^{n} X_i \right) \leq c \tag{6.2}$$

(see Corollary 1.17).

The main problem considered in this chapter is to obtain improved (sharp) VaR bounds (resp. TVaR bounds) under the additional assumption

$$\mathrm{Var}(S) \leq s^2 \tag{6.3}$$

on the variance of the portfolio S. This assumption gives a restriction on the positive dependence on X in particular in the upper part of the distributions and thus is expected to lead to an improvement of the upper risk bounds. We define for $\alpha \in [0, 1]$ the constrained VaR bounds:

$$M_\alpha(s^2) = \overline{\mathrm{VaR}}_\alpha^+(s^2) = \sup\{\mathrm{VaR}_\alpha^+(S); X_i \sim F_i, \mathrm{Var}(S) \leq s^2\} \tag{6.4}$$

and

$$m_\alpha(s^2) = \underline{\mathrm{VaR}}_\alpha(s^2) = \inf\{\mathrm{VaR}_\alpha(S); X_i \sim F_i, \mathrm{Var}(S) \leq s^2\} \tag{6.5}$$

as the sharp upper resp. lower risk bounds with additional variance information.

To investigate these constrained VaR bounds, we introduce a random variable X^* having a two-point distribution:

$$X^* = \begin{cases} A, & \text{with probability } \alpha, \\ B, & \text{with probability } 1 - \alpha, \end{cases} \tag{6.6}$$

where $A = A_\alpha$, $B = B_\alpha$ are unconstrained bounds in Theorem 6.1. Then by definition of A, B,

$$EX^* = ES^c = ES = \mu,$$

and

$$\mathrm{Var}(X^*) = \alpha(A - \mu)^2 + (1 - \alpha)(B - \mu)^2, \tag{6.7}$$

and

$$\mathrm{VaR}_\alpha(X^*) = A, \quad \mathrm{VaR}_\alpha^+(X^*) = B.$$

It is easy to see that X^* has minimum variance among all random variables Y such that $P(Y = A) = \alpha$ and that satisfy $EY = \mu$.

When the variable X^* satisfies the variance constraint, i.e., $\text{Var}(X^*) \leq s^2$, then the unconstrained risk bounds A and B cannot be improved.

However, if $\text{Var}(X^*) > s^2$, then A and B are generally too wide, and better bounds can be constructed. It is then intuitive that better bounds can be found by constructing a random variable Y taking two values a larger than A and b smaller than B with respective probabilities α and $1 - \alpha$ in such a way that the variance constraint of the portfolio sum is satisfied. Note that a two-point distribution taking value $\mu - s\sqrt{\frac{1-\alpha}{\alpha}}$ with probability α and value $\mu + s\sqrt{\frac{\alpha}{1-\alpha}}$ with probability $1 - \alpha$ is the only two-point distribution with mean μ and variance s^2.

The following theorem shows that the construction as outlined above gives bounds on value-at-risk in the presence of a variance constraint.

Theorem 6.3 (Constrained bounds) *Let $\alpha \in (0, 1)$, $X_i \sim F_i$, $i = 1, 2, \ldots, n$, and $S = \sum_{i=1}^{n} X_i$ satisfy $\text{Var}(S) \leq s^2$. Then, we have*

$$a := \max\left(\mu - s\sqrt{\frac{1-\alpha}{\alpha}}, A\right) \leq m \leq \text{VaR}_\alpha(S) \leq \text{VaR}_\alpha^+(S)$$

$$\leq M \leq b := \min\left(\mu + s\sqrt{\frac{\alpha}{1-\alpha}}, B\right). \tag{6.8}$$

In particular, if $s^2 \geq \alpha(A - \mu)^2 + (1 - \alpha)(B - \mu)^2$, then $a = A$ and $b = B$, and the unconstrained bounds are not improved by the constraint on the variance.

Proof Define the function

$$B(t) := \frac{1}{1-\alpha} \int_{\alpha-t}^{1-t} \text{VaR}_u(S^c) \, du \tag{6.9}$$

on the interval $[0, \alpha]$, and note that $B(0) = B$. Let us also define the variable X_t^*, $t \in [0, \alpha]$, which takes the values $A(t)$ and $B(t)$,

$$X_t^* = \begin{cases} A(t) & \text{with probability } \alpha, \\ B(t) & \text{with probability } 1 - \alpha, \end{cases} \tag{6.10}$$

in which $A(\alpha) := \frac{\mu - B(\alpha)(1-\alpha)}{\alpha}$. Note that $A(0) = A$ and that $X_0^* =_d X^*$. Furthermore, as per construction, $EX_t^* = \mu$. The variance function $t \to \text{Var}(X_t^*)$ has the following monotonicity property:

Monotonicity property of $\text{Var}(X_t^*)$. The variance of X_t^* given by $\text{Var}(X_t^*) = \alpha(A(t) - \mu)^2 + (1 - \alpha)(B(t) - \mu)^2$ is a maximum when $t = 0$.

To prove this property, note that by increasing t we decrease $B(t)$ and thus increase $A(t)$. Hence, there exists $0 < \beta < \alpha$ such that $t \to \text{Var}(X_t^*) := \alpha(A(t) - \mu)^2 + (1 - \alpha)(B(t) - \mu)^2$ is continuously decreasing on $[0, \beta]$ with minimum value given by $\text{Var}(X_\beta^*) = 0$ and with maximum value given by $\text{Var}(X_0^*) = \alpha(A - \mu)^2 + (1 - \alpha)(B - \mu)^2$.

We first prove that

$$a := A(t^*) \leq m \leq \text{VaR}_\alpha(S) \leq \text{VaR}_\alpha^+(S) \leq M \leq b := B(t^*), \tag{6.11}$$

in which t^* is defined as

$$t^* := \min \left\{ t \mid 0 \leq t \leq \alpha, \text{Var}(X_t^*) \leq s^2 \right\}. \tag{6.12}$$

Proof of (6.11): If $\text{Var}(X_{t^*}^*) < s^2$, this means that $t^* = 0$. Hence, $A(t^*)$ and $B(t^*)$ correspond to the absolute bounds, and there is nothing to prove. We further assume that $\text{Var}(X_{t^*}^*) = s^2$ and denote by G the distribution of $X_{t^*}^*$. We first prove that b is an absolute upper bound for feasible solutions of (6.4). To this aim, assume there exist $(X_1, X_2, \ldots, X_n)$ such that $\text{VaR}_\alpha^+(S) > b$. One has that, for all $a \leq x < b$, $F_S(x) \leq G(x) = \alpha$. When $b \leq x$, $F_S(x) \leq G(x) = 1$. Since $G(x) = 0$ when $x < a$, this implies that

$$\begin{cases} \text{for all } x < a, & F_S(x) \geq G(x), \\ \text{for all } x \geq a, & F_S(x) \leq G(x). \end{cases} \tag{6.13}$$

In other words, the distribution function F_S crosses G once from above. Since $ES = \mu$, this implies that $X_{t^*}^* \leq_{cx} S$ (see Karlin and Novikoff, 1963; Müller and Stoyan, 2002). Since $\text{Var}(X_{t^*}^*) = s^2$, the feasibility of $(X_1, X_2, \ldots, X_n)$ requires that $\text{Var}(S) = \text{Var}(X_{t^*}^*)$. In view of the convex ordering between S and $X_{t^*}^*$, this is only possible when $S \overset{d}{=} X_{t^*}^*$, which is a contradiction.

The proof that a is an absolute lower bound can be given in a similar way. Now let $(X_1, X_2, \ldots, X_n)$ be such that $\text{VaR}_\alpha(S) < a$. One has that, for all $x \leq a$, $F_S(x) \geq G(x) = 0$. When $a \leq x < b$, $F_S(x) \geq G(x) = \alpha$. Since $G(x) = 1$ when $x \geq b$, this implies that

$$\begin{cases} \text{for all } x < b, & F_S(x) \geq G(x), \\ \text{for all } x \geq b, & F_S(x) \leq G(x). \end{cases} \tag{6.14}$$

In other words, the distribution function F_S crosses G once from above. By symmetry of the argument, the result follows from the first part of the proof.

From the expression (6.11), we can finish the proof of Theorem 6.3. If $s^2 \geq \alpha(A - \mu)^2 + (1 - \alpha)(B - \mu)^2$ then the result is obvious from Theorem 6.1 and the monotonicity property stated above. In the other case, by the monotonicity property of $\text{Var}\, X_t^*$ there exists t^* such that $\text{Var}(X_{t^*}^*) = s^2$. Hence, a and b can be seen as the mass points from a two-point distribution satisfying the mean constraint μ and the variance constraint s^2. This yields the desired expressions for a and b. □

The presence of the variance constraint does not always make it possible to strengthen the bounds A and B. Indeed, the variable X^* taking the values A and B may also satisfy the variance constraint, i.e., $\text{Var}(X^*) \leq s^2$ and, in this case, $a = A$ and $b = B$. Hence, we conclude that if s^2 is not too large, i.e., when $s^2 \leq \alpha(A - \mu)^2 + (1 - \alpha)(B - \mu)^2$, then the bounds a and b that are obtained for the constrained case strictly outperform the bounds in the unconstrained case.

The question thus arises: What is meant by "too large"? This aspect pertains to the characteristics of the problem and the data at hand. However, a few observations are of interest. When all risks are distributed identically, then the bounds A and B grow linearly with the size of the portfolio; however, as the standard deviation of a portfolio is subadditive, the condition $s^2 \geq \alpha(A - \mu)^2 + (1 - \alpha)(B - \mu)^2$ becomes more difficult to satisfy, meaning that it becomes more likely that bounds a and b are better than A and B. For example, when the risks are approximately independent (e.g., in a life insurance context), then bounds a and b will strictly improve upon A and B for moderate portfolio sizes.

Two observations follow:

(i) When no information on the dependence is available, we are still able to conclude that

$$\mu - \sum_{i=1}^{n} \sigma_i \sqrt{\frac{1 - \alpha}{\alpha}} \leq \text{VaR}_\alpha \left(\sum_{i=1}^{n} X_i \right) \tag{6.15}$$

$$\leq \text{VaR}_\alpha^+ \left(\sum_{i=1}^{n} X_i \right) \leq \mu + \sum_{i=1}^{n} \sigma_i \sqrt{\frac{\alpha}{1 - \alpha}},$$

where $\sigma_i^2 = \text{Var}(X_i)$, $1 \leq i \leq n$. Indeed, as the standard deviation is a subadditive risk measure, $\sum_{i=1}^{n} \sigma_i$ is an upper bound for the standard deviation of the portfolio so that (6.15) becomes a consequence of Theorem 6.3. Note that, in fact, the unconstrained bound, B, is also an upper bound for the constrained case and improves upon $\mu + \sum_{i=1}^{n} \sigma_i \sqrt{\frac{1-\alpha}{\alpha}}$.

(ii) Sometimes the correlations between some of the risks X_i are known. This partial information can then be used to provide an upper bound on the variance of the portfolio sum, which could sharpen the unconstrained VaR bounds that we derived in Theorem 6.3. For example, assume that $\text{Var}(X_1 + X_2 + \cdots + X_i) \leq s_1^2$ and $\text{Var}(X_{i+1} + X_2 + \cdots + X_n) \leq s_2^2$. Then, invoking subadditivity of the standard deviation again and using a similar rearrangement as for the proof of Theorem 6.3,

$$\mu - (s_1 + s_2) \sqrt{\frac{1 - \alpha}{\alpha}} \leq \text{VaR}_\alpha \left(\sum_{i=1}^{n} X_i \right) \tag{6.16}$$

$$\leq \text{VaR}_\alpha^+ \left(\sum_{i=1}^{n} X_i \right) \leq \mu + (s_1 + s_2) \sqrt{\frac{\alpha}{1 - \alpha}}.$$

Remark 6.4 (Connection with moment bounds) If we replace "the information on marginal distributions" by "the bounded variance information on the joint portfolio," we can formulate the following moment problem:

$$M(\mu, s^2) = \sup \{ \text{VaR}_\alpha^+(S); ES = \mu \text{ and } \text{Var}(S) \leq s^2 \}. \tag{6.17}$$

For this moment problem we obtain as result (see Part IV) that $M(\mu, s^2) = \mu + s \sqrt{\frac{\alpha}{1-\alpha}}$. This bound corresponds to the Cantelli bound (see also Barrieu and Scandolo, 2015). Thus, if the variance constraint is not "too large," we find that the moment bound $M(\mu, s^2)$ coincides with the analytical bound b. Note that Hürlimann (2002) extends these results to the case of bounded variables; see Chapter 4 for more details. Given that the bound b turns out to be nearly sharp for sufficiently large portfolios, our approach provides evidence that the literature on moment bounds is of significant interest when analyzing VaR bounds. For smaller (or concentrated) portfolios, however, one cannot expect sharpness of the bound b. In Section 6.2 we adapt the RA or BRA algorithm to infer dependence (presented in Section 2.3) to approximate the sharp bounds for any size n and variance bound s^2.   ◊

In the following we discuss sharpness of the analytical bounds and provide a precise connection with convex order. Next we show that the constrained bounds are sharp if and only if the risks $X_1, X_2, \ldots, X_n$ have a mixing property.

Theorem 6.5 (Sharpness of variance-constrained bounds) *Let $X_i = f_i(U) \sim F_i$, $1 \leq i \leq n$, where U is uniformly distributed on $(0, 1)$, and $S = \sum_{i=1}^{n} X_i$ satisfies the variance constraint, i.e., $\mathrm{Var}(S) \leq s^2$. Without loss of generality, let $\{S \geq \mathrm{VaR}_\alpha(S)\} = \{U \geq \alpha\}$. Then, the upper bound b in Theorem 6.3 is attained if and only if the lower bound a is attained and, equivalently, if*

$$S = b \text{ on } \{U \geq \alpha\} \text{ and } S = a \text{ on } \{U < \alpha\}, \tag{6.18}$$

i.e., S is mixing on the upper α-part $\{U \geq \alpha\}$ and on the lower α-part $\{U < \alpha\}$ with mixing constants b and a.

The proof of Theorem 6.5 can be given along similar lines to that of Theorem 4.7.

In general, the risk bounds proposed in Theorem 6.3 are not sharp. Theorem 6.5, however, suggests how to obtain sharp VaR bounds when there is a constraint on the variance of the sum. Roughly speaking, the outcomes of the variables should be rearranged to produce a dependence between the risks such that the outcomes for the sum are as concentrated as possible around the two values a and b, which occur with respective probabilities $1 - \alpha$ and α. This idea is concordant with the aim of finding convex order bounds. Indeed, the improved result in Theorem 6.3 based on convex order that was valid for the unconstrained case extends in a similar way to the variance-constrained case.

Let $Y_i = f_i(U) \sim F_i$, $i = 1, 2, \ldots, n$, and let $S = \sum_{i=1}^{n} Y_i$ with $\mathrm{Var}(S) \leq s^2$ be an admissible solution for the constrained VaR_α upper bound problem. Let the upper α-part of the distribution of S, $\{S \geq \mathrm{VaR}_\alpha(S)\}$, be, without loss of generality, identical to $\{U \geq \alpha\}$.

In what follows, for given random variables X, Y and subset T of Ω, $X|_T \leq_{\mathrm{cx}} Y|_T$ denotes that the conditional distribution of the restriction $X|_T$ is smaller in convex order than $Y|_T$.

Theorem 6.6 (Variance-constrained bounds and convex order) *If $X_i = f_i(U)$, and $\overline{S} = \sum_{i=1}^{n} X_i$ satisfies*

$$\overline{S}|_{U \geq \alpha} \leq_{\mathrm{cx}} S|_{U \geq \alpha} \quad \text{and} \quad \overline{S}|_{U < \alpha} \leq_{\mathrm{cx}} S|_{U < \alpha}, \tag{6.19}$$

then $\overline{S}$ is admissible, $\mathrm{Var}(\overline{S}) \leq s^2$, and $\overline{S}$ is an improvement of S in the sense that

$$\mathrm{VaR}_\alpha^+(\overline{S}) \geq \mathrm{VaR}_\alpha^+(S) \quad \text{and} \quad \mathrm{VaR}_\alpha(\overline{S}) \leq \mathrm{VaR}_\alpha(S).$$

Proof We note that by the convex ordering assumption in (6.19), we obtain for the upper α-part $T = \{U \geq \alpha\}$ of S:

$$E(\overline{S})^2 = E((\overline{S})^2 \mid T)P(T) + E((\overline{S})^2 \mid T^c)P(T^c)$$
$$\leq E(S^2 \mid T)P(T) + E(S^2 \mid T^c)P(T^c)$$
$$= ES^2,$$

and thus $\mathrm{Var}(\overline{S}) \leq \mathrm{Var}(S) \leq s^2$. The argument for the increase of $\mathrm{VaR}_\alpha^+(\overline{S})$ (resp. decrease of $\mathrm{VaR}_\alpha(\overline{S})$) compared to $\mathrm{VaR}_\alpha^+(S)$ (resp. $\mathrm{VaR}_\alpha(S)$) follows from the convex ordering assumption by Theorem 1.15 b. □

Theorem 6.6 says that in order to obtain sharp upper bounds of $\mathrm{VaR}_\alpha^+(S)$ and sharp lower bounds for $\mathrm{VaR}_\alpha(S)$, one should try to rearrange the random variables Y_i on the upper α-part $\{U \geq \alpha\}$ of S and on their lower α-part $T^c = \{U < \alpha\}$ such that the distribution is as flat as possible in convex order on both parts. In particular, this holds true for an optimal solution of the variance-constrained problem. The two flattening steps can be performed separately or simultaneously. In the following section, this idea is used to approximate sharp VaR_α bounds in the variance-constrained case.

6.2 Algorithm to Approximate VaR Bounds

The algorithm requires random variables that are discretely distributed, which is possible since any distribution can be approximated to any degree of accuracy by discrete ones. Let N be the number of points used to discretize the random variables, and assume that each risk X_j, $j = 1, 2, \ldots, n$, is sampled into N equiprobable values.

To avoid confusion, when we compute the bounds A, B and a, b, respectively, for the discretized (sampled) risks, we use the notations A_N, B_N and a_N, b_N, respectively. Hence,

$$B_N = \sum_{j=1}^{n} \frac{1}{N-k} \sum_{i=k+1}^{N} x_{ij}, \quad A_N = \frac{1}{k} \sum_{j=1}^{n} \sum_{i=1}^{N} x_{ij} - \frac{N-k}{k} B_N. \tag{6.20}$$

Denote by μ_N the mean of the sum of the discretized risks. We describe in detail the extended rearrangement algorithm that we propose as a suitable way to compute numerical VaR bounds of portfolios in the presence of a variance constraint. We point out that, unlike the theoretical VaR bounds, the algorithm does not require that the risks have finite mean or variance.

Recall that the standard rearrangement algorithm (RA) was used successfully to compute approximate sharp VaR bounds for the sum of n dependent risks with given marginal distributions in Chapter 2. Here, we show how it can also be used in the variance-constrained case.

Based on Theorems 6.3 and 6.5, it is natural to modify the rearrangement algorithm used by Embrechts et al. (2013) and Puccetti and Rüschendorf (2012b) to construct numerically the minimum and maximum VaR-bounds. This algorithm simultaneously rearranges the upper and the lower part of the distribution of the sum and then moves through the domains in a systematic way to satisfy the variance constraint. While the algorithm proposed in Bernard et al. (2017c) to use the RA iteratively on two submatrices is reasonable and provides accurate results, it is lacking a strong theoretical foundation. In fact, there might exist other partitions of the matrix $\mathbf{X}$ that allow for better approximations of the sharp bounds. In this respect, the following algorithm appears to be useful in selecting these two submatrices. In this, we use the algorithm presented in Section 2.3 to infer dependence using as target distribution a two-point distribution. In the third step, we adapt this two point distribution to meet the variance constraint. The numerical algorithm to obtain VaR bounds with a variance constraint thus involves the following three steps.

Step 1: In a first step, we add an extra column to the initial matrix $\mathbf{X}$ representing a variable $-X_{n+1}$ that takes the values $-a$ and $-b$ with probabilities α and $1 - \alpha$, respectively.

Step 2: Then in a second step we apply the RA or BRA to the full matrix. As a result we obtain the candidate approximate sharp lower bound $m_N := \max_{i \in \mathcal{A}} (x^*_{i1} + \cdots + x^*_{in})$ as well as the candidate approximate sharp upper bound $M_N := \min_{i \in \mathcal{B}} (x^*_{i1} + \cdots + x^*_{in})$, in which $\mathcal{A}$ and $\mathcal{B}$ are defined as $\mathcal{A} = \{i \in \{1, 2, \ldots, N\} \mid x^*_{i,n+1} = a\}$ and $\mathcal{B} = \{1, 2, \ldots, N\} \backslash \mathcal{A}$. In the ideal situation, we observe that after running the RA the row sums are all equal to zero, and thus m_N and M_N are sharp.

However, in general, the row sums will not be equal to zero, and it is not a priori clear whether the sum $X^*_1 + X^*_2 + \cdots + X^*_n$ meets the variance constraint. If it does, then m_N and M_N are our approximations for the sharp bounds.

Step 3: If the variance constraint is not satisfied, then we consider in a third step the variable $-X^\varepsilon_{n+1}$, $\varepsilon > 0$, taking the values $-a_\varepsilon$ and $-b_\varepsilon$ (with respective probabilities α and $1 - \alpha$), in which case

$$a_\varepsilon = \max \left(\mu - (s - \varepsilon) \sqrt{\frac{1 - \alpha}{\alpha}}, A \right)$$

and

$$b_\varepsilon = \min \left(\mu + (s - \varepsilon) \sqrt{\frac{\alpha}{1 - \alpha}}, B \right),$$

and we repeat the above procedure to yield new candidate approximations for the sharp bounds. By gradually increasing ε, one obtains a situation in which $X^*_1 + X^*_2 + \cdots + X^*_n$ meets the variance constraint and approximate sharp VaR bounds are obtained.

Example 6.7

a) **Normally distributed risks**

Assume that $X = (X_1, X_2, \ldots, X_n)$ is a vector of dependent standard normally distributed random variables with a correlation matrix (ϱ_{ij}) such that $\varrho_{ii} = 1$ for $i = 1, \ldots, n$ and $\varrho_{ij} = \varrho$ for all $i \neq j$. Note that the dependence structure of X is only partially specified through the knowledge of the pairwise correlations. The variance of the sum is equal to

$$\text{Var}(S) = \sum_{i=1}^{n} \sum_{j=1}^{n} \varrho_{ij}.$$

However, this information does not allow us to compute the VaR of the portfolio sum $S = X_1 + X_2 + \cdots + X_n$ precisely. Hence, we apply the above procedure to compute numerical upper and lower approximate sharp bounds on VaR satisfying the variance constraint. We then compare these bounds with the theoretical bounds in Theorem 6.1 and Theorem 6.3.

In Table 6.1 we assess the VaRs at 95 %, 99 %, and 99.5 % for three levels of correlation, $\varrho = 0$, $\varrho = 0.15$, and $\varrho = 0.3$, and for portfolio sizes $n = 10$ and $n = 100$. Panel A shows the constrained lower and upper VaR bounds using the RA with discretization level $N = 10,000$. We denote these constrained lower and upper bounds by m_N and M_N, respectively. Panel B shows the corresponding constrained bounds a_N and b_N as an application of Theorem 6.3, whereas Panel C shows the unconstrained bounds A_N and B_N using Theorem 6.1. The last rows in Panels B and C show the values for $a_\infty := a$ and $b_\infty := b$, respectively: $A_\infty := A$ and $B_\infty := B$. These bounds are thus based on the original (nondiscretized) distributions and are explicitly given as

$$A = -n\frac{\varphi(\Phi^{-1}(\alpha))}{\alpha}, \quad B = n\frac{\varphi(\Phi^{-1}(\alpha))}{1-\alpha}$$

and

$$a = \max\left(-s\sqrt{\frac{\alpha}{1-\alpha}}, A\right), \quad b = \min\left(s\sqrt{\frac{1-\alpha}{\alpha}}, B\right),$$

where φ and Φ denote the standard normal density and distribution function and where $s^2 = n + n(n-1)\varrho$ is the variance of S. Finally, in Panel D, X is multivariate normally distributed, and the dependence is thus assumed to be Gaussian, in which case the VaRs of the portfolio can be computed exactly.

Panel A: Approximate sharp bounds obtained by the BRA

(m_N, M_N)		$n = 10$	
	$\varrho = 0$	$\varrho = 0.15$	$\varrho = 0.3$
VaR$_{95\%}$	$(-0.72; 13.77)$	$(-1.08; 20.60)$	$(-1.08; 20.60)$
$N = 10,000$ VaR$_{99\%}$	$(-0.27; 26.56)$	$(-0.27; 26.56)$	$(-0.27; 26.56)$
VaR$_{99.5\%}$	$(-0.14; 28.76)$	$(-0.14; 28.76)$	$(-0.14; 28.76)$

(m_N, M_N)		$n = 100$	
	$\varrho = 0$	$\varrho = 0.15$	$\varrho = 0.3$
VaR$_{95\%}$	$(-2.29; 43.58)$	$(-9.13; 173.30)$	$(-10.84; 206.00)$
$N = 10,000$ VaR$_{99\%}$	$(-1.01; 99.40)$	$(-2.68; 265.60)$	$(-2.68; 265.60)$
VaR$_{99.5\%}$	$(-0.71; 141.00)$	$(-1.44; 287.60)$	$(-1.44; 287.60)$

Panel B: Constrained bounds as in Theorem 6.3

(a_N, b_N)		$\varrho = 0$	$n = 10$ $\varrho = 0.15$	$\varrho = 0.3$
	VaR$_{95\%}$	(−0.72; 13.77)	(−1.08; 20.60)	(−1.08; 20.60)
$N = 10,000$	VaR$_{99\%}$	(−0.27; 26.56)	(−0.27; 26.56)	(−0.27; 26.56)
	VaR$_{99.5\%}$	(−0.14; 28.76)	(−0.14; 28.76)	(−0.14; 28.76)
	VaR$_{95\%}$,	(−0.72; 13.78)	(−1.08; 20.63)	(−1.08; 20.63)
$N = +\infty$	VaR$_{99\%}$	(−0.27; 26.65)	(−0.27; 26.65)	(−0.27; 26.65)
	VaR$_{99.5\%}$	(−0.14; 28.92)	(−0.14; 28.92)	(−0.14; 28.92)

Panel B (cont.): Constrained bounds as in Theorem 6.3

(a_N, b_N)		$\varrho = 0$	$n = 100$ $\varrho = 0.15$	$\varrho = 0.3$
	VaR$_{95\%}$	(−2.29; 43.55)	(−9.13; 173.40)	(−10.84; 206.00)
$N = 10,000$	VaR$_{99\%}$	(−1.01; 99.42)	(−2.68; 265.60)	(−2.68; 265.60)
	VaR$_{99.5\%}$	(−0.71; 141.00)	(−1.44; 287.60)	(−1.44; 287.60)
	VaR$_{95\%}$,	(−2.29; 43.59)	(−9.13; 173.50)	(−10.86; 206.30)
$N = +\infty$	VaR$_{99\%}$	(−1.01; 99.50)	(−2.69; 266.50)	(−2.69; 266.50)
	VaR$_{99.5\%}$	(−0.71; 141.10)	(−1.45; 289.20)	(−1.45; 289.20)

Panel C: Unconstrained bounds as obtained in Theorem 6.1, independent of ϱ

(A_N, B_N)		$n = 10$	$n = 100$
	VaR$_{95\%}$	(−1.08; 20.60)	(−10.84; 206.00)
$N = 10,000$	VaR$_{99\%}$	(−0.27; 26.56)	(−2.68; 265.60)
	VaR$_{99.5\%}$	(−0.14; 28.76)	(−1.45; 287.60)
	VaR$_{95\%}$	(−1.09; 20.63)	(−10.86; 206.30)
$N = +\infty$	VaR$_{99\%}$	(−0.27; 26.65)	(−2.69; 266.50)
	VaR$_{99.5\%}$	(−0.14; 28.92)	(−1.45; 289.20)

Panel D: Exact VaR numbers when the risks are multivariate normally distributed

	$n = 10$ $\varrho = 0$	$\varrho = 0.15$	$\varrho = 0.30$	$n = 100$ $\varrho = 0$	$\varrho = 0.15$	$\varrho = 0.30$
VaR$_{95\%}$	5.20	7.97	10.01	16.45	65.48	91.14
VaR$_{99\%}$	7.36	11.28	14.15	23.26	92.62	128.90
VaR$_{99.5\%}$	8.14	12.49	15.67	25.76	102.51	142.72

Table 6.1 Bounds on value-at-risk of sums of normally distributed risks.

There are several observations. First, when comparing the results of Panel A and Panel B, we observe that the BRA is performing remarkably well. The obtained numerical bounds m_N and M_N are very close to their theoretical counterparts a_N and b_N, showing that the BRA is able to construct the dependence between the risks such that the sum is closely concentrated on two values: a_N and b_N.

Second, as b_N is an upper bound for $\overline{\text{VaR}}_\alpha$ whereas M_N is an approximation for it from below, the observation that they are close to each other implies that both quantities are good approximations for $\overline{\text{VaR}}_\alpha$. A similar observation holds for m_N and a_N.

Third, the distance between the upper and lower bounds, as reported in Panels A, B, and C, is typically significant. For example, Panel B shows that the true 95 %-VaR of a portfolio of 100 uncorrelated but not independent normally distributed risks lies in the interval $(-2.29; 43.58)$. Considering that the given portfolio has zero mean and a standard deviation of 10, this interval appears to be rather wide. Note that when the risks are independent, then the exact 95 %-VaR can be computed and is given by 16.45. In other words, when the risks are known to be independent, the 95 %-VaR is approximately three times smaller than the reported upper bound, i.e., 43.59, which is valid when we only know that the correlations are equal to zero. When we ignore the variance constraint, then the upper bound is as high as 206.3. We also observe that the distance between the bounds becomes wider when the level of the probability α used to assess VaR increases. These observations already suggest that misspecification of models is a significant concern, especially when the VaRs are assessed at high probability levels (which is the case in Solvency frameworks such as Solvency II and Basel III, where $\alpha = 99.5$ %).

Third, when comparing the results shown in Panel B and Panel C, we observe that adding a variance constraint may have a significant impact on the unconstrained bounds. When the portfolio exhibits low to moderate correlation, then the constrained bounds a and b that we propose improve strongly upon the unconstrained ones, A and B. It is straightforward to show that

$$a > A, \ b < B \iff \varrho \leq \widehat{\varrho}_1(n, \alpha) := \frac{n\left(\varphi(\Phi^{-1}(\alpha))\right)^2}{\alpha(1-\alpha)(n-1)} - \frac{1}{n-1}. \tag{6.21}$$

The relation (6.21) allows us to derive the critical correlation values $\widehat{\varrho}_1(n, \alpha)$ as a function of portfolio size n and probability level α. If the correlation ϱ of the portfolio is lower than the critical value, then the constrained bounds improve upon the unconstrained ones. We report some values here: $\widehat{\varrho}_1(10, 0.95) = 0.138$, $\widehat{\varrho}_1(10, 0.995) = -0.0644$, $\widehat{\varrho}_1(100, 0.95) = 0.216$, and $\widehat{\varrho}_1(100, 0.995) = 0.0324$. Note that the critical correlation levels decrease when the probability level that is used for VaR assessment increases, indicating that adding dependence information does not readily allow for improving the unconstrained bounds when going "deep in the tail." In this regard, it is also of interest to compare the constrained upper bound b with the portfolio VaR that one obtains in the case in which all risks are perfectly dependent (comonotonic), i.e., when the portfolio VaR is equal to $n\Phi^{-1}(\alpha)$. We find that

$$b < n\Phi^{-1}(\alpha) \iff \varrho \leq \widehat{\varrho}_2(n, \alpha) := \frac{n(1-\alpha)(\Phi^{-1}(\alpha))^2}{\alpha(n-1)} - \frac{1}{n-1}. \tag{6.22}$$

As critical values for the correlation parameter, we report

$$\widehat{\varrho}_2(10, 0.95) = 0.0471, \qquad \widehat{\varrho}_2(10, 0.995) = -0.0741,$$
$$\widehat{\varrho}_2(100, 0.95) = 0.1337, \quad \text{and} \quad \widehat{\varrho}_2(100, 0.995) = 0.0236,$$

which shows that adding the variance constraint may give rise to a VaR bound that strictly improves upon the comonotonic VaR. When the variance constraint s^2 is "too high" or the probability level α is "too high," there is no improvement.

b) Pareto-distributed risks

We assume that X is an homogeneous portfolio of dependent Pareto-distributed random variables (of type II). Hence, $F_i(x) = 1 - (1 + x)^{-\vartheta}, i = 1, 2, \ldots, n$, with $x > 0$ and with a tail parameter $\vartheta > 0$. The correlation matrix (ϱ_{ij}) is such that $\varrho_{ii} = 1$ for $i = 1, \ldots, n$ and $\varrho_{ij} = \varrho$ for all $i \neq j$.

We first consider the case $\vartheta = 3$ so that the first two moments exist, which allows us to compute the VaR bounds that we discussed in the previous sections. We first calculate

$$EX_i = \frac{1}{\vartheta - 1}, \quad \text{Var}(X_i) = \frac{2}{(\vartheta - 1)(\vartheta - 2)} - \frac{1}{(\vartheta - 1)^2},$$

$$F_{X_i}^{-1}(\alpha) = (1 - \alpha)^{-1/\vartheta} - 1, \quad \text{TVaR}_\alpha(X_i) = \frac{(1 - \alpha)^{-1/\vartheta}}{(1 - \frac{1}{\vartheta})} - 1,$$

$$\text{LTVaR}_\alpha(X_i) = \frac{1}{\alpha}(EX_i - (1 - \alpha)\text{TVaR}_\alpha(X_i)).$$

When applying Theorem 6.1, we find that the unconstrained bounds are

$$B = n\frac{(1 - \alpha)^{-1/\vartheta}}{(1 - \frac{1}{\vartheta})} - n, \quad A = \frac{n}{\alpha}\frac{1}{\vartheta - 1} - B\frac{(1 - \alpha)}{\alpha}.$$

And, from Theorem 6.3,

$$a = \max\left(-s\sqrt{\frac{1 - \alpha}{\alpha}}, A\right), \quad b = \min\left(s\sqrt{\frac{1 - \alpha}{\alpha}}, B\right),$$

where $s^2 = (n + n(n - 1)\varrho)(\frac{2}{(\vartheta-1)(\vartheta-2)} - \frac{1}{(\vartheta-1)^2})$.

We present the results of our calculations in Table 6.2 in a similar way as in the previous example. Panel A shows the numerical sharp lower and upper bounds obtained using the RA. Note that the discretization involves the computation of $x_{ij} = (1 - i/(N + 1))^{-1/\vartheta} - 1$ for $i = 1, \ldots, N; j = 1, 2, \ldots, n$. Panel B shows the corresponding constrained bounds, and Panel C shows the bounds in the unconstrained case. The results are in line with those that we obtain in the case of normally distributed risks. Also, in this case the numerical procedure gives rise to numerical bounds that are close to those that are obtained in Theorem 6.3. In other words, the absolute bounds are "nearly sharp" in this case.

The Pareto distribution has heavy tails, and hence one observes a significant difference between A_N given by (6.20) and A (respectively, B_N and B), as in Table 6.2. The difference between the upper and lower bounds is again significant, confirming that in the case in which there is little or no information on the dependence, the model risk that goes along with a particular model is an issue. Note also that the impact of the variance constraint is more significant than in the normal case.

For example, when $\alpha = 99.5\%$, $n = 100$, and $\varrho = 0.15$, we find a numerical sharp bound $m_{10,000} = 499.1$, close to the bound $b_{10,000} = 500.01$, whereas the unconstrained bound amounts to $B_{10,000} = 741.1$. Similar conclusions are reached for other discretization levels, such as $N = 1000$ and $N = 100,000$.

Panel A: Approximate sharp bounds obtained by the numerical algorithm procedure as presented in Section 6.2

(m_N, M_N)		$n = 10$		
		$\varrho = 0$	$\varrho = 0.15$	$\varrho = 0.3$
	$\text{VaR}_{95\%}$	(4.401; 15.72)	(4.091; 21.85)	(3.863; 26.19)
$N = 10,000$	$\text{VaR}_{99\%}$	(5.486; 28.69)	(4.591; 43.45)	(4.492; 53.22)
	$\text{VaR}_{99.5\%}$	(6.820; 39.48)	(5.471; 59.60)	(4.850; 73.11)

(m_N, M_N)		$n = 100$		
		$\varrho = 0$	$\varrho = 0.15$	$\varrho = 0.3$
	$\text{VaR}_{95\%}$	(47.96; 84.72)	(42.48; 188.9)	(39.61; 243.3)
$N = 10,000$	$\text{VaR}_{99\%}$	(48.99; 129.5)	(46.61; 366.0)	(45.36; 489.5)
	$\text{VaR}_{99.5\%}$	(49.23; 162.8)	(47.54; 499.1)	(46.68; 671.5)

Panel B: Variance-constrained bounds as obtained in Theorem 6.3

(a_N, b_N)		$n = 10$		
		$\varrho = 0$	$\varrho = 0.15$	$\varrho = 0.3$
	$\text{VaR}_{95\%}$	(4.398; 16.03)	(4.089; 21.92)	(3.861; 26.23)
$N = 10,000$	$\text{VaR}_{99\%}$	(4.725; 30.20)	(4.589; 43.64)	(4.490; 53.50)
	$\text{VaR}_{99.5\%}$	(4.800; 40.74)	(4.705; 59.80)	(4.634; 73.77)
	$\text{VaR}_{95\%}$	(4.372; 16.94)	(4.037; 23.30)	(3.791; 27.96)
$N = +\infty$	$\text{VaR}_{99\%}$	(4.725; 32.25)	(4.578; 46.77)	(4.470; 57.41)
	$\text{VaR}_{99.5\%}$	(4.806; 43.63)	(4.702; 64.22)	(4.634; 77.72)

Panel B (cont.): Variance-constrained bounds as obtained in Theorem 6.3

(a_N, b_N)		$n = 100$		
		$\varrho = 0$	$\varrho = 0.15$	$\varrho = 0.3$
	$\text{VaR}_{95\%}$	(47.96; 84.74)	(42.48; 188.9)	(39.61; 243.4)
$N = 10,000$	$\text{VaR}_{99\%}$	(48.99; 129.6)	(46.59; 367.3)	(45.33; 491.7)
	$\text{VaR}_{99.5\%}$	(49.23; 162.9)	(47.54; 500.0)	(46.65; 676.3)
	$\text{VaR}_{95\%}$	(48.01; 87.75)	(42.09; 200.3)	(38.99; 259.2)
$N = +\infty$	$\text{VaR}_{99\%}$	(49.13; 136.2)	(46.53; 393.1)	(45.18; 527.4)
	$\text{VaR}_{99.5\%}$	(49.39; 172.2)	(47.56; 536.4)	(46.60; 726.9)

Panel C: Unconstrained bounds as obtained in Theorem 6.1 independent of ϱ

(A_N, B_N)		$n = 10$	$n = 100$
	VaR$_{95\%}$,	(3.646; 30.33)	(36.46; 303.3)
$N = 10{,}000$	VaR$_{99\%}$	(4.447; 57.76)	(44.47; 577.6)
	VaR$_{99.5\%}$	(4.633; 74.11)	(46.33; 741.1)
	VaR$_{95\%}$	(3.647; 30.72)	(36.47; 307.2)
$N = +\infty$	VaR$_{99\%}$	(4.448; 59.62)	(44.48; 596.2)
	VaR$_{99.5\%}$	(4.635; 77.72)	(46.35; 777.2)

Table 6.2 Bounds on value-at-risk of sums of Pareto distributed risks ($\vartheta = 3$).

Also in this case we find that the constrained bounds a and b that we propose may improve strongly upon the unconstrained ones, A and B. Using the same notation in dependence on ϱ as in the previous example, we find that $\widehat{\varrho}_1(10, 0.95) = 0.2756$, $\widehat{\varrho}_1(10, 0.995) = 0.1842$, $\widehat{\varrho}_1(100, 0.95) = 0.3415$, and $\widehat{\varrho}_1(100, 0.995) = 0.2583$.

Finally, the constrained upper bound b also sharpens the comonotonic VaR bound $n(1-\alpha)^{-1/\vartheta} - n$. As critical values for the correlation parameter that make it possible to improve upon the comonotonic VaR bounds, we find $\widehat{\varrho}_2(10, 0.95) = 0.0608$, $\widehat{\varrho}_2(10, 0.995) = 0.0201$, $\widehat{\varrho}_2(100, 0.95) = 0.1462$, and $\widehat{\varrho}_2(100, 0.995) = 0.1092$.

6.3   Generalization to Constraints on Higher Moments on Default Probabilities

In this section, we consider loan portfolios under the so-called default mode paradigm. Hence, a credit loss occurs if the loan (i.e., the underlying obligor) defaults during the considered time horizon and other value changes (e.g., due to a downgrade) are not recognized. The portfolio is then described by n (dependent) default indicators $\mathbb{I}_i, i = 1, \ldots, n$ taking value 1 in the case of a default with probability p_i and zero otherwise, i.e., $P(\mathbb{I}_i = 1) = p_i$. and $P(\mathbb{I}_i = 0) = 1 - p_i, i = 1, \ldots, n$.

Furthermore, let EAD_i denote the "exposure at default" and LGD_i the "loss given default" of risk i. The "exposure at default" is the maximum amount of loss on the ith loan, provided that there is a default. The "loss given default" is the percentage of the maximum amount that is effectively lost in the event of a default. We assume that all EAD_i and LGD_i are deterministic and known. The portfolio loss S during the reference period is then given by

$$S = \sum_{i=1}^{n} X_i,$$

where $X_i = v_i \mathbb{I}_i$ and $v_i = EAD_i \, LGD_i$. Hence, the credit losses X_i follow a scaled Bernoulli distribution with known scaling factor v_i, i.e., $X_i \sim v_i \mathcal{B}(p_i)$. We denote its distribution by F_i. Without loss of generality, we assume that $v_1 \geq v_2 \geq \cdots \geq v_n > 0$. We aim at computing the worst case outcome, i.e., the VaR of the portfolio loss S at a given confidence level α, $0 < \alpha < 1$. Hence, we are interested in $\mathrm{VaR}_\alpha^+(S)$, defined as

$$\mathrm{VaR}_\alpha^+(S) = \sup\{x \in \mathbb{R} \mid F_S(x) \leq \alpha\},$$

where $F_S(x)$ is the distribution function of S.

A precise computation of VaR of the portfolio loss S can only be obtained if one knows the joint distribution of the default vector $(\mathbb{I}_1, \mathbb{I}_2, \ldots, \mathbb{I}_n)$. However, this joint distribution is hard to obtain. In this regard, we point out that financial institutions typically use models that allow specification of default probabilities and default correlations. While default probabilities and correlations together reveal the level of all pairwise default probabilities, i.e., the specification of the distributions of the pairs $(\mathbb{I}_i, \mathbb{I}_j)$, the probability that three or more loans default together is not known. In fact, lack of sufficient default statistics occurs since joint defaults are rarely observed. This makes it hard, if not impossible, to specify the probabilities that several loans default together so that the joint distribution of $(\mathbb{I}_1, I_2, \ldots, \mathbb{I}_n)$ cannot readily be specified. In other words, all models that assess VaRs of credit risk portfolios strictly require additional ad hoc and hard to justify assumptions to describe the full dependence, and all provide different VaR numbers. In this chapter we aim at quantifying this inherent uncertainty on VaR estimates.

We first assume that besides the information on the net exposures v_i the only information that is available concerns the probabilities of default p_i of each loan, i.e., the distributions of the different default events $\mathbb{I}_i$, $i = 1, 2, \ldots, n$, are known but not their joint distribution. In this context, the maximum and minimum VaR of the portfolio of loans can be obtained as in Theorem 6.1. The unconstrained bounds are very wide, confirming that using dependence information is crucial in improving the bounds.

However, it appears realistic to have a reasonable estimate for the variance and perhaps even the skewness of the portfolio loss S, providing information on the dependence among credit loans. Hence, in this chapter we are interested in the maximum possible VaR of a portfolio of loans in which the loss distributions F_i, $i = 1, 2, \ldots, n$, of the constituent risky loans are known, as well as some higher order moments of the portfolio loss. These moments are typically not precisely known but have to be estimated from available data. To capture the statistical uncertainty on these estimates, we propose a robust approach in the sense that we only assume an upper bound value c_k for each unknown higher-order moment of S for $k = 2, 3, \ldots, K$. Note that using inequality constraints is in the sense that the VaR bounds will be wider as compared to a situation in which the moments are assumed to be known exactly. Typically, c_k is the point estimate of the kth moment, but it can also be an upper bound of it. In summary, we consider the following problem:

$$\sup \{ \mathrm{VaR}_\alpha^+(S); X_j \sim F_j, \ ES^k \leq c_k, k = 2, 3, \ldots, K\}. \tag{6.23}$$

As for the lower bound for VaR, we consider the problem

$$\inf \{ \mathrm{VaR}_\alpha(S); X_j \sim F_j, \ ES^k \leq c_k, k = 2, 3, \ldots, K\}. \tag{6.24}$$

In what follows, we always tacitly assume that the problems (6.23) and (6.24) are well posed in the sense that there exist portfolios that satisfy the constraints. In particular, by denoting $ES := \mu$ and observing that (since $S \geq 0$)

$$\mu^k \leq ES^k, \tag{6.25}$$

it follows that $c_k \geq \mu^k$, $k = 2, 3, \ldots, K$ will hold.

6.3.1   Sharp VaR Bounds for Portfolios with Homogeneous Exposures

In this subsection, we assume that all exposures are equal, that is, for all i, $v_i = EAD_i\, LGD_i = v$. In this case, we are able to give explicit sharp bounds. As each loss X_i takes value zero or v, it is clear that the portfolio sum S can only take values that are multiples of v and between zero (no loss) and nv (all loans default). Therefore, the bounds A and B established in Theorem 6.1 cannot be sharp, i.e., by a potential dependence between the loans, as soon as they are not a multiple of v.

For credit risk portfolios that are homogeneous in the composition of the exposures, the problem of finding a dependence structure that makes it possible to attain the lower and upper VaR of the portfolio of loans can be solved without using a numerical procedure. It is closely related to solving the problem of finding the dependence structure that minimizes the variance. First, we show that the problem

$$(\mathcal{P}) \quad \min\{\mathrm{Var}(Y_1 + Y_2 + \cdots + Y_n); Y_j \sim v\mathcal{B}(p_j), \text{ for all } j = 1, 2, \ldots, n\} \tag{6.26}$$

can be solved exactly. Armed with this result, we provide sharp VaR bounds in Theorem 6.9. In this regard, it is convenient to denote by $\lceil x \rceil$ (resp. $\lfloor x \rfloor$) the smallest (resp. largest) integer number that is larger (resp. smaller) than x.

Lemma 6.8 (Minimum variance portfolio)   *Consider problem $(\mathcal{P})$ in (6.26). Assume that the exposures are identical, i.e., $v = v_j$, $j = 1, 2, \ldots, n$. Define for $j = 1, \ldots, n$,*

$$a_j = \left(\sum_{i=1}^{j} p_i\right) \bmod 1, \quad and \quad I_j = \begin{cases} [a_{j-1}, a_j], & \text{if } a_j > a_{j-1}, \\ [0, a_j] \cup [a_{j-1}, 1], & \text{if } a_j < a_{j-1}, \end{cases}$$

where we define $a_0 = 0$. Then, the solution to $(\mathcal{P})$ in (6.26) is obtained for Y_j^* given as

$$Y_j^* = v\mathbb{1}_{\{U \in I_j\}}, \tag{6.27}$$

where U is a standard uniformly distributed random variable. Furthermore,

$$\mathrm{Var}(Y_1^* + Y_2^* + \cdots + Y_n^*) = v^2 p^*(1 - p^*),$$

where $p^* = \frac{\mu}{v} - \lfloor \frac{\mu}{v} \rfloor$.

Proof   We first observe that $Y_j^* \sim v\mathcal{B}(p_j)$. Furthermore, one easily verifies that $S_n^* = Y_1^* + Y_2^* + \cdots + Y_n^*$ only takes values ℓv with probability $(1 - p^*)$ or $(\ell + 1)v$ with probability p^*. Note that $p^* = 0$ may hold. It is straightforward to show that

$$\mathrm{Var}(S_n^*) = v^2 p^*(1 - p^*),$$

and we then only need to show that any other sum $S_n = Y_1 + Y_2 + \cdots + Y_n$ with $Y_j \sim v\mathcal{B}(p_j)$ has a larger variance. Consider any admissible sum S_n; in particular, S_n takes values in $\{0, v, 2v, \ldots, nv\}$.

It is clear that for all $x \in]0, \ell v[$, $F_{S_n}(x) \geq F_{S_n^*}(x) = 0$, and for all $x \in [(\ell+1)v, +\infty[$, $F_{S_n}(x) \leq F_{S_n^*}(x) = 1$. Since $F_{S_n}(x)$ and $F_{S_n^*}(x)$ are constant on the interval $[\ell v, (\ell + 1)v[$, one has the following:

$$\text{There exists } c \geq 0, \quad \begin{cases} \text{for all } x \in (0, c), & F_{S_n}(x) \geq F_{S_n^*}(x), \\ \text{for all } x \in (c, +\infty), & F_{S_n}(x) \leq F_{S_n^*}(x), \end{cases} \quad (6.28)$$

namely, $c = (\ell + 1)v$ if $F_{S_n}(\ell v) > F_{S_n^*}(x)$ and $c = \ell v$ if $F_{S_n}(\ell) \leq F_{S_n^*}(x)$. In other words, the distribution function F_{S_n} crosses $F_{S_n^*}$ exactly once from above. Since $ES_n = ES_n^*$, this implies the well-known consequence that $Eh(S_n^*) \leq Eh(S_n)$ for all convex functions $h(x)$. Taking $h(x) = x^2$ ends the proof. $\qquad\square$

Note that the minimum variance portfolio $(Y_1^*, Y_2^*, \ldots, Y_n^*)$ has the property that its sum is concentrated on two values around the mean. Specifically,

$$Y_1^* + Y_2^* + \cdots + Y_n^* = \begin{cases} v\lfloor \frac{\mu}{v} \rfloor & \text{with probability } 1 - p^*, \\ v\lceil \frac{\mu}{v} \rceil & \text{with probability } p^*. \end{cases} \quad (6.29)$$

The variance is a traditional measure for comparing variability ("degree of riskiness") among risks. A more general concept to discuss and compare variability of risks is the convex order. From the proof of Lemma 6.8, one can see that the minimum variance portfolio $(Y_1^*, Y_1^*, \ldots, Y_n^*)$ is also a convex minimum among all portfolios with fixed marginal distributions. In other words, if ϱ is a risk measure that is consistent with convex order, then the problem

$$\min\{\varrho(Y_1 + Y_2 + \cdots + Y_n); Y_j \sim v_j\mathcal{B}(p_j), 1 \leq j \leq n\} \quad (6.30)$$

has the same solution as problem $(\mathcal{P})$.

The next theorem gives exact sharp bounds for the VaR of the portfolio sum in (6.23) and (6.24) without moment constraints, i.e., an analog of Theorem 6.1 applied in the case of credit risk. It shows that there exists a dependence structure among the risks $X_1, X_2, \ldots, X_n$ such that these bounds are attainable.

Theorem 6.9 (Unconstrained VaR bounds) *Consider problems* (6.23) *and* (6.24) *without moment constraints, i.e., with* $c_k = \infty$, $k = 2, 3, \ldots, K$. *Assuming that all exposures are identical* $v_i = v$ *for all* $i = 1, \ldots, n$, *then*

$$v\left\lceil \frac{A}{v} \right\rceil \leq \text{VaR}_\alpha(S) \leq \text{VaR}_\alpha^+(S) \leq v\left\lfloor \frac{B}{v} \right\rfloor, \quad (6.31)$$

where $A := \text{LTVaR}_\alpha(S^c)$ *and* $B := \text{TVaR}_\alpha(S^c)$. *Furthermore, these bounds are sharp.*

Proof The proof essentially follows from Lemma 6.8. Consider variables Y_i given as $Y_i = v\mathbb{1}_{\{U \leq \alpha\}}V_i + v\mathbb{1}_{\{U > \alpha\}}W_i$, in which V_i and W_i are Bernoulli distributed random variables that are independent of the uniform random variable U and such that $Y_i \sim vB(p_i)$, $i = 1, 2, \ldots, n$, $E\left(\sum_{i=1}^n V_i\right) = \frac{A}{v}$, $E\left(\sum_{i=1}^n W_i\right) = \frac{B}{v}$.

Applying Lemma 6.8, they can be chosen such that the portfolio sum $\sum_{i=1}^{n} Y_i$ takes four values, namely $v \left\lfloor \frac{A}{v} \right\rfloor$, $v \left\lceil \frac{A}{v} \right\rceil$, $v \left\lfloor \frac{B}{v} \right\rfloor$, and $v \left\lceil \frac{B}{v} \right\rceil$. One observes that for this dependence among V_i and W_i, $\text{VaR}_\alpha^+(Y_1 + Y_2 + \cdots + Y_n) = v \left\lfloor \frac{B}{v} \right\rfloor$ and $\text{VaR}_\alpha(Y_1 + Y_2 + \cdots + Y_n) = v \left\lceil \frac{A}{v} \right\rceil$, which ends the proof.   □

It is possible to improve the approximations for the VaR bounds when the pairwise correlations among the credit losses X_i, $i = 1, 2, \ldots, n$, are available.

6.3.2   VaR Bounds when Default Probabilities and Pairwise Correlations Are Known

In this regard, we can assume that n is even, possibly by adding a risk with zero exposure, i.e., by taking $v_n = 0$. When the correlations between the X_i are known, the distribution of each pair (X_i, X_j) and thus also of each partial sum $S_{i,j} = X_i + X_j$, $i \neq j = 1, 2, \ldots, n$, is known. The reason is that in the case of (scaled) Bernoulli distributions, knowledge of pairwise correlations and single default probabilities implies knowledge of the pairwise joint default probabilities and thus also knowledge of the distribution of partial sums that only involve two components. Note that the probabilities that three or more loans default together cannot be determined based on single default probabilities and default correlations alone.

In particular, when $n = 2$, the VaRs of S can be exactly determined. It is then also clear that VaR bounds on S that are only based on the marginal distributions of the X_i are wider than ones that are based on the marginal distributions of the $S_{i,j}$. As a result we can use this correlation information to reduce the unconstrained VaR bounds.

For every permutation π of $\{1, 2, \ldots, n\}$, we find that for every portfolio sum $S = X_1 + X_2 + \cdots + X_n$,

$$\text{VaR}_\alpha^+(S) \le \sum_{i=1}^{n/2} \text{TVaR}_\alpha(S_{\pi(i),\pi(i+1)}) \le B. \tag{6.32}$$

Any permutation π leads to a reduction of the unconstrained bound B. We denote by Π the set of permutations. As a result of (6.32) we obtain the following improvement of the unconstrained bound in Theorem 6.1.

Theorem 6.10 (Improved bounds with known correlations)
For any portfolio sum $S = X_1 + X_2 + \cdots + X_n$ *it holds that*

$$C \le \text{VaR}_\alpha(S) \le \text{VaR}_\alpha^+(S) \le D \tag{6.33}$$

in which

$$C := \inf_{\pi \in \Pi} \left\{ \sum_{i=1}^{n/2} \text{LTVaR}_\alpha(S_{\pi(i),\pi(i+1)}) \right\}$$

and $$D := \sup_{\pi \in \Pi} \left\{ \sum_{i=1}^{n/2} \text{TVaR}_\alpha(S_{\pi(i),\pi(i+1)}) \right\}.$$

Note that it can be hard to compute the infimum and supremum exactly because of the large number of possible permutations. However, it is possible to obtain some approximate bounds by using specific permutations. Knowledge of pairwise distributions and some related examples on the magnitude of reduction have also been considered in Embrechts and Puccetti (2010) and Puccetti and Rüschendorf (2012a).

Approximations of Sharp VaR Bounds

As before, the bounds C and D from Theorem 6.10 are not attainable, as it is typically not possible to change the dependence among the risks such that the quantile function of the portfolio sum S becomes flat on $[0, \alpha]$ and $[\alpha, 1]$, respectively. However, we can apply the extended RA as described in Section 6.2. In this regard, we make use of the auxiliary (extra) variable S_{n+1},

$$S_{n+1} = \begin{cases} -D & \text{with probability } 1 - \alpha, \\ -C & \text{with probability } \alpha. \end{cases} \tag{6.34}$$

It is convenient to use the shorthand notation S_i to denote $S_{\pi^*(i), \pi^*(i+1)}$. Note that each S_i can take four values, namely 0, $v_{\pi^*(i)}$, $v_{\pi^*(i+1)}$, and $v_{\pi^*(i)} + v_{\pi^*(i+1)}$, occurring with the appropriate probabilities that are derived from the marginal PDs and the correlations. With no loss of practical generality we can assume that all probabilities are rational numbers. As before, we sample each risk S_j ($j = 1, 2, \ldots, \frac{n}{2}$) into N equiprobable values that are ordered from low to high, i.e., we start again with a portfolio that exhibits comonotonic dependence. Hence, every S_j takes N values s_{ij}, $i = 1, 2, \ldots, N$, all occurring with probability $1/N$. The $N \times \frac{n}{2}$ matrix (s_{ij}) can then be seen as a representation of the multivariate vector $(S_1, S_2, \ldots, S_n)$. Next we add a column, reflecting the variable S_{n+1}, to this matrix to obtain the $N \times (\frac{n}{2} + 1)$ matrix $\mathbf{S}$

$$\mathbf{S} := \begin{bmatrix} s_{1,1} & s_{1,2} & \cdots & s_{1,\frac{n}{2}} & s_{1,\frac{n}{2}+1} \\ s_{2,1} & s_{2,2} & \cdots & s_{2,\frac{n}{2}} & s_{2,\frac{n}{2}+1} \\ \vdots & \vdots & & \vdots & \vdots \\ s_{k,1} & s_{k,2} & \cdots & s_{k,\frac{n}{2}} & s_{k,\frac{n}{2}+1} \\ s_{k+1,1} & s_{k+2,2} & \cdots & s_{k+1,\frac{n}{2}} & s_{k+1,\frac{n}{2}+1} \\ \vdots & \vdots & & \vdots & \vdots \\ s_{N,1} & s_{N,2} & \cdots & s_{N,\frac{n}{2}} & s_{N,\frac{n}{2}+1} \end{bmatrix},$$

in which

$$s_{1,\frac{n}{2}+1} = s_{2,\frac{n}{2}+1} = \cdots = s_{k,\frac{n}{2}+1} = -D$$

and

$$s_{k+1,n+1} = s_{k+2,n+1} = \cdots = s_{N,n+1} = -C.$$

Hence, as before we rearrange the values in the columns of $\mathbf{S}$ such that the rearranged matrix $\mathbf{S}^*$ has the property that all columns are antimonotonic with the sum of all other columns. As a result we obtain an approximation of the sharp VaR bounds based on information on the correlations.

6.3.3   VaR Bounds when Default Probabilities and Bounds on Higher Order Moments Are Known

Next, we consider the general *constrained* VaR maximization and minimization problems (6.23) and (6.24) in which $c_k < \infty$ for some $k = 2, 3, \ldots, K$.

It is clear that the unconstrained bounds A and B are also bounds of the general problem if the two-point variable S^* satisfies the moment constraints. If it does not, this means that S^* exhibits too much spread. In this case, in order to obtain VaR bounds, the idea is to construct another (less-dispersed) two-point variable that is still as dispersed as possible while satisfying the constraints. To this end, define $A(t)$ and $B(t), 0 \le t \le \alpha$, as

$$B(t) := \frac{1}{1 - \alpha} \int_{\alpha - t}^{1 - t} \mathrm{VaR}_u^+(S^c) \, du, \quad A(t) := \frac{ES - B(t)(1 - \alpha)}{\alpha}, \tag{6.35}$$

with $B(0) = B$ and $A(0) = A$. Consider variables $X_{n+1}(t), 0 \le t \le \alpha$,

$$X_{n+1}(t) = \begin{cases} A(t) & \text{with probability } 1 - \alpha, \\ B(t) & \text{with probability } \alpha, \end{cases} \tag{6.36}$$

with $X_{n+1}(0) = X_{n+1}$. For ease of exposition, we further denote $X_{n+1}(t)$ by $X(t)$. The moments of $X(t)$ are given by

$$E(X(t))^k = A^k(t)\alpha + B^k(t)(1 - \alpha), \tag{6.37}$$

and note that $EX(t) = ES = \mu$. For each k, the function $\alpha \to EX(\alpha)^k$ is continuous on $[0, \alpha]$. The function first decreases and has minimum value μ^k for $\gamma \in [0, \alpha]$, with the property that $A(\gamma) = B(\gamma)$. Since $\mu^k \le c_k$ (see (6.25)), there exists

$$t^* := \min \left\{ t \in [0, \gamma] \mid E(X(t))^k \le c_k, \ k = 2, 3, \ldots, K \right\}. \tag{6.38}$$

In Theorem 6.11 we show that the variable $X(t^*)$ yields the upper VaR bound $B(t^*)$ and the lower VaR bound $A(t^*)$ for the constrained VaR problems (6.23) and (6.24). Here, we point out that, as expected, adding dependence information makes it possible to strengthen the unconstrained bounds A and B.

First, when adding more dependence constraints, the value for t^* can obviously only increase, which implies that $B(t^*) - A(t^*)$ decreases (see (6.35)). Second, when the dependence constraints c_k take smaller values, we effectively reduce the set of admissible dependence structures among the risks, which translates into an increasing value for t^* and hence a decrease in $B(t^*) - A(t^*)$. In particular, when there are no dependence constraints or when they take very high values, then $t^* = 0$, and we obtain the unconstrained bounds A and B. By contrast, when the c_k are minimum, i.e., $c_k = \mu^k$, we only allow for a dependence that would make the portfolio sum constant. In this case $t^* = \gamma$, and $B(\gamma) = A(\gamma) = ES$.

Theorem 6.11 (Moment-constrained VaR bounds)   *Consider problems (6.23) and (6.24), and let t^* be defined by (6.38). We have*

$$A(t^*) \le \mathrm{VaR}_\alpha(S) \le \mathrm{VaR}_\alpha^+(S) \le B(t^*).$$

Proof We show that $B(t^*)$ is an upper bound of M. To this end, assume that there exists $T = \sum_{i=1}^{n} X_i$ that satisfies all moment constraints, i.e., $ET^k \leq c_k, k = 2, 3, \ldots, K$, and such that $\mathrm{VaR}_\alpha^+(T) > B(t^*)$. Denote the distribution function of $X(t^*)$ by G. Then, for all $a \leq x < b$, $F_T(x) \leq G(x) = \alpha$. When $b \leq x$, we have $F_T(x) \leq G(x) = 1$. Since $G(x) = 0$ for $x < a$, this implies that

$$\begin{cases} \text{for all } x < a, & F_T(x) \geq G(x), \\ \text{for all } x \geq a, & F_T(x) \leq G(x). \end{cases} \tag{6.39}$$

In other words, the distribution function F_T crosses G once from above. Since $ET = EX_{t^*}$, this implies that $X_{t^*}^* \leq_{\mathrm{cx}} T$ by the cut criterion of Karlin and Novikoff (1963).

Moreover, since T and $X^k(t^*)$ are positive, it holds that $EX^k(t^*) \leq ET^k$ for all $k = 2, 3, \ldots, K$. On the other hand, there exists $\widehat{k} \geq 2$ such that $E(X^{\widehat{k}}(t^*)) = c_{\widehat{k}}$ (by definition of t^*). This implies that $ET^{\widehat{k}} = EX_{T*}^{\widehat{k}}$, because $ET^{\widehat{k}} \leq c_{\widehat{k}}$ by the moment constraint. As $\varphi(x) = x^{\widehat{k}}$ ($\widehat{k} \geq 2$) is strictly convex, it follows that $T \overset{d}{=} X(t^*)$ (see Theorem 3.A.43 of Shaked and Shantikumar, 2007). This ends the proof.   □

The proof that $A(t^*)$ is an absolute lower bound can be given in a similar way.

Theorem 6.11 can be seen as a generalization of Theorem 6.3, which considered the case $K = 2$. A two-point distribution provides the best bounds in all cases, which at first view may seem counterintuitive. The reason is that we have inequality constraints on the moments, and, therefore, not all moment constraints are binding. Note that Theorem 6.11 also covers the unconstrained case; in this case $t^* = 0$, so that $A(t^*) = A$ and $B(t^*) = B$.

Remark 6.12 (Approximation of sharp VaR bounds) Theorem 6.11 shows that the best possible sharp upper and lower VaR bounds are obtained if one can construct a dependence among the risks $X_i, i = 1, 2, \ldots, n$, such that $S = X_1 + X_2 + \cdots + X_n$ takes values $A(t^*)$ and $B(t^*)$. Hence, we can also use the algorithm described in Section 6.2 for the approximation of VaR bounds with a variance constraint for approximating VaR bounds with moment constraints of higher orders. The only difference is that the last column in the $N \times (n+1)$ matrix $\mathbf{X} = (x_{ij})$ contains the realizations of the random variable $-X_{n+1}(t^*)$ instead of $-X_{n+1}$, i.e.,

$$x_{1,n+1} = x_{2,n+1} = \cdots = x_{k,n+1} = -B(t^*)$$
$$\text{and} \quad x_{k+1,n+1} = x_{k+2,n+1} = \cdots = x_{N,n+1} = -A(t^*). \tag{6.40}$$

    ◊

6.3.4   Portfolios with a Homogeneous Structure of Exposures

In this section we assume that $v_j := v, j = 1, 2, \ldots, n$. To discuss sharp bounds it is convenient to consider an auxiliary variable Y taking three values and given as

$$Y = \begin{cases} lv & \text{with probability } \alpha z, \\ (l+1)v & \text{with probability } \alpha(1-z), \\ \lfloor \frac{B(t^*)}{v} \rfloor v & \text{with probability } 1-\alpha, \end{cases} \tag{6.41}$$

in which $0 \le z \le 1$ and $l \in \mathbb{N}$ are the unique values such that $EY = ES$. The first moment requirement amounts to a condition of the type $l + (1-\alpha) = \text{constant}$, which implies that l and α are uniquely determined. Furthermore, t^* is defined by (6.38). The following theorem provides a sharp upper VaR bound for a homogeneous portfolio. The proof is similar to the proof in the unconstrained case (Theorem 6.9) and is therefore omitted.

Theorem 6.13   (Sharp moment-constrained bounds for a homogeneous port-folio) *Consider problems (6.23) and (6.24), and define t^* by (6.38). Assume that the variable Y in (6.41) satisfies the moment constraints, i.e., $EY^k \le c_k$); then*

$$\text{VaR}_\alpha^+(S) \le v \left\lfloor \frac{B(t^*)}{v} \right\rfloor. \tag{6.42}$$

Furthermore, this bound is sharp.

Similarly, one gets the lower bound

$$\text{VaR}_\alpha(S) \ge v \left\lceil \frac{A(t^*)}{v} \right\rceil, \tag{6.43}$$

which is also attained by a corresponding three-point distribution assuming that the moment constraints are satisfied.

The variable Y in (6.41) satisfies the moment constraints in particular when $\frac{A(t^*)}{v}$ and $\frac{B(t^*)}{v} \in \mathbb{N}$ (see also the analysis in Section 6.3.3). In general, the moment constraints are approximately satisfied by this construction of Y, and (6.42) gives approximate best bounds.

The worst case VaR value is typically strictly larger than the VaR that is obtained when assuming the risks are comonotonic. The worst case has a VaR larger than the sum of the individual VaRs. In practice, however, it is often believed that comonotonic dependence among the risks should yield the maximum possible VaR in the sense that portfolios that give rise to a VaR that is higher than the comonotonic VaR equal to the sum of individual VaRs are considered to be unrealistic. However, it is not so clear whether comonotonic scenarios are more realistic, i.e., occur more often than other extreme scenarios. In addition, this feature of having risk bounds that go beyond the comonotonic bounds does not occur when using a measure that is consistent with the convex order. Nevertheless, if we go "deep enough in the tail," we still have that the worst case VaR occurs in the case of comonotonicity.

Theorem 6.14   (Maximum VaR = comonotonic VaR in the tail area) *Consider problem (6.23), and assume there exists an admissible portfolio $(X_1, X_2, \ldots X_n)$ that* *satisfies the moment constraints. Then, there exist* $1 > \alpha^* > 0$ *and a portfolio* $(X_1^*, X_2^*, \ldots, X_n^*)$ *satisfying the constraints such that for* $\alpha \in [\alpha^*, 1]$,

$$\mathrm{VaR}_\alpha^+(X_1^* + X_2^* + \cdots + X_n^*)$$
$$= \mathrm{VaR}_\alpha^+(X_1^*) + \mathrm{VaR}_\alpha^+(X_2^*) + \cdots + \mathrm{VaR}_\alpha^+(X_n^*)$$
$$= nv. \tag{6.44}$$

Proof Without loss of generality, we can express X_i as $X_i = F_i^{-1}(V_i)$ for uniformly distributed V_i, $i = 1, 2, \ldots, n$. Next, we consider variables $X_i^* = \mathbb{1}_{\{U \le u\}} F_i^{-1}(uV_i) + \mathbb{1}_{\{U > u\}} F_i^{-1}(U)$ in which $0 < u < 1$ is chosen such that $(X_1^*, X_2^*, \ldots, X_n^*)$ also satisfies the moment constraints. One observes then that for $\alpha > \alpha^* := \max(1 - \min(p_1, p_2, \ldots, p_n), u)$,

$$\mathrm{VaR}_\alpha^+(X_1^* + X_2^* + \cdots + X_n^*) = nv. \qquad \square$$

6.4 Discussion of Consequences for Credit Risk Modeling

In this section, we discuss three main models that are used in the financial industry to assess VaRs of credit risk portfolios, namely the KMV model, the CreditRisk$^+$ model, and the Beta model (see Gordy, 2000; Vanderdorpe et al., 2008). Next, we will analyze to what extent these industry standards are robust with respect to model misspecification.

6.4.1 Credit Risk Portfolio Models

KMV Model (Merton's Model of the Firm)
Many financial institutions, as well as Basel III and Solvency II regulation, rely on "Merton's model of the firm" when computing the VaR of a portfolio of loans. The basic idea is very simple: A default is an event in which *the asset value drops below a threshold value (a liability that is due)*. Formally, after normalization, default of the ith risk occurs when $N_i < c_i$, where N_i is the normalized asset return, and c_i is the threshold value such that $p_i = P(N_i < c_i)$. Merton's model further assumes that the joint asset (log-)returns are multivariate normally distributed. Hence, for a loan portfolio, the loss S is written as

$$S = \sum_{i=1}^n v_i \mathbb{1}_{\{N_i < c_i\}}, \tag{6.45}$$

in which $(N_1, N_2, \ldots, N_n)$ is multivariate normally distributed with some correlation matrix ϱ. Each asset return N_i (driving the randomness of the ith loan) can be expressed as a linear combination of independent factors that are standard normally distributed. Specifically,

$$N_i = \sum_{j=1}^r \sqrt{\varrho_{ij}} M_j + \varepsilon_i \sqrt{1 - \sum_{j=1}^r \varrho_{ij}}, \tag{6.46}$$

in which M_j is the explaining factor of the asset return N_j, and in which ε_i represents the idiosyncratic (individual) risk. The factor weights $\sqrt{\varrho_{ij}}$ can also be interpreted as the correlation between the ith return N_i and the jth factor factor M_j. It is natural to assume that there is always a strictly positive portion of idiosyncratic risk that remains inherent in N_i, i.e., $1 - \sum_{j=1}^{r} \varrho_{ij} > 0$ and $r < n$. Hence, we can also write the portfolio loss S as

$$S = \sum_{i=1}^{n} v_i I_i, \tag{6.47}$$

in which I_i is a Bernoulli random variable with (stochastic) probability $p_i(M_1, M_2, \ldots, M_r)$ given as

$$p_i(M_1, M_2, \ldots, M_r) = \Phi\left(\frac{\Phi^{-1}(p_i) - \sum_{j=1}^{r} \sqrt{\varrho_{ij}} M_j}{\sqrt{1 - \sum_{j=1}^{r} \varrho_{ij}}} \right), \tag{6.48}$$

and where Φ is the distribution of the standard normal random variable. It is then clear that $\mathrm{VaR}_\alpha(S)$ can be obtained, for instance, using Monte Carlo simulations.

When the asset returns are assumed to be driven by one single factor M only, the above formulas further simplify in an obvious way. In this case, an analytic formula for computing the VaR becomes available for "large homogeneous portfolios." Indeed, assuming that $v_i = v$, $p_i = p$, $\varrho_{ij} = \varrho$ and letting the number of loans $n \to \infty$, we find that

$$\lim_{n \to \infty} \mathrm{VaR}_\alpha\left(\frac{S}{nv} \right) = \Phi\left(\frac{\Phi^{-1}(p) + \sqrt{\varrho} \cdot \Phi^{-1}(\alpha)}{\sqrt{1 - \varrho}} \right). \tag{6.49}$$

Note that in this case the (squared) weight ϱ is the correlation between different asset returns N_i and N_j. This model is then an example of a one-factor mixture model in which the default event of the obligor is assumed to be driven by a global economic factor M. It can also be seen as the one-factor version of the KMV model that is highly used in the industry and also appears in regulatory frameworks. For example, the Basel III standard framework relies on formula (6.49) to determine the required capital that banks need to hold for their credit portfolios; see the Basel committee on banking supervision (2010). The Solvency II framework also uses this formula to decide the amount of capital that insurers need to hold as a buffer in case reinsurance or derivative counterparts fail.

CreditRisk$^+$

Starting from the expression of the loss of a portfolio of loans in (6.47), it is clear that other assumptions for the default probabilities can be made (reflecting other choices for dependence among the risks), and (6.48) merely reflects only one such possibility. Note that we can rewrite the portfolio loss (6.47) as

$$S = \sum_{i=1}^{n} \sum_{j=1}^{I_i} v_i \tag{6.50}$$

in which I_i is a Bernoulli random variable with (stochastic) probability $p_i(M_1, M_2, \ldots, M_r)$ given by (6.48). Since the dependent Bernoulli random variables I_i are too difficult to work with, one substitutes them by other dependent random variables N_i that are "close" to the I_i but that are more tractable. Hence, we consider

$$S_* = \sum_{i=1}^{n} \sum_{j=1}^{N_i} v_i \tag{6.51}$$

in which N_i is a Poisson random variable with (stochastic) intensity $p_i(\Gamma_1, \Gamma_2, \ldots, \Gamma_r)$ given as

$$p_i(\Gamma_1, \Gamma_2, \ldots, \Gamma_r) = p_i\left(w_i + \sum_{j=1}^{r} w_{ij}\Gamma_j\right). \tag{6.52}$$

The coefficient $w_i \geq 0$ reflects the portion of idiosyncratic risk that can be attributed to the ith risk, whereas $w_{ij} \geq 0$ reflects its affiliation to the jth common factor. The random variables Γ_i are assumed to be independent Gamma-distributed variables with respective variances σ_i^2. Since for any r and $a > 0$, the random variable $a\Gamma_r$ will be distributed like a Gamma random variable, we can assume without loss of generality that $E(\Gamma_i) = 1$. Assuming that conditionally on $(\Gamma_1 = \gamma_1, \Gamma_2 = \gamma_2, \ldots, \Gamma_r = \gamma_r)$, the random variables N_i are mutually independent, we find after some computations for the moment generating function S_* that

$$m_{S_*}(t) = \exp\left(\sum_{i=1}^{n} w_i p_i(\exp(t v_i) - 1)\right) \tag{6.53}$$

$$-\sum_{j=1}^{r} \frac{1}{\sigma_j^2} \ln\left(1 - \sigma_j^2 \sum_{i=1}^{n} w_{ik} p_i(\exp(t v_i) - 1)\right)\right).$$

Using the fast Fourier transform (FFT), one can easily derive an algorithm that can be used to find the probability distribution function of S_*; see, e.g., Haaf et al. (2004).

Single Factor Model

Similarly as for the KMV model, let us consider the single factor model and assume that there exists a single random variable $\Gamma = \gamma$ representing the "global economy" with variance $\sigma^2 (= 1/\beta)$ such that, conditionally given $\Lambda = \lambda$, the random variables N_i are Poisson-distributed with parameters $p_i \lambda$. Then, the expression of the moment generating function in (6.53) can be simplified to

$$m_{S_*}(t) = \left(\frac{\beta}{\beta - \sum_{i=1}^{n} p_i(\exp(t v_i) - 1)}\right)^{\beta}, \tag{6.54}$$

which is the moment generating function of a compound negative binomially distributed variable S_*, i.e.,

$$S_* = \sum_{i=1}^{N} Y_i \qquad (6.55)$$

with

$$N \sim \text{NB}\left(\beta, \frac{\beta}{\beta + \sum_{i=1}^{n} p_i}\right),$$

and where the $Y_i \overset{d}{=} Y$ are i.i.d. and independent of N, with moment-generating function given as

$$m_Y(t) = \frac{\sum_{i=1}^{n} p_i \exp(tv_i)}{\sum_{i=1}^{n} p_i}.$$

The compound negative binomial distribution can be computed using (6.53) or by using the recursion of Panjer (1981). As compared to the KMV model, the VaR numbers can be computed explicitly, and Monte Carlo simulations are not required. Observe that this model formally allows that a credit loan defaults more than once. However, a realistic model calibration based on the given default probabilities and the default correlations generally ensures that the probability that this occurs is very small. For more details, see Credit Suisse (1997).

Beta Model
The Beta distribution has always been influential in modeling credit portfolio risk. In this case, one assumes that

$$\frac{S}{\sum_{i=1}^{n} v_i} \sim \text{Beta}(a, b),$$

where the parameters $a > 0$ and $b > 0$ are typically determined using moment matching techniques. While the use of the Beta model seems to be ad hoc and not well motivated, it is shown in Dhaene et al. (2003) that this model is closely linked to the single factor CreditRisk$^+$ model. Indeed, by taking the CreditRisk$^+$ specifications and letting $v_i = v$, $p_i = p$, and $n \to \infty$, we find that

$$\frac{S}{nv} \sim \text{Gamma}(\beta, \beta/p),$$

and a further analysis (see Dhaene et al., 2003) shows that the Gamma distribution tends to be very close to a Beta distribution with the same two moments (when the probabilities p are small enough).

6.4.2   Assessing Model Risk

In this section we analyze the robustness of VaR computations for several types of portfolios. We first compute VaR numbers using one or more industry models and next assume that the models are not completely correct, i.e., we only trust the marginal distributions of the risky loans and the first K portfolio moments, $K = 1, 2, 3, 4$. We compare the VaR numbers obtained under the model(s) with the maximum and minimum possible values when $K = 1$ (unconstrained bounds) and when $K = 2, 3$, or 4 (constrained bounds). The bounds are computed using the algorithm presented in Section 6.2.

Credit Risk Portfolio

In large portfolios a correspondence between the theoretical bounds and the (approximate) sharp bounds is to be expected, implying that significant reductions in the unconstrained VaR bounds can be obtained. The following example shows that for smaller portfolios one can still observe strong reductions. We analyze a small concentrated portfolio of 25 exposures as described in Appendix B of Credit Suisse (1997). The exposures range from 0.4 to 20.4, and the total exposure is 130.5 (all numbers mentioned are in million euros). As for the default rates, they range between 1.5 % and 30 % (Credit Suisse, 1997, pg. 61, Table 9). The portfolio expected loss is then equal to 14.2. Also, the portfolio standard deviation is assumed to be known (it is derived from default statistics) and is equal to 12.7; see Credit Suisse (1997, pg. 62). Assuming the single factor version of the CreditRisk$^+$ model, it is then straightforward to compute the value of the remaining parameter β in (6.54) (by simple moment matching). Next, the VaRs can be computed using Panjer's recursion, for instance. The results of the CreditRisk$^+$ model are given in the second column of Table 6.3; see also pg. 63 in Credit Suisse (1997). For example, the 99.5 %-VaR is equal to 62.

However, other modeling assumptions could be made. For instance, if we only trust the marginal distributions and the portfolio variance (moments up to $K = 2$), we observe from Table 6.3 that the true 99.5 %-VaR can actually be any value between 13.6 and 130.5, showing that the 99.5 %-VaR can easily be underestimated by a factor 2. In Table 6.3, we also provide VaR bounds assuming that more information on the higher-order moments[1] is available. This table makes clear that adding higher-order information reduces the gap between the upper and lower bounds for VaR significantly, thus reducing model uncertainty on VaR assessment. When the probability level α that is used to assess the VaR is not too big (e.g., $\alpha = 95$ %) then this has a positive impact on model uncertainty in the sense that the range of possible VaRs becomes "reasonable" as soon as second and/or third moment information is added. At high probability levels (e.g., $\alpha = 99.9$ %), significant model risk is present even when higher-order information is available.

Homogeneous Portfolio

We use this last example to obtain further insight into the VaR when including higher-order information. We consider a homogeneous portfolio presented on pg. 365 in McNeil, Frey, and Embrechts (2015). Using the same parameters as in Table 8.6 on pg. 365 of that book, we set the default probability of all loans equal to $p = 0.049$, and the default correlation is equal to 0.0157. The variance of the portfolio sum of n correlated

---

[1] They have been computed under the specification of the CreditRisk$^+$ model.

VaR assessment of a small portfolio (Appendix B of Credit Suisse, 1997)

$\alpha =$	CreditRisk$^+$	Comon.	Unconstrained
75 %	20.5	21.9	(1.4; 52.5)
95 %	38.9	85.2	(8.8; 117.4)
99 %	55.3	130.5	(13.0; 130.5)
99.5 %	62.0	130.5	(13.6; 130.5)
99.9 %	77.1	130.5	(14.1; 130.5)

$\alpha =$	$K = 2$	$K = 3$	$K = 4$
75 %	(15.4; 31.6)	(15.4; 31.6)	(15.4; 31.6)
95 %	(20.2; 64.6)	(20.2; 55.1)	(20.2; 51.6)
99 %	(20.2; 130.5)	(20.2; 94.9)	(20.2; 80.7)
99.5 %	(20.2; 130.5)	(20.2; 115.1)	(20.2; 94.8)
99.9 %	(20.2; 130.5)	(20.2; 130.5)	(20.2; 130.5)

Table 6.3 Column 2 contains VaRs under the CreditRisk$^+$ model. They can be compared to comonotonic VaRs and the VaR bounds (displayed in brackets) in the unconstrained and the constrained case (K reflects the number of moments of the portfolio sum that are known). The VaR bounds are obtained using the procedure in Section 6.2.

loans (all with exposure that is equal to $1/n$) can be easily calculated. Next, the two parameters of the Beta distribution can be inferred by moment matching, and one can compute higher-order moments. We use the associated higher moments as the moment constraints. We are interested in the VaR of the portfolio of loans at confidence levels 95 %, 99 %, 99.5 %, and 99.9 %.

VaR assessment of a homogeneous portfolio, $n = 10{,}000$

	VaR$_{95\%}$	VaR$_{99\%}$	VaR$_{99.5\%}$	VaR$_{99.9\%}$
Beta	10 %	13.1 %	14.4 %	17.1 %
Comon.	0 %	100 %	100 %	100 %
$K = 2$	16.72 %	31.89 %	43.17 %	90.65 %
$K = 3$	14.95 %	24.29 %	30.24 %	50.95 %
$K = 4$	14.00 %	20.55 %	24.34 %	36.23 %

Table 6.4 We report VaRs as a percentage of total exposure assuming that the portfolio loss follows a beta distribution (Beta). We provide the comonotonic VaRs and note that for the confidence levels $\alpha = 99\%$, $\alpha = 99.5\%$, and $\alpha = 99.9\%$ they coincide with the unconstrained upper bounds. We also report the maximum VaRs when higher-order information is available (obtained using Proposition 6.13).

According to the numerical results, we observe that taking additional moment constraints into account can improve the VaR bounds significantly, especially at the high percentage (e.g., 99.5 % in Table 6.4). We observe that including information on the skewness and kurtosis (the 3rd and 4th moment) reduces the model risk on the VaR assessment, in particular this makes it possible to improve the comonotonic bounds in many instances. However, the bounds on VaR are still very wide, suggesting that inclusion of other sources of dependence information would be useful to obtain further improvements.

6.4.3   Final Remarks

In this chapter, we assess the model error due to incomplete information on dependence. Specifically, we consider the VaR of a portfolio in which the individual marginal distributions and possibly the pairwise correlation or bounds on some moments of the portfolio are known. We develop a numerical procedure to estimate VaR bounds in this given context and provide some explicit results for homogeneous portfolios in an application to credit risk assessment. To assess credit portfolio risk, one needs to model the marginal risks as well as the way they interact. Dependence modeling in a credit context usually focuses on finding the economic dimensions that influence the default behavior of the different loans. Apart from factors that describe the global state of the economy, such as, for example, interest rates, the default drivers typically considered are asset size, industry sector, and geographical situation. As a result, companies of similar size, industry activity, and geographical situation will be grouped together, meaning that they behave similarly, which is akin to saying that they are positively dependent. However, all these dimensions together still do not fully capture all sources of dependence.

Under the internal model approach of Basel III and Solvency II, financial institutions are allowed to use their own model for setting their credit capital requirements. However, in the presence of incomplete information there are several statistically indistinguishable models, and we give in this chapter worst case (yielding the VaR upper bound) and best case models (for the VaR lower bound).

Our numerical illustrations at the end of this chapter show that these two extreme models may provide very different VaR estimates, in particular when these VaR calculations are performed using very high confidence levels (e.g., 99.5 %). In other words, while the inclusion of higher-order moment information helps to reduce the unconstrained VaR bounds at all confidence levels, model risk remains an issue, especially at high confidence levels.

The observation that the VaRs of credit portfolios are hard to estimate raises two concerns: First, it shows that setting capital requirements for credit risk portfolios is a difficult exercise, and it is highly questionable whether a computed regulatory capital charge based on a 99.5 % confidence level actually corresponds to a real 99.5 % confidence level. Second, it casts doubt on the use of internal models to set regulatory capital requirements. When the correct model cannot be identified, two institutions having a similar portfolio can use different, yet statistically indistinguishable, internal models and thus obtain very different capital charges. However, the two institutions should be required to hold a similar amount of regulatory capital because they have the same portfolio. This is also a consequence of the fact that the Basel committee insists that a solvency framework must exhibit comparability between institutions.

7   Distributions Specified on a Subset

In this chapter, we consider the case where, besides marginal information on the risk vector $X = (X_1, \ldots, X_n)$, either the conditional distribution of X on a subset and the probability of the subset S in $\mathbb{R}^n$ is specified or the bounds for the distribution function F_X resp. the copula C_X are given on corresponding subsets. It may be the case that C_X of X is known on a central domain due to availability of sufficient statistical data allowing for precise modeling within this domain. Alternatively, some positive or negative dependence information may be available in the upper or lower tail area. It is assumed that this dependence information is given by one- or two-sided bounds on the distribution function resp. the copula in a particular case that the distribution function resp. the copula is known exactly on this domain.

We first provide several results on improved Hoeffding–Fréchet bounds based on bounds for the distribution function resp. for the copula on a subset S, which allow us to apply the method of improved standard bounds as in Chapter 5. In Section 7.2 we obtain as an application several VaR bounds for the sum based on one-sided bounds on the distribution function F resp. on the copula C on some subdomains and clarify the question of which subdomains are informative vs. uninformative.

In Section 7.3 we consider the case where, due to precise modeling, the conditional distribution of X is known on a central domain. We also assume knowledge of the marginals and of the probability of the domain. Note that there is a relevant difference to the assumptions in Sections 7.1 and 7.2. While knowing the distribution function F on an open domain A implies knowledge of P^X on A, the converse direction is not valid, even for rectangular domains A. So the type of information used in Section 7.3 is typically not comparable to that used in Section 7.2. Based on a numerical approach using the RA, the value of this type of information for improving VaR bounds for the sum is investigated in Section 7.3.4.

7.1   Improved Hoeffding–Fréchet Bounds

Based on the method of improved standard bounds as described in Section 5, a natural procedure to derive risk bounds, in particular VaR-bounds, is first to derive improvements of the Hoeffding–Fréchet bounds using the information of the distribution on a subset and then to apply the method of improved standard bounds as in Chapter 5. As described in Chapter 5, such improved (two-sided) Hoeffding–Fréchet bounds have been developed in the two-dimensional case in Rachev and Rüschendorf (1994), Nelsen et al. (2004), Tankov (2011), Bernard et al. (2014a), and Bernard and Vanduffel (2014). They have been extended to the n-dimensional case in Puccetti et al. (2016) and Lux and Papapantoleon (2019).

Let $G \colon \mathbb{R}^n \to \mathbb{R}$ be an increasing function such that

$$F_-(x) \le G(x) \le F_+(x), \quad x \in \mathbb{R}^n, \quad F_-, F_+ \text{ are the Fréchet bounds.} \tag{7.1}$$

We assume that X is a risk vector with marginals $F_1, \ldots, F_n$ and that the distribution function $F = F_X$ of X satisfies that $F \le G$ or $F \ge G$ on some subset $S \subset \mathbb{R}^n$.

Theorem 7.1 (Improved Hoeffding–Fréchet bounds with subset information)
Let G be an increasing function satisfying (7.1), and define for a given subset $S \subset \mathbb{R}^n$

$$\overline{F}^S(x) = \min \left(\min_{i \le n} F_i(x_i), \inf_{y \in S} \left\{ G(y) + \sum_{i=1}^{n} (F_i(x_i) - F_i(y_i))_+ \right\} \right)$$

and $\underline{F}^S(x) = \max \left(0, \sum_{i=1}^{n} F_i(x_i) - (n-1), \right.$ (7.2)

$$\left. \sup_{y \in S} \left\{ G(y) + \sum_{i=1}^{n} (F_i(y_i) - F_i(x_i))_+ \right\} \right).$$

Then for an n-variate distribution function $F \in \mathcal{F}(F_1, \ldots, F_n)$, it holds that:

(i) If $F(y) \le G(y)$ for all $y \in S$, then $F(x) \le \overline{F}^S(x)$, for all $x \in \mathbb{R}^n$.
(ii) If $F(y) \ge G(y)$ for all $y \in S$, then $F(x) \ge \underline{F}^S(x)$, for all $x \in \mathbb{R}^n$.
(iii) If $F(y) = G(y)$ for all $y \in S$, then

$$\underline{F}^S(x) \le F(x) \le \overline{F}^S(x) \text{ for all } x \in \mathbb{R}^n. \tag{7.3}$$

$\overline{F}^S$ *and* $\underline{F}^S$ *are called improved Hoeffding–Fréchet bounds with subset information.*

Proof (i) Let $X = (X_1, \ldots, X_n)$ be a random vector with distribution function F and without loss of generality, let $y \in S$ satisfy $y_i \le x_i$ for $1 \le i \le n$. Then using the assumption $F(y) \le G(y)$ for $y \in S$, we obtain

$$
\begin{aligned}
F(x) &= P(X_1 \le x_1, \ldots, X_n \le x_n) \\
&= P(X_1 \le y_1, X_2 \le x_2, \ldots, X_n \le x_n) \\
&\quad + P(y_1 < X_1 \le x_1, X_2 \le x_2, \ldots, X_n \le x_n) \\
&= P(X_1 \le y_1, X_2 \le y_2, X_3 \le x_3, \ldots, X_n \le x_n) \\
&\quad + P(y_1 < X_1 \le x_1, X_2 \le x_2, \ldots, X_n \le x_n) \\
&\quad + P(X_1 \le y_1, y_2 < X_2 \le x_2, X_3 \le x_3, \ldots, X_n \le x_n) \\
&= P(X_1 \le y_1, \ldots, X_n \le y_n) + P(y_1 < X_1 \le x_1, X_2 \le x_2, \ldots, X_n \le x_n) \\
&\quad + \cdots + P(X_1 \le y_1, \ldots, X_{n-1} \le y_{n-1}, y_n < X_n \le x_n) \\
&\le G(y) + \sum_{i=1}^{n} (F_i(x_i) - F_i(y_i)).
\end{aligned}
$$

Considering all $y \in S$ and taking the infimum, this implies by the classical Hoeffding–Fréchet bounds that $F \leq (\overline{F})^S(x)$.

(ii) Similarly to (i), for $y \in S$ satisfying $y_i \geq x_i$ for $1 \leq i \leq n$, we obtain that

$$
\begin{aligned}
F(x) &= P(X_1 \leq x_1, \ldots, X_n \leq x_n) \\
&= P(X_1 \leq y_1, \ldots, X_n \leq y_n) - P(x_1 < X_1 \leq y_1, X_2 \leq x_2, \ldots, X_n \leq x_n) \\
&\quad - \cdots - P(X_1 \leq y_1, \ldots, X_{n-1} \leq y_{n-1}, x_n < X_n \leq y_n).
\end{aligned}
$$

Considering that $F(y) \geq G(y)$ is equivalent to $1 - F(y) \leq 1 - G(y)$, it follows that

$$
\begin{aligned}
1 - F(x) &= 1 - P(X_1 \leq y_1, \ldots, X_n \leq y_n) \\
&\quad + P(x_1 < X_1 \leq y_1, X_2 \leq x_2, \ldots, X_n \leq x_n) \\
&\quad + \cdots + P(X_1 \leq y_1, \ldots, X_{n-1} \leq y_{n-1}, x_n < X_n \leq y_n) \\
&\leq 1 - G(y) + \sum_{i=1}^{n} (F_i(y_i) - F_i(x_i)),
\end{aligned}
$$

which implies

$$
F(x) \geq G(y) - \sum_{i=1}^{n} (F_i(y_i) - F_i(x_i)).
$$

Considering all $y \in S$ and taking the supremum, this implies by the classical Hoeffding–Fréchet bounds that $F \geq \underline{F}^S(x)$.

(iii) is a consequence of (i) and (ii).   □

Remark 7.2   a) If X is positive lower orthant dependent (PLOD), i.e.,

$$
F(x) \geq G(x) := \prod_{i=1}^{n} F_i(x_i), \quad x \in S = \mathbb{R}^n, \tag{7.4}
$$

then

$$
\sup_{y \in \mathbb{R}^n} \left\{ \prod_{i=1}^{n} F_i(y_i) - \sum_{i=1}^{n} (F_i(y_i) - F_i(x_i))_+ \right\} = \prod_{i=1}^{n} F_i(x_i) = G(x),
$$

and as a consequence $\underline{F}^S(x) = \prod_{i=1}^{n} F_i(x_i) = G(x)$, i.e., $\underline{F}^S$ coincides with G and is a sharp lower bound. Similarly, if G is a joint distribution function with marginals $F_1, \ldots, F_n$, i.e., $G \in \mathcal{F}(F_1, \ldots, F_n)$, then the improved Hoeffding–Fréchet bounds coincide with G and are sharp, i.e., $\overline{F}^S = G$ under condition (i) and $\underline{F}^S = G$ under condition (ii).

b) **Sharpness of the bounds $\underline{F}^S, \overline{F}^S$**   In general the bounds $\underline{F}^S$ and $\overline{F}^S$ are not distribution functions and are not sharp bounds as can be seen easily in the case where $S = \mathbb{R}^n$.

Some conditions to imply sharpness in the case $n = 2$ are given in Tankov (2011) and Bernard et al. (2012).   ◊

In the particular case where $F_i = U(0, 1)$, $1 \leq i \leq n$, Theorem 7.1 implies the following improved bounds for the copula C of the risk vector.

Corollary 7.3 (Improved copula bounds with subset information) *Let* $S \subset [0, 1]^n$, *and let* Q *be a componentwise increasing function on* $[0, 1]^n$ *such that* $W(u) \le Q(u) \le M(u)$, $u \in [0, 1]^n$. *Define the bounds* $A^{S,Q}, B^{S,Q} : [0, 1]^n \to [0, 1]$ *as*

$$A^{S,Q}(u) = \min \left(M(u), \inf_{a \in S} \left\{ Q(a) + (u_i - a_i)_+ \right\} \right),$$

$$B^{S,Q}(u) = \max \left(W(u), \sup_{a \in S} \left\{ Q(a) - (a_i - u_i)_+ \right\} \right).$$

Then for an n-dimensional copula C, it holds that

(i) *If* $C(u) \le Q(u)$ *for all* $u \in S$, *then* $C(u) \le A^{S,Q}(u)$, *for all* $u \in [0, 1]^n$.
(ii) *If* $C(u) \ge Q(u)$ *for all* $y \in S$, *then* $C(u) \ge B^{S,Q}(u)$, *for all* $u \in [0, 1]^n$.
(iii) *If* $C(u) = Q(u)$ *for all* $y \in S$, *then* $B^{S,Q}(u) \le C(u) \le A^{S,Q}(u)$, *for all* $u \in [0, 1]^n$.

Remark 7.4 (Survival copulas) In the case that lower resp. upper bounds $\overline{Q}_1$ and $\overline{Q}_2$ are available on the survival copula $\overline{Q}$ on a subset $S \subset [0, 1]^n$, Corollary 7.3 implies improved lower resp. upper bounds on the survival copula. If $\overline{Q}_1 \le \overline{Q} \le \overline{Q}_2$ on S, then

$$\overline{B}^{\overline{S}, \overline{Q}_1}(u) \le \overline{Q}(u) \le \overline{A}^{\overline{S}, \overline{Q}_2}(u) \text{ for all } u \in [0, 1]^n, \tag{7.5}$$

where $\overline{S} = \{\underline{1} - u; \ u \in S\}$, $\underline{1} = (1, \dots, 1)$, and $\overline{B}^{\overline{S}, \overline{Q}_1}(u) = B^{\overline{S}, \overline{Q}_1}(\underline{1} - u)$, $\overline{A}^{\overline{S}, \overline{Q}_2}(u) = A^{\overline{S}, \overline{Q}_2}(\underline{1} - u)$. ◊

The improved Hoeffding–Fréchet bounds in Theorem 7.1 (see (7.3) and (7.5)) may be considerable improvements over the classical Hoeffding–Fréchet bounds and thus may lead to strongly improved VaR bounds by the method of one- or two-sided improved standard bounds described in Chapter 5. The degree of improvement depends on the dependence information described by the subset S as well as on the bounds G, Q resp. $\overline{Q}$. In the following section, some effects of this kind are discussed in some applications.

7.2 Distribution Functions or Copulas with Given Bounds on a Subset

In this section, we consider various forms of subset information and their effects on the reduction of VaR bounds.

7.2.1 Positive Dependence in the Tails

In the first considered case, positive dependence of the risk vector is only assumed in the upper tail of its distribution. More precisely, for a risk vector $X = (X_1, \dots, X_n)$ with marginal distribution functions $F_1, \dots, F_n$, we assume that a lower bound $Q = C_\ell$ for the copula C_X is given on the tail domain $S = [\beta, 1]^n$, i.e.,

$$C_X \ge C_\ell \text{ on } [\beta, 1]^n. \tag{7.6}$$

As usual it is assumed that C_ℓ is increasing and satisfies $W \le C_\ell \le M$. Equivalent to (7.6) is to assume a lower bound of the distribution function on a corresponding subset.

Using Theorem 7.1 and Corollary 7.3, we obtain, from the improved Hoeffding–Fréchet bounds with side information.

Theorem 7.5 (Positive dependence in the tails)   *Let the risk vector X have marginals F_i and satisfy the positive dependence condition (7.6) where $0 \leq \beta \leq \alpha < 1$. Then the following estimate holds for the VaR of the joint portfolio $S_n = \sum_{i=1}^{n} X_i$ at level α:*

$$\mathrm{VaR}_\alpha(S_n) \leq \mathrm{VaR}_{\alpha^*}(S_n^c) = \sum_{i=1}^{n} F_i^{-1}(\alpha^*), \qquad (7.7)$$

where $\alpha^* = \delta_{C_\ell}^{-1}(\alpha)$.

Proof   For the determination of the lower bound B^{S,C_ℓ} in (7.4), note that for $u \in [\alpha, 1]^n$ it holds that

$$\sup_{a \in [\beta, 1]^n} \left\{ C_\ell(a) - \sum_{i=1}^{n}(a_i - u_i)_+ \right\} = \sup_{a \in [u, 1]} \left\{ C_\ell(a) - \sum_{i=1}^{n}(a_i - u_i) \right\} = C_\ell(u). \qquad (7.8)$$

The first equality in (7.8) follows, noting that by the assumption of increasingness of C_ℓ we can replace a_j by u_j, $1 \leq j \leq n$, without decreasing the sup. Since the sup is now taken over $[u, 1]$, the second equality follows from the Lipschitz inequality for copulas:

$$C_\ell(a) - C_\ell(u) \leq \sum_{i=1}^{n}(a_i - u_i). \qquad (7.9)$$

Now the statement of (7.7) follows from Corollary 7.3 applied with $Q = B^{S,C_\ell}$, as for $u \in [\alpha, 1]^n$ we have by (7.8) that $\delta_{B^{S,C_\ell}}(u) = \delta_{C_\ell}(u)$.   □

Remark 7.6

a) Theorem 7.5 gives the same VaR bound as Theorem 5.12. While in Theorem 5.12 it is assumed that $C_X \geq C_\ell$ holds on $[0, 1]^n$, in Theorem 7.5 this assumption is only made on the upper tail part $[\alpha, 1]^n$ of the distribution copula of X. Thus the information on the lower part of the distribution does not contribute to establishing the upper bound. A similar effect in the context of sharp VaR bounds without dependence information is described in Puccetti and Rüschendorf (2012b).

b) The computation of the VaR bound $\mathrm{VaR}_{\alpha^*}(S_n^c)$ is straightforward for any number n of marginal distributions and any copula lower bound C_ℓ.

c) There exist several statistical procedures to test the lower bound of a copula on a subset $S \subset [0, 1]^n$. In Scaillet (2005), for example, a Kolmogorov–Smirnov test for quadrant dependence is investigated, which can easily be modified to a test for positive quadrant dependence on S.   ◊

As a concrete, example consider a risk portfolio X with marginals $F_1 = \cdots = F_n = F$ and copula C_X bounded by a Gumbel copula $C_\vartheta^{\mathrm{Gu}}$ on $S = [\alpha, 1]^n$,

$$C_X \geq C_\ell = G_\vartheta^{\mathrm{Gu}} \text{ on } S = [\alpha, 1]^n. \qquad (7.10)$$

In Table 7.1 we give the values of the VaR bound $\mathrm{VaR}_{\alpha^*}(S_n^c)$ for increasing values of ϑ, corresponding to increasing positive dependence assumed in the tail of X.

The VaR bounds $\mathrm{VaR}_{\alpha^*}(S_n^c)$ are compared to the sharp upper bounds without dependence information, i.e., to $\overline{\mathrm{VaR}}_\alpha$, and to the comonotonic VaR, i.e., to $\mathrm{VaR}_\alpha(S_n^c)$. From Table 7.1 one finds that improvements of the sharp unconstrained bounds $\overline{\mathrm{VaR}}_\alpha$

$\mathrm{VaR}_{\alpha^*}(S_n^c)$ α	$\vartheta = 1$	$\vartheta = 3$	$\vartheta = 5$	$\vartheta = 10$	$\overline{\mathrm{VaR}}_\alpha$	$\mathrm{VaR}_\alpha(S_n^c)$
0.990	107.38	60.27	53.68	49.17	84.44	45.00
0.995	152.33	87.38	78.02	71.62	121.49	65.71
0.999	348.20	201.72	180.71	166.36	277.84	153.11

Table 7.1 Table of $\mathrm{VaR}_{\alpha^*}(S_n^c)$, $\overline{\mathrm{VaR}}_\alpha$, and $\mathrm{VaR}_\alpha(S_n^c)$ for $n = 5$ Pareto(2)-risks, $F_i(x) = 1 - (1 + x)^{-2}$, $x > 0$ and Gumbel lower bound on $[\alpha, 1]^n$.

are obtained for sufficiently large ϑ. Note that in the limit as $\vartheta \to \infty$, the Gumbel copula converges to the comonotonic copula,

$$C_\vartheta^{\mathrm{Gu}} \xrightarrow[\vartheta \to \infty]{} M,$$

and, consequently, the VaR-bound $\mathrm{VaR}_{\alpha^*}(S_n^c)$ converges to the comonotonic VaR, $\mathrm{VaR}_\alpha(S_n^c)$. Also notice that in the homogeneous case one gets

$$\tau_{C_\ell}^{-1}(\alpha) = \mathrm{VaR}_{\alpha^*}(S_n^c) = nF^{-1}(\alpha^*). \tag{7.11}$$

Similar conclusions can be drawn when a Gaussian copula is assumed as a lower bound on the portfolio copula, i.e.,

$$C_X \geq C_\ell = C_\varrho^{\mathrm{Ga}} \quad \text{on } S = [\alpha, 1]^n. \tag{7.12}$$

Table 7.2 displays the risk figures as in Table 7.1 for increasing values of the correlation parameter ϱ, chosen as to match the corresponding values of the pairwise Kendall rank correlation of the Gumbel copula in Table 7.1. Given the same strength of dependence, as expected the Gaussian copula yields a smaller relative improvement compared to the Gumbel copula.

Finally we consider an inhomogeneous portfolio in Table 7.3. Since the value α^* in Corollary 7.3 does not depend on the marginals, calculation of the bounds is straightforward. We consider the case $n = 9$ with $F_1 = F_2 = F_3 = \mathrm{Pareto}(2)$, $F_4 = F_5 = F_6 = \mathrm{Log}\,N(0.2, 1)$, and $F_7 = F_8 = F_9 = \mathrm{Gamma}(3,2)$, with a Gumbel lower bound $C_X \geq C_\vartheta^{\mathrm{Gu}}$ on $S = [\alpha, 1]^n$.

$\mathrm{VaR}_{\alpha^*}(S_n^c)$ α	$\varrho =$ 0	0.8660	0.9511	0.9877	$\overline{\mathrm{VaR}}_\alpha$	$\mathrm{VaR}_\alpha(S_n^c)$
0.990	106.58	75.97	63.37	53.96	84.44	45.00
0.995	152.96	112.41	93.89	79.15	121.49	65.71
0.999	348.48	269.65	222.57	187.32	277.84	153.11

Table 7.2 Risk values for a homogeneous Pareto(2) portfolio, $n = 5$, as in Table 7.1 with a Gaussian lower bound $C_X \geq C_\varrho^{\mathrm{Ga}}$ on $S = [\alpha, 1]^n$.

$\mathrm{VaR}_{\alpha^*}(S_n^c)$ $\diagdown$ α	$\vartheta = 1$	$\vartheta = 3$	$\vartheta = 5$	$\vartheta = 10$	$\overline{\mathrm{VaR}}_\alpha$	$\mathrm{VaR}_\alpha(S_n^c)$
0.990	232.76	144.77	132.06	123.24	165.29	114.96
0.995	289.42	180.52	164.53	153.51	206.04	143.23
0.999	510.77	305.86	277.23	257.74	349.62	239.79

Table 7.3 Inhomogeneous portfolio $n = 9$ with Gumbel lower bound $C_X \geq C_\vartheta^{\mathrm{Gu}}$ on $[\alpha, 1]^n$.

7.2.2 Distribution Functions with Bounds on a Central Domain

In this case it is assumed that a lower bound for the copula of the risk vector X is given in the central domain of the distribution or in any subset $S \subset [0, 1]^n$ not intersecting with the upper α-tail area, i.e., $\overline{S} \cap [\alpha, 1]^n = \phi$. By Sklar's theorem this is equivalent to assuming lower bounds for the distribution function in a domain S not intersecting with the upper quantile area. This condition is motivated by the applications treated in Bernard et al. (2013) and Bernard and Vanduffel (2015), where for a risk portfolio it is assumed that the distribution is known by statistical analysis in the central domain of the distribution, while generally only the marginals are known. The following result says that, using the method of standard bounds, the dependence information on the distribution function on a central domain does not imply improvements compared to the unconstrained case with marginal information only. An analysis of this case by an alternative method is given in Section 7.3.

Theorem 7.7 **(Bounds for the distribution function on a central domain)** *Suppose that the risk vector X has marginals $F_1, \ldots, F_n$ and that $C_X \geq C_\ell$ on a subset $S \subset [0, 1]^n$ with $\overline{S} \cap [\alpha, 1]^n = \phi$. Here C_ℓ is increasing and satisfies $W \leq C_\ell \leq M$. Then the improved* VaR *bounds with dependence information yield*

$$\mathrm{VaR}_\alpha(S_n) \leq \sum_{i=1}^n F_i^{-1}(\alpha^*), \quad \alpha^* = \delta_W^{-1}(\alpha). \tag{7.13}$$

Proof For any $u \in S$ there is at least one component i such that $u_i < \alpha$, and thus

$$C_\ell(u) \leq M(u) = \min_{1 \leq j \leq n} u_j < \alpha.$$

This implies for the calculation of the improved lower bound B^{S, C_ℓ}, that

$$\sup_{a \in S} \left\{ C_\ell(a) - \sum_{j=1}^n (a_j - u_j)_+ \right\} \leq \sup_{a \in S} C_\ell(a) < \alpha$$

since $\overline{S} \cap [\alpha, 1]^n = \phi$. As a consequence, the inequality $\delta_B^{S, C_\ell}(u) \geq \alpha$ can only hold for those values of u such that

$$\delta_{B^{S, C_\ell}}(u) = (nu - (n - 1))_+ = \delta_W(u).$$

The statement in (7.13) is now a consequence of the improved standard bounds in Theorem 5.3. $\square$

Remark 7.8

a) Theorem 7.7 implies that dependence information given by bounds on the distribution function on a central part of the distribution does not help to improve the VaR bounds obtained by the method of improved standard bounds as in Section 7.1. For $n > 2$, on the other hand, the standard bounds without additional dependence information are improved by the sharp unconstrained VaR bounds $\overline{\mathrm{VaR}}_\alpha(S_n)$.

b) From the proof of Theorem 7.7 it follows that in order to reach the comonotonic risk value $\mathrm{VaR}_\alpha(S_n^c)$ by the method of improved standard bounds, it is sufficient that $C_X \geq C_\ell = M$ just at the single point set $S = \{(\alpha, \ldots, \alpha)\}$. ◊

Information on a central domain of the risk distribution is relevant in the setting of improved standard bounds only if it partially covers the upper tails of the joint distribution. If $S = [0, \beta]^n$ with β sufficiently large, i.e., when $\delta_{C_\ell}(\beta) \geq \alpha$, then the dependence information concerns a relevant part of the distribution, and the same VaR bounds as in Theorem 7.5 hold. Otherwise, no improvements are obtained, as in the case dealt with in Theorem 7.7.

Theorem 7.9 (Informative domains of the distribution)   *Let X be a risk vector with marginals F_i, $1 \leq i \leq n$, and such that $C \geq C_\ell$ on $S = [0, \beta]^n$ with $\beta \geq \alpha$, C_ℓ increasing, and $W \leq C_\ell \leq M$. Then*

a) *If $\delta_{C_\ell}(\beta) \geq \alpha$, then the VaR bound in (7.7) holds.*
b) *If $\delta_{C_\ell}(\beta) < \alpha$, then only the VaR bound in (7.13) holds.*

Proof   From the proof of Theorem 7.5 it follows that

$$\sup_{a \in [0,\beta]^n} \left\{ C_\ell(a) - \sum_{i=1}^{n} (a_i - u_i)_+ \right\} = C_\ell(u) \text{ for all } u \in [0, \beta]^n.$$

If $\delta_{C_\ell}(\beta) \geq \alpha$, then the same proof as for Theorem 7.5 implies that the bound in (7.7) holds. In the case that $\delta_{C_\ell}(\beta) < \alpha$, we have for all u that

$$\sup_{a \in [0,\beta]^n} \left\{ C_\ell(a) - \sum_{i=1}^{n} (a_i - u_i)_+ \right\} \leq \sup_{a \in [0,\beta]^n} C_\ell(a) \leq C_\ell(\beta, \ldots, \beta) < \alpha.$$

Thus the inequality $\delta_{BS,C_\ell}(u) \geq \alpha$ can only be satisfied for those values of u such that

$$\delta_{BS,C_\ell}(u) = (nu - (n-1))_+ = \delta_W(u).$$

In consequence, as in Theorem 7.7 we reach only the bound

$$\mathrm{VaR}_{\alpha^*}(S_n) \leq \sum_{i=1}^{n} F_i^{-1}(\alpha^*), \quad \alpha^* = \delta_W^{-1}(\alpha). \qquad \square$$

Tables 7.4 and 7.5 display the values of α^* in dependence on the quantile level α and on the copula C_ℓ, which is the Gumbel copula $C_\vartheta^{\mathrm{Gu}}$ in Table 7.4 and the Gaussian copula in Table 7.5. These values can be interpreted by Theorem 7.9 as the minimal value of β at which dependence information given by bounds on the copula on $S = [0, \beta]^n$ becomes significant.

α^*	$n = 5$			$n = 9$		
α	$\vartheta = 3$	5	10	$\vartheta = 3$	5	10
0.990	0.99413	0.99274	0.99148	0.99517	0.99354	0.99196
0.995	0.99707	0.99637	0.99574	0.99759	0.99677	0.99598
0.999	0.99707	0.99637	0.99574	0.99952	0.99936	0.99920

Table 7.4 Values of α^* for a Gumbel copula $C_{\vartheta}^{\mathrm{Gu}}$.

α^*	$n = 5$			$n = 9$		
α	$\varrho = 0.8660$	0.9511	0.9877	$\varrho = 0.8660$	0.9511	0.9877
0.990	0.99618	0.99465	0.99282	0.99720	0.99559	0.99346
0.995	0.99819	0.99744	0.99647	0.99869	0.99791	0.99683
0.999	0.99966	0.99952	0.99934	0.99977	0.99963	0.99940

Table 7.5 Values of α^* for a Gaussian copula $C_{\varrho}^{\mathrm{Gu}}$.

7.3   Distributions Known on a Subset

In this section, we assume that some reference model for a risk $X = (X_1, X_2, \ldots, X_n)$ is available. The reference model, however, is prone to error, e.g., because of lack of data. Specifically, the marginal distributions of the X_i are assumed to be correctly specified, but the joint distribution is only known conditionally on X taking a value in some subset. Typically, the subset contains central observations, i.e., those that are close to the center of mass. We derive risk bounds for risk measures of $\sum_{i=1}^n X_i$. As expected, upper bounds on tail risk measures, such as VaR and TVaR, will only improve the unconstrained upper bounds if the subset is big enough that it includes observations with a sum of components that is high enough (i.e., that contribute to the tail risk). By contrast, lower bounds are typically improved significantly.

7.3.1   Setting and Assumptions

Let $X = (X_1, X_2, \ldots, X_n)$ be some random vector for which we aim to compute $\varrho(\sum_{i=1}^n X_i)$, in which $\varrho(\cdot)$ is some (law-invariant) risk measure. Let $A_f \subset \mathbb{R}^n$ and $A_u = \mathbb{R}^n \setminus A_f$ ("f" stands for fixed and "u" stands for unspecified). We assume that we know

(i) the marginal distribution F_i of X_i on $\mathbb{R}$ for $i = 1, 2, \ldots, n$;
(ii) the distribution of $X \mid \{X \in A_f\}$;
(iii) the probability $p_f := P(X \in A_f)$, as well as $p_u := P(X \in A_u) = 1 - p_f$.

The joint distribution of X is thus not completely specified (unless $A_f = \mathbb{R}^n$ and $A_u = \phi$). Consequently, $\varrho(\sum_{i=1}^n X_i)$ cannot be precisely computed. In what follows, we are interested in finding the extreme possible values of $\varrho(\sum_{i=1}^n X_i)$, i.e., we consider the problems

$$\overline{\varrho} := \sup\left\{\varrho\left(\sum_{i=1}^{n} Y_i\right)\right\}, \qquad \underline{\varrho} := \inf\left\{\varrho\left(\sum_{i=1}^{n} Y_i\right)\right\},$$

where the supremum and the infimum are taken over all joint distributions of random vectors $Y = (Y_1, Y_2, \ldots, Y_n)$ that satisfy (i), (ii), and (iii).

In this respect, it will be useful to consider the indicator variable $\mathbb{I}$ corresponding to the event "$X \in A_f$":

$$\mathbb{I} := \mathbb{1}_{\{X \in A_f\}}, \tag{7.14}$$

which implies that

$$p_f = P(\mathbb{I} = 1) \text{ and } p_u = P(\mathbb{I} = 0). \tag{7.15}$$

Let us also introduce a standard, uniformly distributed random variable U independent of the event "$X \in A_f$" (and thus also independent of $\mathbb{I}$) as well as a random vector $Z = (Z_1, Z_2, \ldots, Z_n)$ defined by

$$Z_i = F^{-1}_{X_i | X \in A_u}(U), \qquad i = 1, 2, \ldots, n, \tag{7.16}$$

where $F^{-1}_{X_i | X \in A_u}$ denotes the (left) inverse of the distribution function

$$F_{X_i | X \in A_u}(x) := P(X_i \le x \mid X \in A_u).$$

Note that $F^{-1}_{X_i | X \in A_u}(x)$ can be computed, as the marginal distribution of X_i is known, and the joint distribution of X is known on A_f (see the properties (i) and (ii)). Further, all Z_i, $i = 1, 2, \ldots, n$, are increasing in the (common) variable U, and thus Z is a comonotonic vector with known joint distribution. Define also

$$T := F^{-1}_{\sum_i X_i | X \in A_f}(U), \tag{7.17}$$

i.e., T is a random variable with distribution $F_{\sum_i X_i | X \in A_f}(x)$.

7.3.2 Bounds on Convex Risk Measures

The next proposition provides easy-to-compute upper and lower bounds for a convex risk measure of the aggregate sum $\sum_{i=1}^{n} X_i$.

Proposition 7.10 (Bounds on a convex risk measure of $\sum_{i=1}^{n} X_i$) *Let $X = (X_1, \ldots, X_n)$ be a random vector that satisfies properties (i), (ii), and (iii), and let $\mathbb{I}$ and $Z = (Z_1, \ldots, Z_n)$ be defined as in (7.14) and (7.16). Let $\varrho(\cdot)$ be some convex risk measure. We have:*

$$\underline{\varrho} \ge \varrho\left(\mathbb{I}\sum_{i=1}^{n} X_i + (1 - \mathbb{I})\sum_{i=1}^{n} E Z_i\right)$$

and

$$\overline{\varrho} = \varrho\left(\mathbb{I}\sum_{i=1}^{n} X_i + (1 - \mathbb{I})\sum_{i=1}^{n} Z_i\right).$$

Proof Recall from Theorem 1.4 that for any vector $(Y_1, Y_2, \ldots, Y_n)$ and any convex function $v(x)$, it holds that

$$\mathrm{E}v\left(\sum_{i=1}^{n} Y_i\right) \leq \mathrm{E}v\left(\sum_{i=1}^{n} F_{Y_i}^{-1}(U)\right), \qquad (7.18)$$

where U is a uniformly distributed random variable on $(0, 1)$. A simple conditioning argument and taking into account Jenssen's inequality then shows that for all convex functions $v(x)$,

$$\mathrm{E}v\left(\sum_{i=1}^{n} (\mathbb{I}X_i + (1 - \mathbb{I})\mathrm{E}(Z_i))\right) \leq \mathrm{E}v\left(\sum_{i=1}^{n} X_i\right)$$

$$\leq \mathrm{E}v\left(\sum_{i=1}^{n} (\mathbb{I}X_i + (1 - \mathbb{I})Z_i)\right).$$

Hence, we have that

$$\underline{\varrho} \geq \varrho\left(\mathbb{I}\sum_{i=1}^{n} X_i + (1 - \mathbb{I})\sum_{i=1}^{n} EZ_i\right)$$

and

$$\overline{\varrho} \leq \varrho\left(\mathbb{I}\sum_{i=1}^{n} X_i + (1 - \mathbb{I})\sum_{i=1}^{n} Z_i\right).$$

Note that the multivariate vector

$$(\mathbb{I}X_1 + (1 - \mathbb{I})Z_1, \ \mathbb{I}X_2 + (1 - \mathbb{I})Z_2, \ldots, \mathbb{I}X_n + (1 - \mathbb{I})Z_n) \qquad (7.19)$$

satisfies conditions (i), (ii), and (iii) so that the stated upper bound for $\overline{\varrho}$ is sharp. $\square$

The intuition for the stated bounds is as follows. The conditional distribution of $\sum_{i=1}^{n} X_i$ is known on the event $\{X \in A_f\}$ but is unknown on the event $\{X \in A_u\}$. On A_u, one then substitutes the sum $\sum_i X_i$ by the constant $\sum_i EZ_i$ to compute the lower bound and by the comonotonic sum $\sum_i Z_i$ to compute the upper bound. Note in particular that when $A_u = \phi$, $\underline{\varrho} = \overline{\varrho}$ and there is no model uncertainty.

While the upper bound stated in Proposition 7.10 is sharp, the stated lower bound may not be sharp because the distribution function of $\mathbb{I}X_i + (1 - \mathbb{I})EZ_i$ is usually not equal to F_i, $i = 1, 2, \ldots, n$. That is, ϱ^- coincides with the stated lower bound if and only if one can construct a dependence among the components of $(Z_1, Z_2, \ldots, Z_n)$ such that $Z_1 + Z_2 + \cdots + Z_n$ becomes constant and thus equal to $EZ_1 + EZ_2 + \cdots + EZ_n$. We use this insight to propose an algorithm below that makes it possible to closely approximate ϱ^-. To this end, we discretize the problem by sampling from a random vector X satisfying the assumptions (i), (ii), and (iii), N realizations $(x_{i1}, x_{i2}, \ldots, x_{in})$. Denote by $\mathbf{X} = (x_{ij})$ the corresponding $N \times n$ assignment matrix. We can interpret $\mathbf{X}$ as a concatenation of two submatrices $\mathbf{X}_f$ and $\mathbf{X}_u$, where $\mathbf{X}_f$ contains all realizations $(x_{i1}, x_{i2} \ldots, x_{in}) \in A_f$, and $\mathbf{X}_u$ consists of the remaining realizations. In what follows we denote by ℓ_f the number of elements in $\mathbf{X}_f$ and by ℓ_u the number of elements in $\mathbf{X}_u$, such that

$$N = \ell_f + \ell_u.$$

Without any loss of generality, we can assume that $\mathbf{X}_f$ corresponds to the ℓ_f first rows of $\mathbf{X}$ and $\mathbf{X}_u$ corresponds to the last ones. Furthermore, we can also assume that in each column of the matrix $\mathbf{X}_u$, the values are sorted in decreasing order, that is, such that $x_{m_1 k} \geq x_{m_2 k} \geq \cdots \geq x_{m_{\ell_u} k}$ for all $k = 1, 2, \ldots, n$. Finally, for $\mathbf{X}_f$ we can assume that the ℓ_f observations $(x_{i_j 1}, x_{i_j 2} \ldots x_{i_j n})$ appear in such a way that for the sums of the components, i.e., $s_j := x_{i_j 1} + x_{i_j 2} + \cdots + x_{i_j n}$, $j = 1, 2, \ldots, \ell_f$, it holds that $s_1 \geq s_2 \geq \cdots \geq s_{\ell_f}$. From now on, the realizations are reported in the following matrix $\mathbf{X}$

$$\mathbf{X} = \begin{bmatrix} x_{i_1 1} & x_{i_1 2} & \cdots & x_{i_1 n} \\ x_{i_2 1} & x_{i_2 2} & \cdots & x_{i_2 n} \\ \vdots & \vdots & \vdots & \vdots \\ x_{i_{\ell_f} 1} & x_{i_{\ell_f} 2} & \cdots & x_{i_{\ell_f} n} \\ x_{m_1 1} & x_{m_1 2} & \cdots & x_{m_1 n} \\ x_{m_2 1} & x_{m_2 2} & \cdots & x_{m_2 n} \\ \vdots & \vdots & \vdots & \vdots \\ x_{m_{\ell_u} 1} & x_{m_{\ell_u} 2} & \cdots & x_{m_{\ell_u} n} \end{bmatrix}, \tag{7.20}$$

where the gray area reflects $\mathbf{X}_f$ and the white area reflects $\mathbf{X}_u$. We define the vectors s_N^f and s_N^u as

$$\begin{bmatrix} s_N^f \\ s_N^u \end{bmatrix} = \begin{bmatrix} s_1 \\ s_2 \\ \vdots \\ s_{\ell_f} \\ \tilde{s}_1 := x_{m_1 1} + x_{m_1 2} + \cdots + x_{m_1 n} \\ \tilde{s}_2 := x_{m_2 1} + x_{m_2 2} + \cdots + x_{m_2 n} \\ \vdots \\ \tilde{s}_{\ell_u} := x_{m_{\ell_u} 1} + x_{m_{\ell_u} 2} + \cdots + x_{m_{\ell_u} n} \end{bmatrix}. \tag{7.21}$$

To approximate the minimum bound ϱ^-, the values contained in the vector s_N^u must be close to each other. To make this the case, we apply the RA introduced in Chapter 2 to the matrix $\mathbf{X}_u$. Denote by $\tilde{s}_i^m$ the corresponding values of s_N^u after applying the RA. For instance, if the risk measure at hand is the variance, we approximate the minimum variance as follows:

$$\frac{1}{N} \left(\sum_{i=1}^{\ell_f} (s_i - \bar{s})^2 + \sum_{i=1}^{\ell_u} (\tilde{s}_i^m - \bar{s})^2 \right), \tag{7.22}$$

where $\bar{s}$ is computed as

$$\bar{s} = \frac{1}{N} \sum_{i=1}^{N} \sum_{j=1}^{n} x_{ij} = \frac{1}{N} \left(\sum_{i=1}^{\ell_f} s_i + \sum_{i=1}^{\ell_u} \tilde{s}_i \right). \tag{7.23}$$

7.3.3   Bounds on Value-at-Risk

The following proposition provides bounds on VaR.

Proposition 7.11 (VaR bounds for $\sum_{i=1}^{n} X_i$) *Let $\varrho = \mathrm{VaR}_\alpha$ for some $0 < \alpha < 1$, and let $X = (X_1, X_2, \ldots, X_n)$ be a random vector that satisfies properties (i), (ii), and (iii). Let $\mathbb{I}$, $Z = (Z_1, Z_2, \ldots, Z_n)$, and U be defined as in (7.14) and (7.16). Define the variables L_i and H_i as*

$$L_i = \mathrm{LTVaR}_U(Z_i) \quad and \quad H_i = \mathrm{TVaR}_U(Z_i),$$

and let

$$V_\alpha := \mathrm{VaR}_\alpha \left(\mathbb{I} \sum_{i=1}^{n} X_i + (1 - \mathbb{I}) \sum_{i=1}^{n} H_i \right),$$

$$v_\alpha := \mathrm{VaR}_\alpha \left(\mathbb{I} \sum_{i=1}^{n} X_i + (1 - \mathbb{I}) \sum_{i=1}^{n} L_i \right).$$

Then bounds on the value-at-risk of the aggregate risk are given by

$$v_\alpha \le \underline{\varrho} \quad and \quad \overline{\varrho} \le V_\alpha. \tag{7.24}$$

Proof For a given random vector $\mathbf{X}$ satisfying properties (i), (ii), and (iii), there exists a vector $(Y_1, Y_2, \ldots, Y_n)$ with marginals Y_i that have the same distribution as Z_i such that

$$\sum_{i=1}^{n} X_i =_d \left(\mathbb{I} \sum_{i=1}^{n} X_i + (1 - \mathbb{I}) \sum_{i=1}^{n} Y_i \right).$$

As per the definition of the VaR, it follows for all $\alpha \in (0, 1)$ that

$$\mathrm{VaR}_\alpha \left(\sum_{i=1}^{n} X_i \right) \tag{7.25}$$

$$= \inf \left\{ x \in \mathbb{R} \mid p_f F_{\left(\sum_{i=1}^{n} X_i \mid \mathbb{I}=1 \right)}(x) + (1 - p_f) F_{\sum_{i=1}^{n} Y_i}(x) \ge \alpha \right\},$$

where $p_f := P(\mathbb{I}=1)$. Note that for all $\alpha \in (0, 1)$,

$$F_{\sum_{i=1}^{n} Y_i}^{-1}(\alpha) \le \mathrm{TVaR}_\alpha \left(\sum_{i=1}^{n} Y_i \right) \le \mathrm{TVaR}_\alpha \left(\sum_{i=1}^{n} Z_i \right),$$

where the second inequality follows from the fact that the Z_i are comonotonic (while having the same distribution as the Y_i). Thus,

$$F_{\sum_{i=1}^{n} Y_i}^{-1}(U) \le R := \mathrm{TVaR}_U \left(\sum_{i=1}^{n} Z_i \right) \text{ a.s.,}$$

which can be also written in terms of their distribution functions. Therefore, for all $x \in \mathbb{R}$, $F_{\sum_{i=1}^{d} Y_i}(x) \ge F_R(x)$. Thus,

$$\text{VaR}_{\alpha}\left(\sum_{i=1}^{n} X_i\right) \le \inf\left\{x \in \mathbb{R} \mid p_f F_{\left(\sum_{i=1}^{n} X_i | \mathbb{I}=1\right)}(x) + (1 - p_f)F_R(x) \ge \alpha\right\}.$$

We observe that the right-hand side of the above equation is by definition the VaR of a sum of mutually exclusive variables; it follows that

$$\text{VaR}_{\alpha}\left(\sum_{i=1}^{n} X_i\right) \le \text{VaR}_{\alpha}\left(\mathbb{I}\sum_{i=1}^{n} X_i + (1 - \mathbb{I})\,\text{TVaR}_U\left(\sum_{i=1}^{n} Z_i\right)\right)$$

$$= \text{VaR}_{\alpha}\left(\mathbb{I}\sum_{i=1}^{n} X_i + (1 - \mathbb{I})\sum_{i=1}^{n} H_i\right),$$

where the last equality follows from the fact that TVaR is additive for the comonotonic sum $\sum_{i=1}^{n} Z_i$. The proof for the lower bound is similar and is therefore omitted.   □

Initially, the appearance of variables L_i and H_i may seem somewhat odd. However, note that the variables Z_i, which played a crucial role in Proposition 7.10, can also be expressed as $Z_i = \text{VaR}_U(Z_i)$, and here we merely use $\text{TVaR}_U(Z_i)$ and $\text{LTVaR}_U(Z_i)$ instead. Thus, Proposition 7.11 has a similar form to that of Proposition 7.10, but the bounds proposed are usually not sharp.[1] We observe that in the case of no uncertainty (i.e., $A_u = \phi$), there is no model risk, as $\mathbb{I} = 1$. When there is full uncertainty, i.e., $A_u = \mathbb{R}^n$, then $\mathbb{I} = 0$, and we are returned to the unconstrained lower bound on the VaR of a portfolio given in Proposition 1.8.

For practical calculations it might be convenient to use an alternative formulation of the stated VaR bounds. The proof of Proposition 7.12 is provided in the appendix of Bernard and Vanduffel (2015).

Proposition 7.12 (Alternative formulation of the VaR bounds)
Let $\varrho = \text{VaR}_{\alpha}$ for some $0 < \alpha < 1$, and let X be a random vector that satisfies properties (i), (ii), and (iii). Let $\mathbb{I}$, $(Z_1, Z_2, \ldots, Z_n)$, and T be defined as in (7.14), (7.16), and (7.17). Recall that $p_f = P(\mathbb{I} = 1)$. Define

$$q_* := \inf\left\{q \in (q_1, q_2) \mid \text{VaR}_q(T) \ge \text{TVaR}_{\frac{\alpha - p_f q}{1 - p_f}}\left(\sum_{i=1}^{n} Z_i\right)\right\},$$

where $q_1 = \max\left\{0, \frac{\alpha + p_f - 1}{p_f}\right\}$ and $q_2 = \min\left\{1, \frac{\alpha}{p_f}\right\}$. Then, the upper bound V_{α} in Proposition 7.11 is given by:

$$V_{\alpha} = \begin{cases} \text{TVaR}_{\frac{\alpha - p_f q_*}{1 - p_f}}\left(\sum_{i=1}^{n} Z_i\right), & \text{if } \frac{\alpha + p_f - 1}{p_f} < q_* < \frac{\alpha}{p_f}, \\ \text{VaR}_{q_*}(T), & \text{if } q_* = \frac{\alpha}{p_f}, \\ \max\left\{\text{VaR}_{q_*}(T), \text{TVaR}_{\frac{\alpha - p_f q_*}{1 - p_f}}\left(\sum_{i=1}^{n} Z_i\right)\right\}, & \text{if } q_* = \frac{\alpha + p_f - 1}{p_f}. \end{cases} \qquad (7.26)$$

---

[1] Note, indeed, that the variables H_i and L_i are not distributed as $(X_i \mid \mathbb{I} = 0)$.

The expressions for the lower bound v_α are obtained by replacing "TVaR" with "LTVaR" in the above statements.

Recall from the discussion of Proposition 7.11 that the stated upper and lower VaR bounds are not sharp in general. To approximate $\underline{\varrho}$ and $\overline{\varrho}$, we present an algorithm that can be applied directly to the matrix **X**. To compute the VaR at level $\alpha \in (0, 1)$, we define

$$k := N(1 - \alpha), \tag{7.27}$$

where we assume that k is an integer. The algorithm is based on Proposition 7.12 and on the following motivation:

Observe that almost surely

$$\mathbb{I} \sum_{i=1}^{n} X_i + (1 - \mathbb{I}) \sum_{i=1}^{n} Z_i \leq \mathbb{I} \sum_{i=1}^{n} X_i + (1 - \mathbb{I}) \sum_{i=1}^{n} H_i.$$

In particular,[2] for all q, r in $[0, 1]$ such that $p_f q + (1 - p_f)r = \alpha$,

$$\max \left\{ \mathrm{VaR}_q(T), \mathrm{VaR}_r \left(\sum_{i=1}^{n} Z_i \right) \right\}$$

$$\leq V_\alpha = \max \left\{ \mathrm{VaR}_{q_*}(T), \mathrm{TVaR}_{r_*} \left(\sum_{i=1}^{n} Z_i \right) \right\}, \tag{7.28}$$

where q_* is defined as in Proposition 7.12, and $r_* = \frac{\alpha - p_f q_*}{1 - p_f}$. The critical issue is to choose q and r, as well as a dependence between the components of the (comonotonic) vector $(Z_1, Z_2, \ldots, Z_n)$, such that the inequality (7.28) turns into an equality. Such an equality is clearly obtained when taking $r = r_*$ (thus $q_* = q$) and a dependence of the vector $(Z_1, Z_2, \ldots, Z_n)$ such that

$$\mathrm{VaR}_{r_*} \left(\sum_{i=1}^{n} Z_i \right) = \mathrm{TVaR}_{r_*} \left(\sum_{i=1}^{n} Z_i \right). \tag{7.29}$$

The best approximation for the sharp upper bound for $\mathrm{VaR}_\alpha \left(\sum_{i=1}^{n} X_i \right)$ is thus likely to occur when the quantile (VaR) function of the $\sum_{i=1}^{n} Z_i$ can be made (nearly) flat on $[r_*, 1]$. In cases in which this property cannot be (nearly) obtained, it cannot be excluded that better approximations can be found (for example, if the quantile function $\sum_{i=1}^{n} Z_i$ can be made flat on another interval $[r, 1]$, in which r is close to r_*). Similar reasoning shows that to reach the stated lower bound as closely as possible, one should make the quantile function of the portfolio sum as flat as possible on the interval $[0, r_*]$. In the algorithm hereafter, we build on these considerations to approximate sharp bounds. Here, s_N^f is interpreted as a discretely distributed random variable (it can be interpreted as the discrete counterpart of T) and similar for s_N^u as the discrete counterpart of T.

---

[2] See Bernard and Vanduffel (2015) for more details and intuition on the proof of this result.

Algorithm for computing the maximum VaR

1. Recall that $p_f = \frac{\ell_f}{N}$. Compute $m_1 := \max\{0, \ell_f - k\}$ (so that $q_1 = \frac{m_1}{\ell_f} = \max\{0, \frac{p+p_f-1}{p_f}\}$) and $m_2 := \min\{\ell_f, N - k\}$ (then $q_2 = \frac{m_2}{\ell_f} = \min\{1, \frac{p}{p_f}\}$).

2. Compute q_*, where

$$q_* := \inf\left\{ q \in (q_1, q_2) \mid \text{VaR}_q(s_N^f) \geq \text{TVaR}_{\frac{\alpha-p_f q}{1-p_f}}(s_N^u) \right\}.$$

3. Apply the RA to the first $\lfloor (1 - r_*)\ell_u \rfloor$ rows of the untrusted part $\mathbf{X}_u$ of the matrix $\mathbf{M}$, where $r_* = \frac{\alpha-p_f q_*}{1-p_f}$ and where $\lfloor \cdot \rfloor$ denotes the floor of a number. Observe that $\lfloor (1 - r_*)\ell_u \rfloor = k + m_* - \ell_f$, where $m_* := \lfloor q_*\ell_f \rfloor$, and note that $m_1 \leq m_* \leq m_2$.

4. To compute this maximum possible VaR, calculate all (row) sums for $\mathbf{X}_u$ and $\mathbf{X}_f$, and sort them from maximum to minimum value, $\widetilde{s}_1 \geq \widetilde{s}_2 \geq \cdots \geq \widetilde{s}_k \geq \cdots \geq \widetilde{s}_N$. Then, the maximum possible VaR is $\widetilde{s}_k$.

The above algorithm for approximating sharp VaR bounds requires running the rearrangement algorithm once only. However, as the RA will rarely generate a perfectly constant sum on the area where it is applied, it is possible that a better approximation might be obtained by applying step 3 to the first $k + m - \ell_f$ rows of the $\mathbf{X}_u$ for some other m, $m_1 \leq m \leq m_2$. The algorithm for computing the minimum VaR is similar to that for the maximum, where TVaR is replaced by LTVaR to compute α_*. Details can be found in Bernard and Vanduffel (2015).

7.3.4 Examples

Normally distributed risks

We assume that $X = (X_1, \ldots, X_n)$ is a random vector with standard normally distributed components. Furthermore, the joint distribution of X is assumed to be a multivariate standard normal distribution with correlation parameter[3] ϱ on the subset $A_f := [q_\beta, q_{1-\beta}]^n \subset \mathbb{R}^n$ (for some $\beta < 50\%$), where q_γ denotes the quantile of the standard normal random variable at level γ. We assume that $n = 20$.

In Table 7.6, we provide the upper and lower bounds for the standard deviation of the portfolio sum for various confidence levels β and correlation levels ϱ. The first column ($\beta = 0$) provides results for cases in which there is no uncertainty on the multivariate distribution of X; as such, it provides a benchmark for assessing model risk. The last column ($\beta = 50$) provides bounds for cases in which there is full uncertainty on the dependence; as such, it corresponds to the situation studied in Part I.

One observes from Table 7.6 that the impact of model risk on the standard deviation can be substantial even when the joint distribution of X is almost perfectly known, i.e., when β is close to zero (p_u is close to 0). Consider, for instance, $\beta = 0.05$ and $\varrho = 0$. In this case, $p_u = 1 - 0.999^{20} \approx 0.02$, and we find that using a multivariate normal assumption (as the benchmark) might underestimate the standard deviation

---

[3] A multivariate standard normal distribution with correlation coefficient ϱ is such that the pairwise correlation is ϱ for all pairs (X_i, X_j) with $i \neq j$.

$A_f =$ $[q_\beta, q_{1-\beta}]^n$	$A_u = \phi$ $\beta = 0$	$\beta = 0.05$	$\beta = 0.5$	$\beta = 5$	$A_u = \mathbb{R}^n$ $\beta = 50$
$\varrho = 0$	4.47	(4.40, 5.65)	(3.89, 10.60)	(1.23, 19.30)	(0.00, 20.00)
$\varrho = 0.1$	7.62	(7.41, 8.26)	(6.23, 11.70)	(1.69, 19.20)	(0.00, 20.00)
$\varrho = 0.5$	14.50	(13.80, 14.60)	(11.10, 15.40)	(3.74, 18.60)	(0.00, 20.00)

Table 7.6 In the first column we report the **standard deviation** of $\sum_{i=1}^{20} X_i$ under the assumption of multivariate normality (no dependence uncertainty, i.e., $A_u = \phi$). Lower and upper bounds of the standard deviation of $\sum_{i=1}^{20} X_i$ are reported as pairs $(\underline{\varrho}, \overline{\varrho})$ for various confidence levels β. We use 3,000,000 simulations. All digits reported in the table are significant.

by $(5.65 - 4.47)/4.47 = 26.4\%$ and overestimate it by $(4.47 - 4.40)/4.40 = 1.6\%$. It thus seems that the assumption of multivariate normality is not particularly robust against misspecification. Here, in fact, it clearly gives rise to a situation in which one is more likely to underestimate risk than to overestimate it. These observations are also intuitive, as the standard deviation is sensitive to high outcomes, and these scenarios occur frequently when considering the upper bound (as the tail events are then assumed to be fully correlated). Furthermore, the example shows that adding some partial information on the dependence (i.e., when $\beta < 50$) can change the unconstrained bounds (the case in which $\beta = 50$) and confirms that dependence is important when assessing the risk of a portfolio. For instance, when $\beta = 0.5$ and $\varrho = 0$, one has that $p_u = 1 - 0.99^{20} \approx 0.18$, and the unconstrained upper bound for the standard deviation shrinks by approximately 50% (from 20 to 10.6).

p_u	$A_u = \phi$ $\beta = 0$	$\beta = 0.05$	$\beta = 0.5$	$\beta = 5$	$A_u = \mathbb{R}^d$ $\beta = 50$
$\varrho = 0$	0	0.02	0.18	0.88	1
$\varrho = 0.1$	0	0.02	0.18	0.87	1
$\varrho = 0.5$	0	0.016	0.12	0.66	1

Table 7.7 Probability p_u that X takes values outside the n-cube $[q_\beta, q_{1-\beta}]^n$, $n = 20$, for a confidence level β and a correlation coefficient ϱ. We use $N = 3,000,000$ simulations.

In Table 7.7 we report, for the levels of correlation ϱ and confidence levels β used in Table 7.6, the probability p_u that $X = (X_1, \ldots, X_n)$ takes values outside the n-cube $A_f = [q_\beta, q_{1-\beta}]^n$. Doing so allows us to better interpret the results of Table 7.6 and will also be useful in understanding the effect of the choice of A_f.

In the above example, the choice for A_f as an n-cube is based solely on the use of the $N(0, 1)$ marginal distributions. Specifically, we can also express A_f as

$$A_f := \left\{(x_{i1}, \ldots, x_{in}) \in \mathbb{R}^n \mid \text{ for all } j \in \{1, 2, \ldots, n\}, f_j(x_{ij}) \geq \varepsilon\right\}, \tag{7.30}$$

where ε is suitably chosen; that is, rare events (i.e., observations that lie in A_u) correspond then to either the largest or the smallest outcomes of the risks, and where f_j denotes the density of X_j. Another natural criterion by which to determine A_f consists in starting from a given fitted multivariate density $\widehat{f}$ (coming, for instance, from a multivariate Gaussian model, a multivariate Student model, or a pair-copula construction model: see Aas et al., 2009; Czado, 2010). The trusted area A_f is then based on the contour levels of the density. We refer the reader to Bernard and Vanduffel (2015) for more details on this point.

Table 7.8 provides for various levels of probability α, confidence level β, and correlation ϱ of the bounds on TVaR. The results are in line with those of the previous example. Model risk is already present for small levels of β, but at the same time the availability of dependence information ($\beta < 50$) allows for strengthening the unconstrained bounds ($\beta = 50$) significantly. Interestingly, the degree of model risk also depends on the interplay between the level α used to assess the TVaR and the degree of uncertainty on the dependence as measured by β. When α is large (e.g., $\alpha = 99.5\,\%$), a small proportion of model uncertainty (e.g., $\beta = 0.05$) appears to have a tremendous effect on the model risk of underestimation. We can explain this observation as follows. The TVaR is essentially measuring the average of all upper VaRs, and its level is thus driven mainly by scenarios in which one or more outcomes of the risks involved are high. These scenarios, however, are not considered as trustworthy for depicting the (tail) dependence, with negative impact on the level of the TVaR. In fact, for a given level of α, the model risk of underestimation increases sharply with an increase in the level of β and already approaches its maximum for small to moderate values of β. This effect is further emphasized when the level of α increases. In other words, the TVaR is highly vulnerable to model misspecification, especially when it is assessed at high probability levels.

$A_f = [q_\beta, q_{1-\beta}]^n$		$A_u = \phi$ $\beta = 0$	$\beta = 0.05$	$\beta = 0.5$	$\beta = 5$	$A_u = \mathbb{R}^n$ $\beta = 50$
$\alpha = 95\,\%$	$\varrho = 0$	9.2	(9.1, 11.6)	(8.5, 27.5)	(3.4, 41.3)	(−0.002, 41.3)
	$\varrho = 0.1$	15.7	(15.4, 17.3)	(13.5, 28.4)	(4.7, 41.3)	(0.004, 41.3)
	$\varrho = 0.5$	29.9	(28.1, 30.5)	(22.9, 34.0)	(10.0, 41.3)	(−0.002, 41.3)
$\alpha = 99.5\,\%$	$\varrho = 0$	12.9	(12.8, 30.4)	(12.1, 57.9)	(7.5, 57.9)	(−0.004, 57.9)
	$\varrho = 0.1$	22	(21.5, 33.3)	(19.0, 57.8)	(10.0, 57.9)	(−0.002, 57.9)
	$\varrho = 0.5$	42	(37.4, 47.6)	(29.6, 57.9)	(15.2, 57.9)	(0.019, 57.9)

Table 7.8 TVaR$_{95\,\%}$ and TVaR$_{99.5\,\%}$ of $\sum_{i=1}^{20} X_i$ are reported in the absence of uncertainty (multivariate standard normal model with $A_u = \phi$). Lower and upper bounds for the TVaR of $\sum_{i=1}^{20} X_i$ are reported as pairs $(\underline{\varrho}, \overline{\varrho})$ for various levels of confidence β, correlation ϱ, and level, α. Bounds are obtained based on $N = 3,000,000$ simulations. All digits reported are significant.

The VaR bounds reported in Table 7.9 were obtained within a few minutes, using 3,000,000 Monte Carlo simulations. We make the following observations. First, model risk is clearly present even when the dependence is "mostly" known (i.e., β is small). Furthermore, the precise degree of model error depends strongly on the level α that is used to assess the VaR. Let us consider the benchmark model with $\varrho = 0$ (the risks are independent and standard normally distributed) and $\beta = 0$ (no uncertainty). We find that VaR$_{95\,\%}$ ($\sum_{i=1}^{20} X_i$) $= \sqrt{20}\Phi^{-1}(95\,\%) = 7.35$ and, similarly, VaR$_{99.5\,\%}$ ($\sum_{i=1}^{20} X_i$) $= 11.5$, VaR$_{99.95\,\%}$ ($\sum_{i=1}^{20} X_i$) $= 14.7$. However, if $\beta = 0.05$, then $p_u \approx 0.02$, and the benchmark model might underestimate the 95 %-VaR by

$(8.08 - 7.36)/8.08 = 8.9\%$ or overestimate it by $(7.36 - 7.27)/7.27 = 1.24\%$. However, when using the 99.5%-VaR, the degree of underestimation may rise to $(30.4 - 11.5)/30.4 = 62.2\%$, whereas the degree of overestimation is equal only to $(11.5 - 11.4)/11.4 = 0.9\%$. Hence, the risk of underestimation increases sharply with the probability level that is used to assess VaR.

| $A_f = [q_\beta, q_{1-\beta}]^n$ | | $A_u = \phi$ | | | | $A_u = \mathbb{R}^n$ |
		$\beta = 0$	$\beta = 0.05$	$\beta = 0.5$	$\beta = 5$	$\beta = 0.5$
$\alpha = 95\%$	$\varrho = 0$	7.4	(7.3, 8.1)	(6.7, 27.5)	(0.79, 41.3)	(−2.2, 41.3)
	$\varrho = 0.1$	12.5	(12.2, 13.3)	(10.7, 27.7)	(1.5, 41.2)	(−2.2, 41.2)
	$\varrho = 0.5$	23.8	(22.9, 24.2)	(18.9, 30.9)	(7.0, 41.2)	(−2.2, 41.2)
$\alpha = 99.5\%$	$\varrho = 0$	11.5	(11.4, 30.4)	(10.8, 57.8)	(6.1, 57.8)	(−0.3, 57.8)
	$\varrho = 0.1$	19.6	(19.1, 31.4)	(16.9, 57.8)	(8.2, 57.8)	(−0.3, 57.8)
	$\varrho = 0.5$	37.4	(34.3, 45.1)	(27.4, 57.8)	(13.5, 57.8)	(−0.3, 57.8)
$\alpha = 99.95\%$	$\varrho = 0$	14.7	(14.6, 71.0)	(13.8, 71.1)	(9.3, 71.1)	(−0.04, 71.1)
	$\varrho = 0.1$	25.1	(24.2, 71.1)	(21.5, 71.1)	(12.1, 71.1)	(−0.0, 71.1)
	$\varrho = 0.5$	47.7	(41.3, 71.1)	(32.3, 71.1)	(17.2, 71.1)	(−0.04, 71.1)

Table 7.9 $\text{VaR}_{95\%}$, $\text{VaR}_{99.5\%}$ and $\text{VaR}_{99.95\%}$ of $\sum_{i=1}^{20} X_i$ are reported in the absence of uncertainty (multivariate standard normal model with $A_u = \phi$). Lower and upper bounds for the VaR of $\sum_{i=1}^{20} X_i$ are reported as pairs $(\underline{\varrho}, \overline{\varrho})$ for various levels of confidence β, correlation ϱ, and probability α. We use $N = 3{,}000{,}000$ simulations, and all digits reported are significant.

Finally, note that when very high probability levels are used in VaR calculations ($\alpha = 99.95\%$; see the last three rows in Table 7.9), the constrained upper bounds are very close to the unconstrained upper bound, even when there is almost no uncertainty on the dependence ($\beta = 0.05$). The bounds computed by application of the RA, as in Section 2.2, are thus almost the best possible bounds, even though it seems that the multivariate model is known at a very high confidence level. This implies that any effort to accurately fit a multivariate model will not reduce the model risk on VaR when assessed at high confidence level.

Note that when no information on the dependence is available ($\beta = 50$), the upper and lower bounds stated in Proposition 7.11 simply reduce to the expressions $\sum_{i=1}^{n} \text{TVaR}_\alpha(X_i)$ and $\sum_{i=1}^{n} \text{LTVaR}_\alpha(X_i)$, respectively and thus coincide with the bounds mentioned in Section 2.2. Using these formulas, we find that the bounds on the VaR_α of sums of 20 independent $N(0, 1)$ risks are

$$A = -20\frac{\varphi(\Phi^{-1}(\alpha))}{\alpha}, \qquad B = 20\frac{\varphi(\Phi^{-1}(\alpha))}{1 - \alpha},$$

and we observe that one obtains consistency with the bounds reported in Table 7.9. For example, when $\alpha = 95\%$, we find that $(A, B) = (-2.17, 41.25)$, which conforms with the numbers in Table 7.9.

Pareto distributed risks

We assume that $n = 20$ individual risks X_i are all Pareto distributed with parameter $\vartheta = 3$ and that their dependence is modeled by a Gaussian copula with parameter ϱ (pairwise correlation). The distribution function of each X_i is given as

$$F(x) = 1 - (1 + x)^{-\vartheta},$$

$x > 0$, and their value-at-risk at level $\alpha \in (0, 1)$ is then given by

$$F_X^{-1}(\alpha) = (1 - \alpha)^{-1/\vartheta} - 1.$$

Assuming that A_f is based on each marginal being between the quantile of level $1 - \beta$ and β, respectively,

$$A_f = \left\{ (x_{i1}, \ldots, x_{in}) \mid \text{for all } k \in \{1, 2, \ldots, n\}, q_\beta \le x_{ik} \le q_{1-\beta} \right\},$$

where

$$q_\beta = (1 - \beta)^{-1/\vartheta} - 1.$$

We find the results in Table 7.10 for the bounds on value-at-risk from the simulation of $N = 3,000,000$ simulations of $n = 20$ Pareto variables.

$A_f = [q_\beta, q_{1-\beta}]^n$		$A_u = \phi$ $\beta = 0$	$\beta = 0.05$	$\beta = 0.5$	$\beta = 5$	$A_u = \mathbb{R}^n$ $\beta = 50$
	$\varrho = 0$	16.6	(16.0, 18.4)	(13.8, 37.4)	(8.6, 61.4)	(7.3, 61.4)
$\alpha = 95\%$	$\varrho = 0.1$	19.7	(18.3, 20.6)	(15.9, 37.8)	(8.8, 61.4)	(7.3, 61.4)
	$\varrho = 0.5$	28	(26.5, 33.5)	(20.6, 43.2)	(10.3, 61.4)	(7.3, 61.4)
	$\varrho = 0$	25.8	(21.5, 60.7)	(17.5, 156.0)	(10.7, 156.0)	(9.3, 156.0)
$\alpha = 99.5\%$	$\varrho = 0.1$	32.5	(27.9, 63.0)	(21.8, 156.0)	(11.6, 156.0)	(9.3, 156.0)
	$\varrho = 0.5$	61.1	(49.0, 94.7)	(31.6, 155.0)	(14.0, 155.0)	(9.3, 155.0)
	$\varrho = 0$	43.5	(26.5, 359.0)	(20.5, 360.0)	(12.4, 360.0)	(9.8, 359.0)
$\alpha = 99.95\%$	$\varrho = 0.1$	51.9	(36.3, 357.0)	(26.8, 359.0)	(13.9, 358)	(9.8, 357.0)
	$\varrho = 0.5$	116	(69.6, 361.0)	(40.1, 361.0)	(16.8, 359.0)	(9.8, 359.0)

Table 7.10 VaR$_{95\%}$, VaR$_{99.5\%}$, and VaR$_{99.95\%}$ of $\sum_{i=1}^{20} X_i$ are reported in the absence of uncertainty (Pareto marginals with Gaussian dependence). Lower and upper bounds for the VaR of $\sum_{i=1}^{20} X_i$ are reported as pairs $(\underline{\varrho}, \overline{\varrho})$ for various levels of confidence β, correlation ϱ, and level α. We use $N = 3,000,000$ simulations, and all digits reported are significant.

The examples show that accounting for the available information coming from a multivariate fitted model may reduce the risk bounds. However, model uncertainty remains a significant concern. For instance, we observe from the numerical experiments that the portfolio VaR at a very high confidence level (as used in the current Basel regulation) might be prone to such a high level of model risk that, even if one knows the multivariate distribution nearly perfectly, its range of possible values remains wide. Hence, we recommend caution regarding regulation based on value-at-risk at a very high level.

Part III

Additional Information on the Structure

The risk bounds and range of dependence uncertainty obtained in the unconstrained case can be significantly improved by including relevant structural information on the underlying class of models. This additional structural information may lead to a positive or a negative dependence restriction and thus, as shown in Part II induce improvements of the upper resp. the lower risk bounds, as shown in Part II.

We deal in this part with several types of structural information, investigate their impact, and show in several applications their potential to reduce dependence uncertainty. In some applications it is possible to combine these structural assumptions with dependence assumptions, as in Part II, leading to particular useful reduction results.

In Chapter 8 we consider first in Section 8.1 the case when, in addition to the one-dimensional marginals, some higher dimensional marginals are known. This assumption allows us to derive improvements of the classical Fréchet bounds by a duality theorem. For general higher-dimensional marginal systems, the "reduction method" allows us to derive good upper and lower bounds by an associated reduced risk model with simple marginals. This reduced problem can be solved by the RA. If the higher-dimensional marginals exhibit strong positive dependence, this leads to improvements of the lower bounds; in comparison to the unconstrained case if they exhibit negative dependence, it leads to improvements of the upper bounds.

In Section 8.2 we consider a general case of additional constraints given by the distribution or the expectations of a class of functionals. This includes in principle the constraints due to higher-dimensional marginals (infinite set of restrictions) but also the case of variance constraints, as in Part II. We give several improved lower and upper dual bounds for such classes of constraints. Using martingale constraints (infinite set of restrictions), this method can also be used to derive improved price bounds for options.

Chapter 9 gives a detailed discussion on partially specified factor models (PSFM). This model assumption is based on an underlying factor model $X_i = f_i(Z, \varepsilon_i)$ with a systematic risk factor Z and individual risks $\varepsilon_1, \ldots, \varepsilon_n$. In comparison to the usual assumption of completely specified factor models, which is in general a hard to verify assumption, in PSFMs only the joint distributions of (ε_i, Z) are specified, which is a much simpler to verify model assumption. We show that the assumption of PSFM may lead to strongly reduced risk bounds, and it can be combined in a particular effective and flexible way with other dependence assumptions, like variance bounds for the aggregated portfolio.

Chapter 10 deals with a systematic investigation of the assumption that the risk vector X is split into k subgroups. For a comparison vector Y, conditions are given for the comparison of the subgroup sums of X and Y and further on the comparison of the copulas of the vectors of subgroup sums to imply a relevant comparison theorem between the aggregated portfolios $\sum_{i=1}^{n} X_i$ and $\sum_{i=1}^{n} Y_i$. This criterion also allows flexible and effective applications and can be combined in a useful way with further constraints, for example with the assumption of PSFMs within the subgroups.

8 Additional Information on Functionals of the Risk Vector

Besides the marginal information on the risk $X = (X_1, \ldots, X_n)$, a useful additional source for reducing dependence uncertainty and risk bounds is to use additional information on some functionals of the risk vector. This information may be available in insurance-type hierarchical models on the aggregation of some branches of the company evaluated by statistical analysis. For example, the distribution of some subgroup sums $\sum_{j \in I_i} X_j \sim Q_i$ might be given, where $I_i \subset \{1, \ldots, n\}, 1 \leq i \leq m$, or the worst case distribution of some subgroups $\max_{j \in I_i} X_j \sim Q_i$ might be known. Alternatively, information on (some) correlations $\tau_{ij} = \text{Cor}(X_i, X_j)$ or covariances $\sigma_{ij} = \text{Cov}(X_i, X_j)$ might be available.

In more general terms, let $T_i \colon \mathbb{R}^n \to \mathbb{R}^1, i \in K$, be a class of measurable real functions. Assume that the distribution Q_i of $T_i(X)$ is known for $i \in K$. Under this information our aim is to derive upper resp. lower risk bounds for the value-at-risk (VaR) of the aggregated risk $\sum_{i=1}^n X_i$. This formulation includes for the case $T_i(X) = \sum_{j \in I_i} X_j$ resp. $T_i(X) = \max_{j \in I_i} X_j$, the situation described above. If we also allow higher dimensional functions $T_i \colon \mathbb{R}^n \to \mathbb{R}^{n_i}$, then we can in this way also describe the case where higher-dimensional marginals are known, considering $T_i(X) = (X_j)_{j \in I_i}$, where $\mathcal{E} = \{I_i, i \in K\}$ is a higher-order marginal system, and $Q_i = P^{X_{I_i}}$ are the corresponding higher-dimensional marginal distributions.

8.1 Higher-Dimensional Marginals

The class of all possible dependence structures given the marginals can be restricted if some higher-dimensional marginals are known. Let $\mathcal{E}$ be a system of subsets J of $\{1, \ldots, n\}$ with $\bigcup_{J \in \mathcal{E}} J = \{1, \ldots, n\}$, and assume that for $J \in \mathcal{E}, F_{X_J} = F_J$ is known.

The class

$$\mathcal{F}_{\mathcal{E}} = \mathcal{F}(F_J; \ J \in \mathcal{E}) \subset \mathcal{F}(F_1, \ldots, F_n) \tag{8.1}$$

resp. the corresponding class of distributions $\mathcal{M}_{\mathcal{E}}$ with marginals $P_J, J \in \mathcal{E}$, is called a *generalized Fréchet class*. In some applications, for example, some two-dimensional marginals might be known in addition to the one-dimensional marginals. The relevant tail risk bounds then are given by

$$M_{\mathcal{E}}(s) = \sup\{P(S \geq s); \ F_X \in \mathcal{F}_{\mathcal{E}}\}$$

$$\text{and} \qquad m_{\mathcal{E}}(s) = \inf\{P(S \geq s); \ F_X \in \mathcal{F}_{\mathcal{E}}\}. \tag{8.2}$$

Under some conditions, a duality result corresponding to the simple marginal case has been established under the assumptions $M_{\mathcal{E}} \neq \phi$ for various classes of functions φ as, for example, for upper-semi-continuous functions (see Rüschendorf, 1984, 1991; Kellerer, 1988). As in Chapter 3, we state the duality theorem as a duality principle taking the following form:

Duality Theorem:

$$M_{\mathcal{E}}(\varphi) = \sup\left\{\int \varphi\,d\mu; \ \mu \in M_{\mathcal{E}}\right\}$$

$$= \inf\left\{\sum_{J \in \mathcal{E}}\int f_J\,dP_J; \ \sum_{J \in \mathcal{E}} f_J \circ \pi_J \geq \varphi\right\}. \tag{8.3}$$

By definition $M_{\mathcal{E}}(s) = M_{\mathcal{E}}(\varphi_s)$, where $\varphi_s(x) = \mathbb{1}_{[s,\infty)}(\sum_{i=1}^{n} x_i)$. For specific classes of indicator functions, one can use the duality result in (8.3) to connect up with Bonferroni type bounds.

Let $(E_i, \mathcal{A}_i)$, $1 \leq i \leq n$ be measure spaces, and let for $J \in \mathcal{E}$, $P_J \in M^1(E_J, \mathcal{A}_J)$ be a marginal system, i.e., P_J are probability measures on the product space $(E_J, \mathcal{A}_J) = (\prod_{j \in J} E_j, \otimes_{j \in J}\mathcal{A}_j)$. The following class of improved Fréchet bounds was given in Rüschendorf (1991).

Proposition 8.1 (Bonferroni type bounds) *Let* $(E_j, \mathcal{A}_j)$, $1 \leq j \leq n$, *and let* $P_J, J \in \mathcal{E})$ *be a marginal system with* $\bigcup_{J \in \mathcal{E}} J = \{1, \dots, n\}$. *For* $A_j \in \mathcal{A}_j$ *and* $A_J := \prod_{j \in J} A_j$, *the following estimates hold:*

a)
$$M_{\mathcal{E}}(A_1 \times \cdots \times A_n) \leq \min_{J \in \mathcal{E}} P_J(A_J). \tag{8.4}$$

b) *In the case that* $\mathcal{E} = J_2^n = \{(i, j); \ i, j \leq n\}$ *and with* $q_i := P_i(A_i^c)$, $q_{ij} = P_{ij}(A_i^c \times A_j^c)$, *it holds that*

$$M_{\mathcal{E}}(A_1 \times \cdots \times A_n) \leq 1 - \sum_{i=1}^{n} q_i + \sum_{i<j} q_{ij}, \tag{8.5}$$

$$m_{\mathcal{E}}(A_1 \times \cdots \times A_n) \geq 1 - \sum_{i=1}^{n} q_i + \sup_{\tau \in \mathcal{T}} \sum_{(i,j) \in \tau} q_{ij}, \tag{8.6}$$

where $\mathcal{T}$ *is the class of all spanning trees of* G_n, *the complete graph of* $\{1, \dots, n\}$.

Remark 8.2

a) In the case of $E_i = \mathbb{R}^1$, (8.5) implies an improvement over the upper Fréchet bound for $F \in \mathcal{F}_{\mathcal{E}}$:

$$F(x) \leq \min_{1 \leq i \leq n} F_i(x_i) \tag{8.7}$$

to

$$F(x) \leq \sum_{i=1}^{n} F_i(x_i) - (n-1) + \sum_{i<j} \overline{F}_{ij}(x_i, x_j). \tag{8.8}$$

Equation (8.6) gives the improved lower bound

$$F(x) \geq \sum_{i=1}^{n} F_i(x_i) - (n-1) + \sup_{\tau \in \mathcal{T}} \sum_{(i,j) \in \tau} \overline{F}_{ij}(x_i, x_j). \tag{8.9}$$

The difference between the upper bound in (8.8) and the lower bound is small when the terms $\overline{F}_{ij}(x_i, x_j)$ outside a specific tree τ are small. Note that the estimates in (8.8) and (8.9) can be applied by the method of improved standard bounds as described in Chapter 5 to obtain improved tail risk bounds for the sum.

b) Decomposable marginal systems are build up by simple configurations, namely the simple marginals $\mathcal{M}(P_1, \ldots, P_n)$, the series system $\mathcal{M}(P_{1,2}, P_{2,3}, \ldots, P_{n-1,n})$, and the star-like system $\mathcal{M}(P_{1,j}; 2 \leq j \leq n)$. For these basic systems, the method of conditioning allows one to derive improved Fréchet bounds, which imply by the method of improved standard bounds improved tail risk estimates for the sum (see Rüschendorf, 1991 and Section 5.2.2 of Rüschendorf, 2013). ◊

For a non-overlapping system $\mathcal{E} = \{I_1, \ldots, I_m\}$ with $I_k \cap I_l = \phi$ for $k \neq l$, define

$$Y_r := \sum_{i \in I_r} X_i, \quad H_r = F_{Y_r}, \quad r = 1, \ldots, m, \quad \text{and} \quad \mathcal{H} = \mathcal{F}(H_1, \ldots, H_m). \tag{8.10}$$

Then consider the *reduced tail risk bounds*

$$M_{\mathcal{H}}(s) := \sup \left\{ P\left(\sum_{i=1}^{m} Y_i \geq s \right); \ F_Y \in \mathcal{H} \right\}$$

$$\text{and} \quad m_{\mathcal{H}}(s) := \inf \left\{ P\left(\sum_{i=1}^{m} Y_i \geq s \right); \ F_Y \in \mathcal{H} \right\}, \tag{8.11}$$

where F_Y is the distribution function of $Y = (Y_1, \ldots, Y_m)$. $M_{\mathcal{H}}$ and $m_{\mathcal{H}}$ are tail risk bounds corresponding to a simple marginal system with marginals H_i. It turns out that $M_{\mathcal{H}} = M_{\mathcal{E}}$ and $m_{\mathcal{H}} = m_{\mathcal{E}}$ and thus the problem of sharp risk bounds can be reduced for non-overlapping marginal systems to the case of simple marginals.

Theorem 8.3 (Non-overlapping marginal systems) *The case of a non-overlapping marginal system $\mathcal{E}$ can be reduced to the simple marginal system $\mathcal{H}$, i.e.,*

$$M_{\mathcal{E}}(s) = M_{\mathcal{H}}(s) \quad \text{and} \quad m_{\mathcal{E}} = m_{\mathcal{H}}(s). \tag{8.12}$$

Proof For the proof, note that for $F_X \in \mathcal{F}_{\mathcal{E}}$ we have that $F_{Y_r} = F_{\sum_{i \in J_r} X_i} = H_r$, and thus $F_Y \in \mathcal{H}$ for $Y = (Y_1, \ldots, Y_m)$. This implies that

$$M_{\mathcal{E}}(s) \leq M_{\mathcal{H}}(s).$$

Conversely, if $Y = (Y_1, \ldots, Y_m)$ is any vector with distribution function $F_Y \in \mathcal{H}$, then by a classical result on stochastic equations, there exist $X_{J_m} \sim F_{J_r}$ such that $\sum_{j \in J_i} X_j = Y_i$ a.s., $1 \le i \le m$. This implies the converse inequality

$$M_{\mathcal{H}}(s) \le M_{\mathcal{E}}(s).$$

The argument for the equality $m_{\mathcal{E}} = m_{\mathcal{H}}$ is similar. $\qquad\square$

Similar exact reduction results are given in Puccetti and Rüschendorf (2012a) also for basic series systems and for the star-like system by the conditioning method. In general, however, one obtains no exact results by the reduction method for general marginal systems, but one obtains reasonably good bounds for any consistent system of marginal information that can be determined numerically.

Let $\mathcal{E} = \{J_1, \ldots, J_m\}$ be any marginal system with corresponding Fréchet class $\mathcal{F}_{\mathcal{E}} = \mathcal{F}(F_J, J \in \mathcal{E})$. Let $\eta_i := \#\{I_r \in \mathcal{E}; i \in J_r\}$ denote the number of sets $J \in \mathcal{E}$ in which the index i appears. For a risk vector X with $F_X \in \mathcal{F}_{\mathcal{E}}$, define the weighted sum

$$Y_r := \sum_{i \in J_r} \frac{X_i}{\eta_i} \quad \text{and} \quad H_r := F_{Y_r}, r = 1, \ldots, m. \tag{8.13}$$

Let $\mathcal{H} = \mathcal{F}(H_1, \ldots, H_m)$ denote the corresponding Fréchet class with simple marginals.

Theorem 8.4 (Reduced Fréchet bounds) *Let $\mathcal{F}_{\mathcal{E}} \ne \phi$ be a consistent marginal system such that $M_{\mathcal{E}} \ne \phi$. Then the following reduced Fréchet bounds hold true:*

$$M_{\mathcal{E}}(s) \le M_{\mathcal{H}}(s) \quad \text{and} \quad m_{\mathcal{E}}(s) \ge m_{\mathcal{H}}(s). \tag{8.14}$$

Proof Since the distribution function F_X of the risk vector X belongs to $\mathcal{F}_{\mathcal{E}} = \mathcal{F}(F_J; \ J \in \mathcal{E})$, the distribution function F_Y of Y belongs to the simple marginal class $\mathcal{H} = \mathcal{F}(H_1, \ldots, H_m)$. Now using that

$$\sum_{i=1}^{n} X_i = \sum_{r=1}^{m} \sum_{i \in J_r} \frac{X_i}{\eta_i} = \sum_{r=1}^{m} Y_r, \tag{8.15}$$

we obtain

$$\begin{aligned} M_{\mathcal{E}}(s) &= \sup\{P(X_1 + \cdots + X_n \ge s; \ F_X \in \mathcal{F}_{\mathcal{E}}\} \\ &= \sup\{P(Y_1 + \cdots + Y_m \ge s); \ F_X \in \mathcal{F}_{\mathcal{E}}\} \\ &\le \sup\{P(Y_1 + \cdots + Y_m \ge s); \ F_Y \in \mathcal{H}\} = M_{\mathcal{H}}(s), \end{aligned}$$

and, similarly, $m_{\mathcal{E}}(s) \ge m_{\mathcal{H}}(s)$. $\qquad\square$

By this reduction result, the RA-algorithm can be used to calculate the reduced Fréchet bounds $M_{\mathcal{H}}$ and $m_{\mathcal{H}}$ to obtain upper resp. lower bounds for $M_{\mathcal{E}}$ resp. $m_{\mathcal{E}}$. It also allows us to apply the method of standard bounds and the method of dual bounds to obtain upper bounds for $M_{\mathcal{H}}$ by the dual bounds $D_{\mathcal{H}}$ resp. by the standard bounds $S_{\mathcal{H}}$,

$$M_{\mathcal{E}} \le M_{\mathcal{H}} \le D_{\mathcal{H}} \le S_{\mathcal{H}}, \tag{8.16}$$

and, similarly, lower bounds for $m_{\mathcal{E}}$:

$$s_{\mathcal{H}} \le d_{\mathcal{H}} \le m_{\mathcal{H}} \le m_{\mathcal{E}}. \tag{8.17}$$

In Puccetti and Rüschendorf (2012a), a weighting scheme generalizing (8.13) was introduced. Consider a weighting vector $a = \{(a_i^r); \ 1 \leq i \leq n, 1 \leq r \leq m\}$ with $a_i^r \in [0, 1]$, $a_i^r > 0$ only if $i \in J_r$ and

$$\sum_{r=1}^{m} a_i^r = 1. \tag{8.18}$$

The idea is to choose an "optimal" system of weights for the reduction to simple marginals, putting different weights on the components within their groups J_i. Defining

$$Y_r^a := \sum_{i=1}^{n} a_i^r X_i, \tag{8.19}$$

we obtain

$$\sum_{r=1}^{m} Y_r^a = \sum_{r=1}^{m} \sum_{i=1}^{n} a_i^r X_i = \sum_{i=1}^{n} \sum_{r=1}^{m} a_i^r X_i$$

$$= \sum_{i=1}^{n} X_i \sum_{r=1}^{m} a_i^r = \sum_{i=1}^{n} X_i. \tag{8.20}$$

Since $a_i^r > 0$ only if $i \in J_r$, the random variables Y_r^a only depend on those components X_j with $j \in J_r$. Thus the distribution function H_r^a of Y_r^a is uniquely determined by the marginal assumption $X_{J_r} \sim F_{J_r}$. In consequence

$$F_{Y^a} \in \mathcal{F}(H_1^a, \ldots, H_m^a) = \mathcal{H}^a, \tag{8.21}$$

where $Y^a = (A_r^a)_{r=1,\ldots,m}$, and we obtain a reduction result analogous to Theorem 8.4.

Theorem 8.5 (Weighting scheme bounds)   *For a consistent marginal system $\mathcal{F}_\mathcal{E} \neq \phi$ and a weighting scheme a satisfying (8.18), define the simple Fréchet bounds $M_{\mathcal{H}^a}(s)$ and $m_{\mathcal{H}^a}$ with $\mathcal{H}^a$ as in (8.21). Then*

$$M_\mathcal{E}(s) \leq M_{\mathcal{H}^a}(s) \quad and \quad m_\mathcal{E}(s) \geq m_{\mathcal{H}^a}(s). \tag{8.22}$$

The reduced VaR bounds with higher-order marginals information given by $M_\mathcal{E}$ corresponding to the tail risk bounds in (8.14) and (8.22) are denoted for fixed weighting vector a by $\overline{\text{VaR}}_\alpha^r$ and $\underline{\text{VaR}}_\alpha^r$. The magnitude of reduction of these reduced VaR bounds depends on the structure of the higher-order marginals.

Example 8.6 (Two-dimensional marginals)   We consider as a marginal system the series system $\mathcal{E} = \{\{2j - 1, 2j\}, \ 1 \leq j \leq 300\}$, a system of two-dimensional marginals $F_{2j-1,2j}$ with Pareto(2) risks and $n = 600$. In scenario A) we assume that all two-dimensional marginals are comonotonic; in scenario B) we assume that they are independent. We apply the standard weights in (8.13). As expected, in scenario A) the improvement of the VaR bounds by the two-dimensional marginals is of minor order compared to the sharp VaR bounds $\overline{\text{VaR}}_\alpha$; however, in scenario B) the improvement is of considerable magnitude (see Figure 8.1 and Table 8.1).

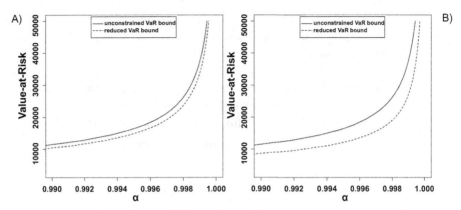

Figure 8.1 Reduced bounds, $n = 600$ Pareto(2) variables, A) ∼ comonotone marginals $F_{2j-1,2j}$, B) ∼ independent variables $F_{2j-1,2j}$.

α	VaR_α^+	$\overline{\mathrm{VaR}}_\alpha^{r,\mathrm{A}}$	$\overline{\mathrm{VaR}}_\alpha^{r,\mathrm{B}}$	$\overline{\mathrm{VaR}}_\alpha(L)$
0.990	5400.00	10309.14	8496.13	11390.00
0.995	7885.28	14788.71	12015.04	16356.42
0.999	18373.67	33710.3	26832.2	37315.70

Table 8.1 Reduced bounds in scenario A) and B), VaR_α^+ = comonotonic VaR.

As a result it is found that higher-order marginals may lead to a considerable reduction of the upper VaR bounds when the known higher-dimensional marginals do not specify strong positive dependence. Similarly, the lower VaR bounds are improved considerably if the marginals specify some kind of negative dependence.

8.2 Functionals of the Risk Vector

Knowing higher-order marginals P_J, $J \in \mathcal{E}$ as assumed in Section 8.1 also implies that the distribution of all functionals $T(X) = T_J(X_J)$ is known. This assumption is often too strong in applications. In this section we assume that for some real functionals $T_1, \ldots, T_m$, the distribution of $T_i(X)$, $1 \le i \le m$ is known.

We call this the **(DF)-assumption** for functionals T_i:

$$T_i(X) \sim Q_i, \quad 1 \le i \le m. \tag{8.23}$$

We also consider a weaker form of this assumption, the **(EF)-assumption**

$$ET_i(X) = a_i, \quad 1 \le i \le m \tag{8.24}$$

and the corresponding **(EF$^\le$)-assumption**

$$ET_i(X) \le a_i, \quad 1 \le i \le m. \tag{8.25}$$

The (DF)- and (EF)-assumptions are quite flexible and widely applicable and allow a unified derivation of upper and lower tail-risk bounds and, therefore, of reduced upper and lower VaR-bounds.

Several methods used in previous chapters for VaR-reduction can be subsumed under these kinds of functional assumptions. The assumption that a variance bound $\text{Var}(S_n) \le \sigma^2$ as in Section 11.1 or a related higher moment assumption $ES_n^k \le c_k$ is exactly of the form (EF$^\le$) from the definition in (8.25) for one functional $T_1(X) = \left(\sum_{i=1}^n X_i\right)^2$ resp. $\left(\sum_{i=1}^n X_i\right)^k$. The assumption that the distributions G_r of the weighted subgroup sums

$$Y_r = \sum_{j \in I_r} \frac{1}{\eta_j} X_j = T_r(X), \quad 1 \le r \le m, \tag{8.26}$$

with $\eta_j := |\{r; \; j \in I_r\}|$ and $\mathcal{E} = \{I_1, \ldots, I_m\}$ known, is of the form (DF) as in (8.23) and is a weakening of the assumption of knowing the multivariate distributions F_I of X_I for $I \in \mathcal{E}$ made in Section 8.1.

Similarly we can consider the case

$$T_r(X) = \max_{j \in I_r} X_j, \quad 1 \le r \le m \tag{8.27}$$

of knowing the distributions of the subgroup maxima. Finally, the assumption that the covariances or correlations among the $X_i, i = 1, \ldots, n$, are known corresponds to the assumption (EF) as in (8.24) when taking

$$T_{i,j}(X) = X_i X_j, \quad 1 \le i < j \le n. \tag{8.28}$$

We denote by $M^{\text{DF}}, m^{\text{DF}}$ resp. $M^{\text{EF}}, m^{\text{EF}}$ resp. $M^{\text{EF}^\le}, m^{\text{EF}^\le}$ the corresponding maximal resp. minimal tail risk probabilities and the corresponding VaR bounds by $\text{Var}_\alpha^{\text{MF}}$, etc. Under the (DF)-assumption, we introduce dual problems of the form

$$U^{\text{DF}}(s) = \inf \left\{ \sum_{i=1}^n \int f_i \, dF_i + \sum_{j=1}^m \int g_j \, dQ_j; \; (f_i, g_j) \in \overline{A}^{\text{DF}}(s) \right\}$$

$$\text{and} \quad I^{\text{DF}}(s) = \sup \left\{ \sum_{i=1}^n \int f_i \, dF_i + \sum_{j=1}^m \int g_j \, dQ_j; \; (f_i, g_i) \in \underline{A}^{\text{DF}}(s) \right\}, \tag{8.29}$$

where $\overline{A}^{\text{DF}}(s) = \left\{ (f_i, g_j); \; \sum_{i=1}^n f_i(x_i) + \sum_{j=1}^m g_j \circ T_j(x) \ge 1_{[s,\infty)} \left(\sum_{j=1}^n X_j \right) \right\}$

and $\underline{A}^{\text{DF}}(s) = \left\{ (f_i, g_j); \; \sum_{i=1}^n f_i(x_i) + \sum_{j=1}^m g_j \circ T_j(x) \le 1_{[s,\infty)} \left(\sum_{j=1}^n X_j \right) \right\}$.

Under several regularity conditions, strong duality theorems for these functionals can be proved. For some examples, see for instance Rüschendorf (2017a). We next formulate the simple-to-verify upper and lower bounds properties of these dual functionals and refrain from discussing the more involved strong duality results.

Proposition 8.7 (Upper and lower bounds under (DF)) *Assume that the risk vector satisfies assumption (DF) in (8.23) for functionals $T_1, \ldots, T_m$. Then the following improved tail risk bounds hold:*

$$M^{\text{DF}}(s) \le U^{\text{DF}}(s) \quad \text{and} \quad m^{\text{DF}}(s) \ge I^{\text{DF}}(s). \tag{8.30}$$

Proof By assumption, (DF) holds for any $(f_i, g_j) \in \overline{A}^{DF}(s)$, i.e.,

$$\sum_{i=1}^{n} f_i(x_i) + \sum_{j=1}^{m} g_j \circ T_j(x) \geq 1_{[s,\infty)}\left(\sum_{j=1}^{n} x_j\right),$$

$$P\left(\sum_{i=1}^{n} X_i \geq s\right) \leq E \sum_{i=1}^{n} f_i(X_i) + \sum_{j=1}^{m} E g_j \circ T_j(X)$$

$$= \sum_{i=1}^{n} \int f_i \, dF_i + \sum_{j=1}^{m} \int g_j \, dQ_j.$$

Taking sup on the left-hand side and inf on the right-hand side, this implies

$$M^{DF}(s) \leq U^{DF}(s).$$

The inequality $m^{DF} \geq I^{DF}$ follows similarly. □

In some cases it may be useful to relax the dual problems by omitting the simple marginal information, i.e., considering admissible dual functions of the form $(0, g_j)$. Define

$$\widetilde{U}^{DF}(s) = \inf\left\{\sum_{j=1}^{m} \int g_j \, dQ_j; \; \sum_{j=1}^{m} g_j \circ T_j(x) \geq 1_{[s,\infty)}\left(\sum_{j=1}^{n} x_j\right)\right\}$$

$$\text{and} \quad \widetilde{I}^{DF}(s) = \sup\left\{\sum_{j=1}^{m} \int g_j \, dQ_j; \; \sum_{j=1}^{m} g_j \circ T_j(x) \leq 1_{[s,\infty)}\left(\sum_{j=1}^{n} x_j\right)\right\}. \tag{8.31}$$

Corollary 8.8 (Relaxed upper and lower bounds under (DF))
Under assumption (DF) for $T_1, \ldots, T_m$, it holds that

$$M^{DF}(s) \leq \widetilde{U}^{DF}(s) \quad \text{and} \quad m^{DF}(s) \geq \widetilde{I}^{DF}(s). \tag{8.32}$$

Proof Corollary 8.8 follows from Proposition 8.7, noting that by definition of $\widetilde{U}^{DF}$ and $\widetilde{I}^{DF}$ it holds that

$$U^{DF}(s) \leq \widetilde{U}^{DF}(s) \quad \text{and} \quad I^{DF}(s) \geq \widetilde{I}^{DF}(s). □$$

In some cases the sharpness of the bounds can be seen directly, and the dual bounds in (8.30) and (8.32) can be reduced to simple marginal bounds. We consider the subgroup sum case in (8.26) for a marginal system $\mathcal{E} = \{I_1, \ldots, I_m\}$ and the weighted sum

$$T_r(x) = \sum_{j \in I_r} \frac{1}{\eta_j} X_j =: Y_r. \tag{8.33}$$

Here η_j counts the number of subsets that have j, as an element. In the case of non-overlapping sets $\{I_r\}$ with $\bigcup_{r=1}^{m} I_r = \{1, \ldots, n\}$, it holds that $\eta_j = 1$ for all j, and the weighted sum Y_r is identical to the partial sum over subgroup I_r.

Denote $H_r = F_{Y_r}$ the partial sum distribution of Y_r and $\mathcal{H} = \mathcal{F}(H_1, \ldots, H_m)$ the corresponding simple marginal systems (see (8.10)). Under assumption (DF), the distributions H_r of Y_r are known, and we obtain:

Theorem 8.9   (Upper and lower bounds with partial sum information) *Let the risk vector X satisfy assumption* (DF) *for the partial sum functionals in* (8.33). *Then:*

a)
$$U^{\mathrm{DF}}(s) \le \tilde{U}^{\mathrm{DF}}(s) \quad \text{and} \quad I^{\mathrm{DF}}(s) \ge \tilde{I}^{\mathrm{DF}}(s);$$

b)
$$M_{\mathcal{E}}(s) \le M^{\mathrm{DF}}(s) \le M_{\mathcal{H}}(s) = \tilde{U}^{\mathrm{DF}}(s)$$

$$\text{and} \quad m_{\mathcal{E}}(s) \ge m^{\mathrm{DF}}(s) \ge m_{\mathcal{H}}(s) = \tilde{I}^{\mathrm{DF}}(s); \tag{8.34}$$

c) *if the marginal system is non-overlapping, then*

$$M_{\mathcal{E}}(s) = M^{\mathrm{DF}}(s) = M_{\mathcal{H}}(s) = \tilde{U}^{\mathrm{DF}}(s)$$

$$\text{and} \quad m_{\mathcal{E}}(s) = m^{\mathrm{DF}}(s) = m_{\mathcal{H}}(s) = \tilde{I}^{\mathrm{DF}}(s). \tag{8.35}$$

Proof   a), b) The proof of a) and b) follows by combining Proposition 8.7 and Corollary 8.8 with the arguments used in the proof of Theorem 8.4 for the inequality $M_{\mathcal{E}}(s) \le M_{\mathcal{H}}(s)$. In particular note that $\sum_{i=1}^{n} X_i = \sum_{r=1}^{m} Y_r$ and that by assumption (DF) we have that $F_{Y_r} = H_r$. The equality $M_{\mathcal{H}}(s) = \tilde{U}^{\mathrm{DF}}(s)$ is the classical strong duality for simple marginal systems.
c) From b) we have the inequality

$$M_{\mathcal{E}}(s) \le M_{\mathcal{H}}(s).$$

Conversely, if $Y = (Y_1, \ldots, Y_m)$ is any vector with distribution function $F_Y \in \mathcal{H}$, then by a classical result on stochastic equations there exist $X_{I_r} \sim F_{I_r}$ such that $\sum_{j \in I_r} X_j = Y_r$ a.s., $1 \le r \le m$. This implies the converse inequality

$$M_{\mathcal{H}}(s) \le M_{\mathcal{E}}(s).$$

Since for the simple marginal system the strong duality theorem holds, we obtain

$$M_{\mathcal{H}}(s) = \tilde{U}^{\mathrm{DF}}(s),$$

and thus the equalities in (8.34) are obtained. The case of the lower bounds is similar. □

As a consequence of Theorem 8.9, we only need the weaker assumption (DF) of knowledge of the distribution of the subgroup sums in order to derive the same bounds as in Theorems 8.3 and 8.4 given there under the stronger assumption of knowledge of the higher-order marginals.

Under the (EF)- resp. (EF$^{\le}$)-assumption for functionals $T_1, \ldots, T_m$, similarly to the (DF)-case we introduce dual functionals

$$U^{\mathrm{EF}}(s) = \inf \left\{ \sum_{i=1}^{n} \int f_i \, \mathrm{d}F_i + \sum_{j=1}^{m} \lambda_j a_j; \; (f_i, \lambda_j) \in \overline{A}^{\mathrm{EF}}(s) \right\} \tag{8.36}$$

$$\text{and} \quad I^{\mathrm{EF}}(s) = \sup \left\{ \sum_{i=1}^{n} \int f_i \, \mathrm{d}F_i + \sum_{j=1}^{m} \lambda_j a_j; \; (f_i, \lambda_j) \in \underline{A}^{\mathrm{EF}}(s) \right\} \tag{8.37}$$

and, similarly, in the inequality case $U^{\mathrm{EF}^{\leq}}, I^{\mathrm{EF}^{\leq}}$, where

$$\overline{A}^{\mathrm{EF}}(s) = \left\{(f_i, \lambda_j); \ f_i \in L^1(F_i), \lambda_j \in \mathbb{R},\right.$$

$$\left. \sum_{i=1}^{n} f_i(x_i) + \sum_{j=1}^{m} \lambda_j T_j(x) \geq 1_{[s,\infty)}\left(\sum_{j=1}^{n} x_j\right)\right\},$$

$$\overline{A}^{\mathrm{EF}^{\leq}}(s) = \left\{(f_i, \lambda_j); \ f_i \in L^1(F_i), \lambda_j \in \mathbb{R}_+,\right.$$

$$\left. \sum_{i=1}^{n} f_i(x_i) + \sum_{j=1}^{m} \lambda_j T_j(x) \leq 1_{[s,\infty)}\left(\sum_{j=1}^{n} x_j\right)\right\},$$

and $\underline{A}^{\mathrm{EF}}(s)$ and $\underline{A}^{\mathrm{EF}^{\leq}}(s)$ are defined similarly.

Proposition 8.10 (Upper and lower bounds under (EF) resp. (EF$^{\leq}$)) *Assume that the risk vector satisfies:*

a) *Assumption (EF) for the functionals $T_1, \ldots, T_m$, then*

$$M^{\mathrm{EF}}(s) \leq U^{\mathrm{EF}}(s) \quad and \quad m^{\mathrm{EF}}(s) \geq I^{\mathrm{EF}}(s). \tag{8.38}$$

b) *Assumption (EF$^{\leq}$) for $T_1, \ldots, T_m$, then*

$$M^{\mathrm{EF}^{\leq}}(s) \leq U^{\mathrm{EF}^{\leq}}(s) \quad and \quad m^{\mathrm{EF}^{\leq}}(s) \geq I^{\mathrm{EF}^{\leq}}(s). \tag{8.39}$$

Proof
a) This follows as in Proposition 8.7 using that for $(f_i, \lambda_j) \in \overline{A}^{\mathrm{EF}}(s)$,

$$\sum_{i=1}^{n} f_i(x_i) + \sum_{j=1}^{m} \lambda_j T_j(x) \geq 1_{[s,\infty)}\left(\sum_{i=1}^{n} x_i\right). \tag{8.40}$$

b) Under (EF$^{\leq}$) we have $ET_j(X) \leq a_j$, and, therefore, since $\lambda_j \geq 0$ we have $E\lambda_j T_j(X) \leq \lambda_j a_j$, and thus the inequality (8.40) implies

$$P\left(\sum_{j=1}^{n} X_j \geq s\right) \leq \sum_{i=1}^{n} \int f_i \, dF_i + \sum_{j=1}^{m} \lambda_j a_j. \qquad \square$$

In some cases it may be useful to omit the marginal information and to use relaxed duals as in (8.31) and Corollary 8.8. Define

$$\widetilde{M}^{\mathrm{EF}}(s) = \inf\left\{\sum_{j=1}^{m} \lambda_j a_j; \ \lambda_j \in \mathbb{R}, \sum_{j=1}^{m} \lambda_j T_j(x) \geq 1_{[s,\infty)}\left(\sum_{i=1}^{n} X_i\right)\right\} \tag{8.41}$$

and, similarly, $\widetilde{I}^{\mathrm{EF}}(s)$, $\widetilde{U}^{\mathrm{EF}^{\leq}}(s)$, and $\widetilde{I}^{\mathrm{EF}^{\leq}}(s)$.

Corollary 8.11 (Relaxed upper and lower bounds under (EF) resp. (EF$^{\leq}$))

a) *Under assumption (EF) for $T_1, \ldots, T_m$ it holds that*

$$M^{\mathrm{EF}}(s) \leq \widetilde{U}^{\mathrm{EF}}(s) \quad and \quad m^{\mathrm{EF}}(s) \geq \widetilde{I}^{\mathrm{EF}}(s). \tag{8.42}$$

b) *Under assumption* (EF$^\leq$) *for* $T_1, \ldots, T_m$ *it holds that*

$$M^{\mathrm{EF}^\leq}(s) \leq \tilde{U}^{\mathrm{EF}^\leq}(s) \quad and \quad m^{\mathrm{EF}^\leq}(s) \geq \tilde{I}^{\mathrm{EF}^\leq}(s). \tag{8.43}$$

Remark 8.12   The dual method for deriving upper and lower bounds for the tail risks under the assumptions (EF), (EF$^\leq$), and (DF) has immediate extensions to deriving upper and lower bounds for the expectation $E\varphi(X)$ of a function φ of the risk vector. We just have to change the admissible class of dual functions: for example, change $\overline{A}^{\mathrm{DF}}(s)$ to

$$\overline{A}^{\mathrm{DF}}(\varphi) := \Big\{ (f_i, g_j); \; \sum_{i=1}^{n} f_i(x_i) + \sum_{j=1}^{m} g_j(T_j(x)) \geq \varphi(x) \Big\}. \tag{8.44}$$

Denoting the maximal expectation of φ under (DF) by $\mathrm{M}^{\mathrm{DF}}(\varphi)$, we obtain similarly to Proposition 8.7:

$$M^{\mathrm{DF}}(\varphi) \leq U^{\mathrm{DF}}(\varphi) \quad and \quad m^{\mathrm{DF}}(\varphi) \geq I^{\mathrm{DF}}(\varphi). \tag{8.45}$$

Similar bounds are valid under (EF) and (EF$^\leq$).   ◇

We consider some applications of assumption (DF) with maxima or with subgroup maxima information. Let $\mathcal{E} = \{I_1, \ldots, I_m\}$ be a non-overlapping system with $\bigcup_{j=1}^{m} I_j = \{1, \ldots, n\}$, and consider the subgroup maxima

$$T_r(X) := \max_{j \in I_r} X_j. \tag{8.46}$$

Under assumption (DF) for $T_1, \ldots, T_m$, i.e., knowing the distribution G_r of the subgroup maxima

$$G_r \sim T_r(X), \tag{8.47}$$

we obtain improved bounds for the distribution function or the survival function (tail risk) of the max in comparison to the case of marginal information only, dealt with in Rüschendorf (1980) (see also Corollary 2.20 in Rüschendorf, 2013).

Theorem 8.13   **(Tail risk of max under subgroup max information)** *Let the risk vector X satisfy condition* (DF) *for the subgroup maxima* $T_1, \ldots, T_m$ *in* (8.46); *then*

$$\Big(\sum_{r=1}^{m} G_r(t) - (m - 1) \Big)_+ \leq F_{\max_{1 \leq i \leq n} X_i}(t) \leq \min_{r \leq m} G_r(t), \tag{8.48}$$

and the bounds in (8.48) *are sharp.*

Proof   By definition of the subgroup maxima T_r we have the basic equality

$$\max_{1 \leq i \leq n} X_i = \max_{r = 1, \ldots, m} T_r(X). \tag{8.49}$$

Since $T_r(X) \sim G_r$, we obtain from the Hoeffding–Fréchet bounds that the bounds in (8.49) are valid. Defining $(Y_1, \ldots, Y_m)$ as a maximally dependent vector with marginal distributions $G_1, \ldots, G_m$, we obtain (see Rüschendorf, 1980)

$$\max_{i \leq m} Y_i \sim \Big(\sum_{r=1}^{m} G_r - (m - 1) \Big)_+. \tag{8.50}$$

Similarly, considering $Z_1, \ldots, Z_m$, the comonotonic vector with marginals G_i, the upper bound is attained:

$$\max_{i \le m} Z_i \sim \min_{r \le m} G_r. \tag{8.51}$$

By a well-known result on stochastic equations, it is possible to construct a vector X with marginals F_i such that a.s. $\max_{j \in I_r} X_j = Y_r$, and similarly it is possible to construct a vector X with marginals F_i such that a.s. $\max_{j \in I_r} X_j = Z_r$. This implies that the bounds in (8.48) are sharp.   □

Remark 8.14

a) By (8.48) it holds for the maximal tail risk $\overline{F}_{\max\limits_{i \le n} X_i}(t)$ that

$$\max_{j \le m} \overline{G}_r(t) \le \overline{F}_{\max\limits_{i \le n} X_i}(t) \le \min \left\{ \left(m - \sum_{r=1}^{m} G_r(t) \right)_+, 1 \right\}. \tag{8.52}$$

This is an improvement over the sharp simple marginal bound resulting from Corollary 3.7:

$$\max_{1 \le i \le n} \overline{F}_i(t) \le \overline{F}_{\max\limits_{1 \le n} X_i}(t) \le \min \left(n - \sum_{i=1}^{n} F_i(t), 1 \right). \tag{8.53}$$

b) Theorem 8.13 also results from an application of the relaxed dual bounds as in Corollary 8.11, based on the inequality

$$\max_{1 \le i \le n} X_i \le \inf_{v \in \mathbb{R}^m} \sum_{r=1}^{m} (v_r + (T_r(X) - v_r)_+) \tag{8.54}$$

and noting that the upper bound is attained for the maximally dependent vector $T_r(X) = Y_r$, $1 \le r \le m$. This gives the lower bound in (8.55), while the upper bound is a direct consequence of the stochastic ordering result

$$T_r(X) \ge_{\text{st}} Z_r, \quad 1 \le r \le m. \tag{8.55}$$

c) For the upper tail risk of the aggregated sum given the assumption (EF) for the subgroup max functionals $T_r(X) = \max_{j \in I_r} X_j$ by the reduced form of the dual functionals, it is indicated to consider inequalities of the form

$$\sum_{i=1}^{n} x_i \le \sum_{r=1}^{m} \alpha_r \max_{j \in I_r} x_j. \tag{8.56}$$

Assuming that $x_i \ge 0$, this inequality implies $\alpha_r \ge n_r = |I_r|$ and therefore this implies the tail risk bound for the sum

$$M^{\text{EF}}(s) \le M_{\mathcal{H}}(s), \tag{8.57}$$

where $\mathcal{H} = \{H_1, \ldots, H_m\}$, $H_r(t) = G_r\left(\frac{t}{n_r}\right)$. The upper bound in (11.51) can be evaluated by the RA-algorithm, but it seems typically to be too rough based on the rough inequality in (8.56).   ◊

The tail risks problem in Remark 8.14 c) with maximal subgroup information seems to be better able to be dealt with by the method based on knowledge of a distribution function on a subset, as considered in Puccetti et al. (2016) and Lux and Papapantoleon (2017). Note that knowing the distribution G_r of $Y_r = \max_{j \in I_r} X_j$ amounts to knowing the distribution function $F_{X_{I_r}} = F^{(r)}$ on the subset $S = \{(t, \ldots, t); t \in \mathbb{R}\}$ since

$$F_{X_{I_r}}((t, \ldots, t)) = P(X_j \le t; \ j \in I_r) = P(\max_{j \in I_r} X_j \le t) = G_r(t). \tag{8.58}$$

This implies by the improved Hoeffding–Fréchet bounds in Puccetti et al. (2016) and Lux and Papapantoleon (2017) that

$$\underline{F}_r^S(x) \le F^{(r)}(x) \le \overline{F}_r^S(x), \quad x \in \mathbb{R}^{I_r}, \tag{8.59}$$

where

$$\overline{F}_r^S(x) = \min \left(\min_{j \in I_r} F_j(x_j), \inf_{t \in \mathbb{R}} \left\{ G_r(t) + \sum_{j \in I_r} (F_j(x_j) - F_j(t))_+ \right\} \right)$$

and $\underline{F}_r^S(x) = \max \left(0, \sum_{j \in I_r} F_j(x_j) - (n_r - 1), \right.$ (8.60)

$$\left. \sup_t \left\{ G_r(t) + \sum_{j \in I_r} (F_j(t) - F_j(x_j))_+ \right\} \right).$$

The bounds in (8.59) imply that the distribution function $F = F_X$ of the risk vector is bounded by

$$\underline{F}^S(x) := \left(\sum_{r=1}^m \underline{F}_r^S(x_{I_r}) - (n - 1) \right)_+ \le F(x) \le \min_{r=1,\ldots,m} \overline{F}_r^S(x_{I_r}) =: \overline{F}^S(x). \tag{8.61}$$

These estimates allow one to apply the method of improved standard bounds to obtain:

Theorem 8.15   (Tail risk of sum under subgroup max information) *Let the risk vector X satisfy condition* (DF) *for the subgroup maxima* $T_1, \ldots, T_m$ *in* (8.46); *then*

$$P\left(\sum_{i=1}^n X_i \le s \right) \ge \bigvee \underline{F}^S(s), \tag{8.62}$$

where $\bigvee$ *denotes the sup-convolution.*

Remark 8.16   Similarly we get a lower bound for the tail risk of the sum. Denoting the survival function of $F^{(r)}$ by $\widehat{F}^r$ and defining

$$\widehat{F}_r^S(x) = \max \left(0, \sum_{j \in I_r} \overline{F}_j(x_j) - (n_r - 1), \right.$$

$$\left. \sup_t \left(G_r(t) - \sum_{j \in I_r} (\overline{F}_j(t) - \overline{F}_j(x_j))_+ \right) \right),$$

we obtain

$$P\left(\sum_{i=1}^n X_i \ge s \right) \ge \bigvee \widehat{F}_r^S(s). \tag{8.63}$$

◇

We next consider an example for the application of the (DF) condition to derive an upper bound for an optimization problem for stop loss premia.

Example 8.17 (**Stop loss premia for a portfolio with additional sum information**) Assume that $X = (X_1, X_2)$ is a risk vector with marginals $X_1 \sim F_1$, $X_2 \sim F_2$, and assume that the distribution of $T(X) = X_1 + X_2$ is known to be G_1, i.e., we make the (DF) assumption for $T_1 = T$. Our aim is to determine under this condition the maximum stop loss premium for the portfolio $2X_1 + X_2$, i.e., to determine $M^{DF}(\varphi_s)$ for $\varphi_s(x) = (2x_1 + x_2 - s)_+$. With the mean excess functions defined for a random variable X by $\pi_X(t) = E(X - t)_+$, the problem can be written in the form

$$\pi_{2X_1+X_2}(t) = \max \tag{8.64}$$

under assumption (DF), as specified above.

Note that for all $a_1, a_2, a_3 \geq 0$ with

$$a_1 + a_3 = 2 \text{ and } a_2 + a_3 = 1 \text{ and for all } u = (u_i) \text{ with } u_1 + u_2 + u_3 = s, \tag{8.65}$$

we have

$$(2x_1 + x_1 - s)_+ \leq (a_1 x_1 - u_1)_+ + (a_2 x_2 - u_2)_+ + (a_3(x_1 + x_2) - u_3)_+. \tag{8.66}$$

This inequality implies taking expectations and infima:

$$M^{DF}(\varphi_s) = \sup\{E(2X_1 + X_2 - s)_+; \ X \text{ satisfies (DF)}\}$$
$$\leq \inf\left\{a_1 \pi_{X_1}\left(\frac{u_1}{a_1}\right) + a_2 \pi_{X_2}\left(\frac{u_2}{a_2}\right) + a_3 \pi_{X_1+X_2}\left(\frac{u_3}{a_3}\right); \tag{8.67}\right.$$
$$\left. a, u \text{ satisfy (8.65)}\right\}.$$

By assumption DF for T, the excess functions are known. As a consequence this dual problem can be solved for distributions of the mean excess functions involved having analytical form. ◇

In the following application of the method of dual bounds, we consider the case where in addition to the marginals, the covariances

$$\sigma_{ij} = \text{Cov}(X_i, X_j) = EX_i X_j - \mu_i \mu_j, \quad \mu_i = EX_i, \tag{8.68}$$

are also specified. This corresponds to assumption (EF) for $T_{ij}(X) = X_i X_j$ with $s_{ij} = EX_i X_j = \sigma_{ij} + \mu_i \mu_j$. From Proposition 8.10 we obtain the improved upper bounds formulated here for a function φ of the risk vector (as in Remark 8.2).

Theorem 8.18 (Risk bound with covariance information)

a) *Let the risk vector X satisfy additionally the (EF) condition $EX_i X_j = s_{ij}$, $1 \leq i \leq j \leq n$; then for a risk function φ, it holds that*

$$M^{\mathrm{EF}}(\varphi) \leq U^{\mathrm{EF}}(\varphi)$$

$$= \inf\left\{\sum_{i=1}^{n}\int f_i\,\mathrm{d}F_i + \sum_{i,j=1}^{n}\alpha_{ij}s_{ij};\quad f_i \in L^1(F_i),\right. \tag{8.69}$$

$$\left.\alpha_{ij}\in\mathbb{R}, \varphi \leq \sum_{i=1}^{n}f_i(x_i) + \sum\alpha_{ij}x_ix_j\right\}.$$

b) *Under* $(\mathrm{EF}^{\leq})$, *inequality* (8.69) *holds with* $\alpha_{ij}\in\mathbb{R}_+$.

Remark 8.19

a) For certain classes of functions φ, the exact duality in (8.69) is stated in Rüschendorf (2017a).

b) Considering, as in Bernard et al. (2017c), $\varphi = 1_{\{\sum_{i=1}^{n}x_i \geq s\}}$, the tail risk of the sum functional, and assuming that besides the marginals F_i it is known that

$$ES_n^2 \leq s^2 \tag{8.70}$$

or, equivalently, $\mathrm{Var}(S_n) \leq \alpha^2 = s^2 - \mu^2$, $\mu = ES_n$, then the dual corresponding to (8.69) simplifies to the form

$$U^{\mathrm{EF}^{\leq}}(s) = \inf\left\{\sum_{i=1}^{n}\int f_i\,\mathrm{d}F_i + \alpha s^2;\quad \alpha \geq 0, f_i \in L^1(F_i),\right. \tag{8.71}$$

$$\left.1_{\{\sum_{i=1}^{n}x_i \geq s\}} \leq \sum_{i=1}^{n}f_i(x_i) + \alpha\left(\sum_{i=1}^{n}x_i\right)^2\right\}.$$

In Bernard et al. (2017b) (see also Chapter 6), good upper bounds for this case are given. In comparison, (8.71) gives theoretical sharp upper bounds that can be evaluated, however only in strongly relaxed form.

c) **Model-independent price bounds.** In a similar way, the method of dual bounds in this chapter also applies to various other types of constraints. For robust model-independent price bounds, dual representations with martingale constraints have been investigated (see Beiglböck et al., 2013 and Acciaio et al., 2016). These constraints are posed due to the fact that reasonable pricing measures have the martingale property. The dual method in this chapter can be extended to infinitely many constraints to deal with the problem of determining robust model-independent price bounds, for example for the case of models based solely on the martingale constraint additional to the marginal structure over time.   ◊

9 Partially Specified Risk Factor Models

In this chapter, we consider the case where, besides the marginal distributions, some structural information is available. This additional information is given by a partially specified factor model in which each risk X_i has a known joint distribution with the common risk factor Z, but we dispense with the conditional independence assumption that is typically made in fully specified factor models. The chapter is based on Bernard et al. (2017b). The assumption that the risks are conditionally independent given the factor is challenging and often appears to be made in an ad-hoc fashion and not grounded in data or statistics (Connor and Korajczyk, 1993). In the context of asset pricing, Chamberlain and Rothschild (1983) and Ingersoll (1984) relax the conditional independence assumption slightly and develop so-called approximate factor models.

We derive easy-to-compute bounds on risk measures such as value-at-risk (VaR) and law-invariant convex risk measures (e.g., tail value-at-risk (TVaR)) and demonstrate their asymptotic sharpness. We show that the dependence uncertainty spread is typically reduced substantially and that, contrary to the case in which only marginal information is used, it is *not* necessarily larger for VaR than for TVaR.

In more detail, this chapter is organized as follows. In Section 9.1, we introduce "partially specified factor models." By representing the distribution of X as a mixture, we obtain sharp upper and lower bounds for the tail probabilities and thus, by inversion, for the VaR as well. The evaluation of these bounds typically poses a considerable challenge. Hence, we derive a more explicit mixture representation for the sharp VaR bound that will be the basis for obtaining VaR bounds in Section 9.4 that are asymptotically sharp and that can be practically evaluated. First, however, in Section 9.3, we study sharp upper and lower bounds for law-invariant convex risk measures, including the tail value-at-risk (TVaR) as a special case. These bounds follow from the availability of largest and smallest elements with respect to convex order for the distribution of S. The largest elements are attained by the dependence structure of comonotonicity conditionally on Z (see Dhaene et al., 2006 for an overview on comonotonicity), and, when the distributions satisfy suitable assumptions, the smallest elements are attained by a dependence structure of joint mixes conditionally on Z (see Chapter 1). We obtain explicit convex lower bounds for S in the context of a mean-variance mixture model.

In Section 9.4, based on the mixing type formula for the sharp VaR bounds, we obtain approximations of the VaR bounds by means of TVaR-based estimates. This procedure leads to greatly simplified formulas that are well suited for numerical evaluation.

We demonstrate that the TVaR-based approximations are asymptotically sharp. These results extend those of Puccetti and Rüschendorf (2014) and Embrechts et al. (2015), where only the marginal information is used. Furthermore, Embrechts et al. (2015) show that in this setting the VaR of large portfolios (asymptotically) exhibits a larger dependence uncertainty spread than the TVaR. These authors use this feature as an argument in support of the use of TVaR in risk management. We provide an example showing that such a result does not hold for a partially specified factor model. By supplementing the partially specified factor model with (conditional) variance information, we derive further improved bounds, and we discuss their asymptotic sharpness.

Finally, in Section 9.5, we assess the model uncertainty of a credit risk portfolio that is modeled using a Bernoulli mixture model (KMV model). The reduction in the dependence uncertainty spread that we observe depends on the magnitude of the common risk factor in comparison to the idiosyncratic factors. All in all, the results of the chapter show that the assumption of a partially specified factor model is a flexible tool with a wide range of possible applications and with a promising capability of reducing risk bounds that are based solely on knowledge of marginal information.

9.1   Partially Specified Factor Models

Factor models offer a useful device for modeling multivariate distributions in various disciplines, including statistics, econometrics, and finance. In particular, they play a central role in asset pricing (Fama and French, 1993; Engle et al., 1990) and are used in monitoring mutual fund performance (Carhart, 1997) and in portfolio optimization (Santos et al., 2013). In risk management, for the derivation of regulatory capital requirements (Gordy, 2003), they constitute the industry standard for the evaluation of credit risk (Gordy, 2000). Specifically, the multivariate normal mean-variance mixture model can be seen as a factor model that generates many of the standard and well-established distributions in quantitative finance, such as the variance gamma, hyperbolic, and normal inverse Gaussian distributions. Importantly, economic theories such as as the arbitrage pricing theory (APT) (Ross, 1976), the capital asset pricing model (CAPM) (Sharpe, 1964), and the rank theory of consumer demands (Lewbel, 1991) are based explicitly on factor models.

In a factor model, the risks X_i are expressed in a functional form as

$$X_i = f_i(Z, \varepsilon_i), \quad 1 \le i \le n, \tag{9.1}$$

where ε_i are idiosyncratic risk components, and Z is a common risk factor taking values in a set $D \subset \mathbb{R}^d$. The typical assumptions for factor models, as in (9.1), are that the factor Z has known distribution and that, conditionally on $Z = z \in D$, the risks X_i are independent (i.e., the ε_i are independent) with known (conditional) distribution $F_{i|z}$. Clearly, there might be further possible model risk due to misspecification of the law of the factor or of the conditional laws of the risks given the factors, or as a result of further assumptions, such as the number of factors. Here, however, we concentrate on the risk contribution arising from possible departures from the conditional independence assumption.

When we dispense with the assumption of conditional independence among the individual risks $X_i, i = 1, \ldots, n$, their joint distribution is no longer specified. However, the joint distributions H_i of (X_i, Z), and thus also the marginal distribution F_i of X_i, $i = 1, \ldots, n$, are known. This setting is denoted as *a partially specified factor model*. As compared to considering only the information on the marginal distributions of $X_1, \ldots, X_n$, using the additional information on the common risk factor Z leads to improved risk bounds when assessing the risk of the aggregated portfolio $S = \sum_{i=1}^{n} X_i$.

Let $(\Omega, \mathcal{A}, P)$ be an atomless probability space and X be a set of real-valued random variables on $(\Omega, \mathcal{A}, P)$. We take $X = L^0$, the set of all random variables in this section and $X = L^1$ in the sections that follow. In this chapter we consider only law-invariant risk measures ϱ from X to $(-\infty, \infty]$. Let $Z \in X^d$ be a random vector with essential support $D \subset \mathbb{R}^d$. We refer to Z as a *risk factor* and denote its distribution by G.

Let $H = (H_i)_{1 \leq i \leq n}$ be a vector of $(1 + d)$-variate distributions and define the *partially specified factor model*

$$A(H) = \{X \in X^n : (X_i, Z) \sim H_i, \ 1 \leq i \leq n\}$$

as the set of random vectors $X = (X_i)_{1 \leq i \leq n}$ such that for each $i = 1, \ldots, n$, (X_i, Z) has joint distribution (function) H_i. For each $i = 1, \ldots, n$, X_i has distribution F_i and conditional distribution $F_{i|z}$ given $Z = z, z \in D$. We aim at determining (sharp) upper and lower bounds on $\varrho(S)$ where ϱ is some risk measure, $S = \sum_{i=1}^{n} X_i$, and $X \in A(H)$. Specifically, we consider the problems

$$\overline{\varrho}^f = \sup\{\varrho(S) : X \in A(H)\} \text{ and } \underline{\varrho}^f = \inf\{\varrho(S) : X \in A(H)\}. \tag{9.2}$$

In this chapter, we refer to the partially specified risk factor model as *the constrained setting*.

Write $F = (F_i)_{1 \leq i \leq n}$. In comparison to the partially specified factor model, the *model with marginal information* (only) is defined as

$$A_1(F) = \{X \in X^n : X_i \sim F_i, \ 1 \leq i \leq n\},$$

and in this setting one considers the problems

$$\overline{\varrho} = \sup\{\varrho(S) : X \in A_1(F)\} \text{ and } \underline{\varrho} = \inf\{\varrho(S) : X \in A_1(F)\}.$$

We refer to this setting as the *unconstrained setting*.

By definition, the admissible class $A(H)$ of risk vectors X with information on the risk factor Z is contained in $A_1(F)$, i.e., $A(H) \subset A_1(F)$. Hence,

$$\underline{\varrho} \leq \underline{\varrho}^f \text{ and } \overline{\varrho}^f \leq \overline{\varrho}.$$

Note that when the risk factor Z is independent of $X_1, \ldots, X_n$, i.e., $H_{i|z} = F_i, 1 \leq i \leq n$, only the information on the marginal distributions is useful, and the study of the constrained bounds $\overline{\varrho}^f$ and $\underline{\varrho}^f$ reduces to the unconstrained bounds $\overline{\varrho}$ and $\underline{\varrho}$.

Following Embrechts et al. (2015), in the unconstrained setting we define the dependence uncertainty spread of a risk measure ϱ as the difference $\overline{\varrho} - \underline{\varrho}$, and in the constrained setting we define it as $\overline{\varrho}^f - \underline{\varrho}^f$. To measure the improvement of dependence uncertainty we obtain through the factor information, the measure of improvement Δ_ϱ is defined as

$$\Delta_\varrho = 1 - \frac{\overline{\varrho}^f - \underline{\varrho}^f}{\overline{\varrho} - \underline{\varrho}},$$

in which we assume by convention that $\Delta_\varrho = 1$ when $\overline{\varrho} = \underline{\varrho}$.

9.2   Risk Bounds

Specifically, we study the risk bound problems in (9.2) using as risk measure ϱ the tail probability. That is, we consider $P(S \geq t)$ for $t \in \mathbb{R}$, but we also use risk measures such as VaR and TVaR.

When the risk measure ϱ is the tail probability, the VaR, or the TVaR, we denote in the partially specified risk factor model the upper bounds $\overline{\varrho}^f$ in (9.2) by $M^f(t), t \in \mathbb{R}$, $\overline{\text{VaR}}_\alpha^f$ or $\overline{\text{TVaR}}_\alpha^f$, $\alpha \in (0,1)$, respectively. The corresponding risk infima are denoted by $m^f(t)$, $\underline{\text{VaR}}_\alpha^f$, and $\underline{\text{TVaR}}_\alpha^f$, respectively. In the model with marginal information only, $\overline{\varrho}$ is specifically denoted as $M(t), t \in \mathbb{R}$, $\overline{\text{VaR}}_\alpha$, and $\overline{\text{TVaR}}_\alpha$, $\alpha \in (0,1)$, and similarly for other quantities.

It turns out to be useful to describe the risk vector $X = (X_i)_{1 \leq i \leq n} \in A(H)$ through a **mixture representation**:

$$X =_d X_Z \quad \text{with} \quad X_z = (X_{i,z})_{1 \leq i \leq n} \in A_1(F_z), \quad z \in D.$$

Here, $F_z = (F_{i|z})_{1 \leq i \leq n}$ is the vector of conditional distributions of X_i given $Z = z$, $i = 1, \ldots, n$. By conditioning, the distribution F_S of S satisfies

$$F_S = \int F_{S_z} \, dG(z), \tag{9.3}$$

where (and throughout) $S_z = \sum_{i=1}^n X_{i,z}$ is the sum of the conditional variables $(X_{i,z})_{1 \leq i \leq n}$, and the above integral, without further specification, is taken over its natural region D. The random variables $X_{i,z}, i = 1, 2, \ldots, n, z \in D$ can be constructed as

$$X_{i,z} = F_{i|z}^{-1}(U_{i,z}), \quad 1 \leq i \leq n, \tag{9.4}$$

where $U_z = (U_{1,z}, \ldots, U_{n,z})$ is some random vector with $U(0,1)$ marginal distributions, and $(U_z)_{z \in D}$ is independent of Z. Of course, in a similar way as for $A_1(F)$ and $A(H)$, risk bounds can also be defined for the admissible class $A_1(F_z)$. In this chapter, the notation $M_z(t)$ is used to denote the sharp tail probability bound for the class $A_1(F_z)$.

The mixture representation in (9.3) implies the following sharp tail probability bounds. In this section we do not impose any assumptions on G and $F_z, z \in D$.

Proposition 9.1 (Sharp tail probability bounds)   *The sharp upper and lower tail probability bounds for the partially specified risk factor model are given by*

$$M^f(t) = \int M_z(t)\,dG(z), \ \ and \ \ m^f(t) = \int m_z(t)\,dG(z), \ \ t \in \mathbb{R}. \tag{9.5}$$

Proof   For any admissible risk vector $X \in A(H)$, we have that the conditional distribution of $X_i \mid Z = z$ is given by $F_{i\mid z}$. Therefore, conditionally under $Z = z$, the random vector X has marginal distributions $F_{i\mid z}$, $1 \le i \le n$. As a consequence, we obtain, by conditioning,

$$P\left(\sum_{i=1}^{n} X_i \ge t\right) = \int P\left(\sum_{i=1}^{n} X_i \ge t \mid Z = z\right)dG(z) \le \int M_z(t)\,dG(z),$$

and thus $M^f(t) \le \int M_z(t)\,dG(t)$.

Conversely, let $\tilde{X}_z = (X_{i,z})$ be random vectors with marginal distributions $F_{i\mid z}$ such that, for given $\varepsilon > 0$,

$$P\left(\sum_{i=1}^{n} X_{i,z} \ge t\right) \ge M_z(t) - \varepsilon. \tag{9.6}$$

The risk vector X has a representation as a mixture model: $X = X_Z$, where Z is a random variable with distribution G, independent of $(X_{i,z})$. Then, by conditioning, we obtain that (X, Z) is admissible, i.e., $X \in A(H)$ and

$$P\left(\sum_{i=1}^{n} X_i \ge t\right) \ge \int M_z(t)\,dG(z) - \varepsilon. \tag{9.7}$$

As a result, (9.6) and (9.7) establish equality in (9.5). The case of the lower bound is proved in a similar way.   □

Remark 9.2 (Existence of worst case distributions)   By a measurable selection result as in Rüschendorf (1985), a worst case distribution for M^f exists, and thus the ε-argument in the proof of Proposition 9.1 could be avoided in the case of the upper bound. However, the lower bounds m^f and $M_z(t)$ are only attainable when we modify the definition of the value-at-risk slightly (see Bernard et al., 2014b; Bernard et al., 2017c).   ◊

As a corollary to Proposition 9.1, we obtain the following sharp VaR bounds.

Corollary 9.3 (Sharp VaR bounds)   *The sharp upper and lower VaR bounds in the partially specified risk factor model are given by*

$$\overline{\mathrm{VaR}}_\alpha^f = (M^f)^{-1}(1 - \alpha) \ \ and \ \ \underline{\mathrm{VaR}}_\alpha^f = (m^f)^{-1}(1 - \alpha), \ \alpha \in (0, 1), \tag{9.8}$$

where, for $\alpha \in (0, 1)$, respectively,

$$(M^f)^{-1}(1 - \alpha) := \inf\left\{t \in \mathbb{R}: M^f(t) \le 1 - \alpha\right\},$$
$$(m^f)^{-1}(1 - \alpha) := \inf\left\{t \in \mathbb{R}: m^f(t) \le 1 - \alpha\right\}.$$

The representation result in (9.5) shows that when the risk measure at hand is the tail probability or the VaR, the problem of determining sharp bounds in the constrained setting essentially reduces to the aggregation of bounds that are derived using information on conditional distributions $(F_{i|z})$ only. Hence, we can build on the results that have been derived in this unconstrained setting as in Part I. We apply some of these results to the following two-dimensional example with normally distributed risks and compare the dependence uncertainty spread in the unconstrained setting with the one in the constrained setting.

Example 9.4 (VaR bounds for normally distributed risks) Assume that X_1 and X_2 have $N(0, 1)$ distributions and that Z is a risk factor such that (X_i, Z) has a bivariate normal distribution with correlation parameter $r_i \in (-1, 1)$, $i = 1, 2$. A stochastic representation is given by $X_i = r_i Z + \sqrt{1 - r_i^2} \varepsilon_i$, $i = 1, 2$, where ε_1 and ε_2 are $N(0, 1)$-distributed and are independent of Z.

As for the case of *unconstrained bounds* with information on marginal distributions only, we obtain from Theorem 1.9:

$$\overline{\text{VaR}}_\alpha = \text{VaR}_0(\Phi^{-1}(U) + \Phi^{-1}(1 + \alpha - U)) = 2\Phi^{-1}\left(\frac{1+\alpha}{2}\right), \quad \alpha \in (0, 1) \quad (9.9)$$

and

$$\underline{\text{VaR}}_\alpha = \text{VaR}_1(\Phi^{-1}(V) + \Phi^{-1}(\alpha - V)) = 2\Phi^{-1}\left(\frac{\alpha}{2}\right), \quad \alpha \in (0, 1), \quad (9.10)$$

where $U \sim U[\alpha, 1]$ and $V \sim U(0, \alpha)$.

As for the case of *constrained bounds*, in the marginal models in the case of the partially specified factor model, we first consider $r_1 = r_2$. Observe that the $X_{i|z}$ have an $N(r_i z, 1 - r_i^2)$ distribution, $i = 1, 2$. Hence, from (9.9)–(9.10) we obtain sharp upper bounds and lower bounds on $M_z(t)$ and $m_z(t)$, $t \in \mathbb{R}$. Using formula (9.5), we find, after a numerical inversion, the values of $\overline{\text{VaR}}_\alpha^f$ and $\underline{\text{VaR}}_\alpha^f$. Here, the values inside the integrals in (9.5) are known explicitly, and the integral is evaluated numerically.

Next, we consider $r_1 = -r_2$. We obtain that $\overline{\text{VaR}}^f = \sqrt{1 - r_1^2} \overline{\text{VaR}}_\alpha$ and $\underline{\text{VaR}}^f = \sqrt{1 - r_1^2} \underline{\text{VaR}}_\alpha$. Table 9.1 displays the bounds for different values of r_i and α as well as the measure of improvement Δ_{VaR} obtained by using factor information. In Panel A, where $r_1 = r_2$, we observe that the upper bounds do not improve, whereas the lower bounds show essential improvements. In Panel B, where $r_1 = -r_2$, we find the opposite picture: The upper bounds improve significantly, whereas the lower bounds remain essentially the same.

9.2.1 A Mixture Representation of VaR Bounds

In general, there are two main challenges when evaluating the bounds $\overline{\text{VaR}}_\alpha^f$ and $\underline{\text{VaR}}_\alpha^f$. First, the representation result (9.8) requires, for a given probability level $\alpha \in (0, 1)$, the function $t \to M^f(t)$, $t \in \mathbb{R}$, to be established in which each $M^f(t)$ requires aggregation of the (marginal) tail probability bounds $M_z(t)$, $z \in D$; see also Example

Panel A	r_1	VaR_α	$(\underline{\mathrm{VaR}}_\alpha, \overline{\mathrm{VaR}}_\alpha)$	$(\underline{\mathrm{VaR}}_\alpha^f, \overline{\mathrm{VaR}}_\alpha^f)$	Δ_{VaR}
	0	2.326	$(-0.125, 3.920)$	$(-0.125, 3.920)$	0%
$\alpha = 0.95$	0.5	2.601	$(-0.125, 3.920)$	$(0.822, 3.920)$	23.44%
	0.8	2.979	$(-0.125, 3.920)$	$(1.894, 3.880)$	50.92%
	1	3.290	$(-0.125, 3.920)$	$(3.290, 3.290)$	100%
	0	3.643	$(-0.0125, 5.614)$	$(-0.013, 5.614)$	0%
$\alpha = 0.995$	0.5	4.073	$(-0.0125, 5.614)$	$(1.893, 5.614)$	33.87%
	0.8	4.665	$(-0.0125, 5.614)$	$(3.464, 5.606)$	61.93%
	1	5.152	$(-0.0125, 5.614)$	$(5.152, 5.152)$	100%

Panel B	r_1	VaR_α	$(\underline{\mathrm{VaR}}_\alpha, \overline{\mathrm{VaR}}_\alpha)$	$(\underline{\mathrm{VaR}}_\alpha^f, \overline{\mathrm{VaR}}_\alpha^f)$	Δ_{VaR}
	0	2.326	$(-0.125, 3.920)$	$(-0.125, 3.920)$	0%
$\alpha = 0.95$	0.5	2.015	$(-0.125, 3.920)$	$(-0.109, 3.395)$	13.4%
	0.8	1.396	$(-0.125, 3.920)$	$(-0.075, 2.352)$	40%
	1	0.000	$(-0.125, 3.920)$	$(0.000, 0.000)$	100%
	0	3.643	$(-0.0125, 5.614)$	$(-0.0125, 5.614)$	0%
$\alpha = 0.995$	0.5	3.155	$(-0.0125, 5.614)$	$(-0.011, 4.862)$	13.4%
	0.8	2.186	$(-0.0125, 5.614)$	$(-0.007, 3.368)$	40%
	1	0.000	$(-0.0125, 5.614)$	$(0.000, 0.000)$	100%

Table 9.1 VaR bounds in the normal case. Panel A: $r_1 = r_2$. Panel B: $r_1 = -r_2$. VaR_α corresponds to the case in which (X_1, X_2) is bivariate normally distributed with correlation r_1^2 (panel A) resp $-r_1^2$ (panel B).

9.4. Second, even in the unconstrained setting, obtaining sharp VaR bounds is an open problem in general, and analytical results are available only for small portfolios, $n = 2$, some classes of homogeneous portfolios, and asymptotically large portfolios, $n \to \infty$ (see Embrechts et al., 2014). A practical approach to approximating the unconstrained bounds is the rearrangement algorithm (RA) (Chapter 2). In response to these issues, we proceed by expressing the VaR bounds $\overline{\mathrm{VaR}}_\alpha^f$ and $\underline{\mathrm{VaR}}_\alpha^f$ directly in terms of marginal VaR bounds in the remainder of this section. These expressions, combined with the use of some results and ideas that are valid in the unconstrained setting, provide the basis for obtaining bounds in Section 9.4 that can be practically evaluated and that are asymptotically sharp.

To evaluate

$$\overline{\mathrm{VaR}}_\alpha^f = \sup\{\mathrm{VaR}_\alpha(S_Z) \colon X_Z \in A(H)\},$$

we first provide two explicit representations of $\mathrm{VaR}_\alpha(S_Z)$. Define, for $\gamma \in \mathbb{R}$ and $z \in D$,

$$p_z^\gamma = F_{S_z}(\gamma) \text{ and } b^\gamma = \text{ess sup}_{z \in D} \mathrm{VaR}_{p_z^\gamma}(S_z), \tag{9.11}$$

where ess $\sup_{z \in D}$ in (9.11) is taken with respect to G (the distribution of Z). Note that for distributions with positive densities on its support, b^γ is equal to γ.

Proposition 9.5 (VaR representation of mixtures)   *For $\alpha \in (0, 1)$, the VaR at level α of the mixture S_Z has the following representations:*

$$\mathrm{VaR}_\alpha(S_Z) = \inf\left\{\gamma \in \mathbb{R}: \int p_z^\gamma \, dG(z) \geq \alpha\right\}, \tag{9.12}$$

$$\mathrm{VaR}_\alpha(S_Z) = \inf\left\{b^\gamma : \gamma \in \mathbb{R}, \int p_z^\gamma \, dG(z) \geq \alpha\right\}. \tag{9.13}$$

Proof   We first prove (9.12). For $\alpha \in (0, 1)$, we have, by definition,

$$\mathrm{VaR}_\alpha(S_Z) = \inf\{\gamma \in \mathbb{R}: F_{S_Z}(\gamma) \geq \alpha\}$$

$$= \inf\left\{\gamma \in \mathbb{R}: \int F_{S_z}(\gamma) \, dG(z) \geq \alpha\right\}$$

$$= \inf\left\{\gamma \in \mathbb{R}: \int p_z^\gamma \, dG(z) \geq \alpha\right\}. \tag{9.14}$$

To prove (9.13), observe that for all z, $\mathrm{VaR}_{p_z^\gamma}(S_z) = F_{S_z}^{-1}(p_z^\gamma) \leq \gamma$. Hence,

$$b^\gamma = \mathrm{ess\,sup}_{z \in D} \, \mathrm{VaR}_{p_z^\gamma}(S_z) \leq \gamma,$$

and, therefore, $\mathrm{VaR}_\alpha(S_Z) \geq \inf\{b^\gamma : \gamma \in \mathbb{R}, \int p_z^\gamma \, dG(z) \geq \alpha\}$.
Conversely, for any $\gamma \in \mathbb{R}$ with $\int p_z^\gamma \, dG(z) \geq \alpha$ it holds that

$$F_{S_z}(b^\gamma) \geq F_{S_z} \circ F_{S_z}^{-1}(p_z^\gamma) \geq p_z^\gamma.$$

This implies that

$$\int F_{S_z}(b^\gamma) \, dG(z) \geq \int p_z^\gamma \, dG(z) \geq \alpha,$$

i.e., b^γ is also an admissible constant in (9.14), and, therefore,

$$\mathrm{VaR}_\alpha(S_Z) \leq \inf\left\{b^\gamma : \gamma \in \mathbb{R}, \int p_z^\gamma \, dG(z) \geq \alpha\right\},$$

and we obtain equality. □

The second representation for $\mathrm{VaR}_\alpha(S_Z)$ in Proposition 9.5 is of some independent interest and may provide the intuition to develop a convenient expression for $\overline{\mathrm{VaR}}_\alpha^f$ in terms of (marginal) VaR bounds. In the formal derivation of this expression, however, we solely build on the first (basic) representation for $\mathrm{VaR}_\alpha(S_Z)$. Using the shorthand notation $q_z(\beta)$ for $\mathrm{VaR}_\beta(S_z)$ $z \in D$, $\beta \in (0, 1)$ with right-continuous generalized inverse

$$q_z^{-1}(\gamma) = \sup\{x \in [0, 1]: q_z(x) \leq \gamma\},$$

it holds that

$$p_z^\gamma = q_z^{-1}(\gamma).$$

Hence, we obtain that

$$\mathrm{VaR}_\alpha(S_Z) = \inf\left\{\gamma \in \mathbb{R}: \int q_z^{-1}(\gamma) \, dG(z) \geq \alpha\right\} =: b^*(\alpha). \tag{9.15}$$

In order to obtain a corresponding representation of the sharp VaR bound $\overline{\text{VaR}}_\alpha^f$, we define the worst (conditional) VaRs for the conditional sum S_z by

$$\overline{q}_z(\beta) := \overline{\text{VaR}}_\beta(S_z) = \sup\{\text{VaR}_\beta(S_z): X_z \in A_1(F_z)\}, \quad z \in D, \ \beta \in (0,1),$$

with the right-continuous generalized inverse denoted by $(\overline{q}_z)^{-1}(\gamma)$, and, finally,

$$\overline{b}^*(\alpha) := \inf\left\{\gamma \in \mathbb{R}: \int (\overline{q}_z)^{-1}(\gamma)\, dG(z) \geq \alpha\right\}. \tag{9.16}$$

The next proposition shows that $\overline{b}^*(\alpha)$ is the sharp upper bound of the VaR in the presence of information on a risk factor.

Proposition 9.6   (Mixture representation for the sharp VaR upper bound) *For $\alpha \in (0,1)$, the sharp upper bound of VaR in the partially specified risk factor model is given by*

$$\overline{\text{VaR}}_\alpha^f = \overline{b}^*(\alpha) = \text{VaR}_\alpha(\overline{q}_Z(V)),$$

where V is a $U(0,1)$-distributed random variable independent of Z.

Proof We have by definition $\overline{\text{VaR}}_\alpha^f = \sup\{\text{VaR}_\alpha(S_Z),\ X_Z \in A(H)\}$. Since $q_z^{-1}(\gamma) \geq \overline{q}_z^{-1}(\gamma)$ for $z \in D$, (9.15) implies that

$$\overline{\text{VaR}}_\alpha^f = \sup_{X_Z \in A(H)} \inf\left\{\gamma \in \mathbb{R}: \int q_z^{-1}(\gamma)\, dG(z) \geq \alpha\right\} \tag{9.17}$$

$$\leq \inf\left\{\gamma \in \mathbb{R}: \int \overline{q}_z^{-1}(\gamma)\, dG(z) \geq \alpha\right\} \tag{9.18}$$

$$= \overline{b}^*(\alpha).$$

On the other hand, for $z \in D$, let $\overline{X}_z \sim F_z$ be a solution to $\overline{q}_z^{-1}(\alpha) = \text{VaR}_\alpha(\overline{S}_z)$, where $\overline{S}_z = \sum_{i=1}^n \overline{X}_{i,z}$. Then $\text{VaR}_\alpha(\overline{S}_Z) = \overline{b}^*(\alpha)$, and thus equality in (9.17) holds. The equality $\overline{b}^*(\alpha) = \text{VaR}_\alpha(\overline{q}_Z(V))$ follows from

$$\text{VaR}_\alpha(\overline{q}_Z(V)) = \inf\{\gamma \in \mathbb{R}: P(\overline{q}_Z(V) \leq \gamma) \geq \alpha\}$$

$$= \inf\left\{\gamma \in \mathbb{R}: \int \overline{q}_z^{-1}(\gamma)\, dG(z) \geq \alpha\right\} = \overline{b}^*(\alpha). \qquad \square$$

While formula (9.16) for the VaR bound $\overline{b}^*(\cdot)$ is explicit, in general it is still not straightforward to evaluate it. Indeed, we need to obtain the (conditional) VaR bounds $\overline{q}_z(v)$ for $z \in D$, $v \in (0,1)$. Few explicit results exist, and a practical evaluation of $\overline{b}^*(\cdot)$ thus appears to require a repeated use of the RA (for approximating all $\overline{q}_z(v)$). In Section 9.4, however, we show that the sharp upper bound $\overline{b}^*(\alpha)$ can be approximated (from above) by easy-to-compute upper bounds that are defined in terms of the TVaR. Furthermore, these approximations are asymptotically sharp.

Remark 9.7   A mixture representation of $\underline{\text{VaR}}_\alpha^f$ can be obtained in a similar way by replacing the upper bound quantities $\overline{q}_z(\beta) = \overline{\text{VaR}}_\beta(S_z)$ in (9.18) with the corresponding lower bound quantities $\underline{q}_z(\beta) = \underline{\text{VaR}}_\beta(S_z)$, in which $\underline{\text{VaR}}_\beta(S_z) = \inf\{\text{VaR}_\beta(S_z): X_z \in A_1(F_z)\}$, $z \in D$, $\beta \in (0, 1)$. $\diamond$

9.3   Bounds for Convex Risk Measures

It is well known that a law-invariant convex risk measure ϱ is consistent with respect to convex order on proper probability spaces such as L^1 (integrable random variables); see Chapter 4 of Föllmer and Schied (2004), Jouini et al. (2006), Bäuerle and Müller (2006), and Burgert and Rüschendorf (2006). From this section on, we let $\mathcal{X} = L^1$, and all marginal distributions of F and F_z, $z \in D$ are assumed to have finite first moment.

Under this assumption, the study of $\overline{\varrho}^f$ (resp. $\underline{\varrho}^f$) is closely connected to finding $X \in A(H)$ such that $S = X_1 + \cdots + X_n$ becomes the largest (resp. smallest) element with respect to convex order. We define the admissible class of sums in the partially specified risk factor model

$$S(H) = \{X_1 + \cdots + X_n: X \in A(H)\}$$

and note that the upper and lower bounds $\overline{\varrho}^f$ and $\underline{\varrho}^f$ can be equivalently defined in terms of $S(H)$ rather than $A(H)$, i.e.,

$$\overline{\varrho}^f = \sup\{\varrho(S): S \in S(H)\} = \sup\{\varrho(S): X \in A(H)\}$$

and

$$\underline{\varrho}^f = \inf\{\varrho(S): S \in S(H)\} = \inf\{\varrho(S): X \in A(H)\}.$$

9.3.1   Upper Bound

We first focus on $\overline{\varrho}^f$ and thus aim at find an element in $S(H)$ that is largest with respect to convex order. To this end, we recall a classical result of Meilijson and Nadas (1979), who established that the comonotonic sum $S^c = \sum_{i=1}^n F_i^{-1}(U)$, $U \sim U(0, 1)$ is larger in the sense of convex order than any other sum $X_1 + \cdots + X_n$, $X \in A_1(F)$ (see also Chapter 1). This result suggests that on $S(H)$, the conditionally comonotonic sum

$$S_Z^c = \sum_{i=1}^n F_{i|Z}^{-1}(U) \tag{9.19}$$

is a largest element with respect to convex order and thus leads to sharp upper bounds for TVaR and for other law-invariant convex risk measures. In the following, let $U \sim U(0, 1)$ be independent of Z. The notation for (conditionally) comonotonic sums, $S_z^c = \sum_{i=1}^n F_{i|z}^{-1}(U)$, $S_Z^c = \sum_{i=1}^n F_{i|Z}^{-1}(U)$, and $S^c = \sum_{i=1}^n F_i^{-1}(U)$, will be used repeatedly.

Proposition 9.8 (Sharp upper bounds for convex risk measures)   *The following statements hold:*

a) For all $S \in \mathcal{S}(H)$ it holds that $S \leq_{cx} S_Z^c \in \mathcal{S}(H)$.

b) For a law-invariant convex risk measure ϱ, we have $\overline{\varrho}^f = \varrho(S_Z^c)$.

c) We have

$$S_Z^c \leq_{cx} S^c. \tag{9.20}$$

Proof

a) Consider the vector X_Z^c having components $F_{i|Z}^{-1}(U)$, and observe that their conditional distribution functions are $F_{i|z}$ and that their marginal distribution functions are F_i. Hence, $X_Z^c \in A(H)$ and $S_Z^c \in \mathcal{S}(H)$. Furthermore, for any $X \in A(H)$ we can use the mixture representation X_Z for X with $X_{i,z} = F_{i|z}^{-1}(U_{i,z})$, as in Section 9.2. From the convex ordering result in (9.19), it follows that

$$S_z = \sum_i X_{i|z} \leq_{cx} \sum_i F_{i|z}^{-1}(U).$$

This implies, by conditioning, $S_Z \leq_{cx} \sum_i F_{i|Z}^{-1}(U) = S_Z^c$.

b) Since ϱ is consistent with convex order, the result follows from a).

c) The summands of S_Z^c having distribution functions F_i, the result follows from (9.19).   □

Remark 9.9   An improvement of the form $S \leq_{cx} S_Z^c$ can also be found in Kaas et al. (2000). Formula (9.20) has been applied there to obtain improved upper risk bounds, for example for basket options and Asian options – mainly, however, in a lognormal context with conditional distributions that are easy to evaluate (see Vanmaele et al., 2006; Deelstra et al., 2008; Vanduffel et al., 2008).   ◊

9.3.2   Lower Bound

As for the study of $\underline{\varrho}^f$, we notice that obtaining a lower bound with respect to convex order in $\mathcal{S}(H)$ is a more difficult task than obtaining an upper bound. In fact, lower bounds with respect to convex order for sums $S = \sum_{i=1}^n X_i$, $X \in A_1(F)$ are generally not available; however, some analytical cases can be found in Chapter 1.

For any $X = X_Z \in A(H)$ and $z \in D$, let $\mu_z = ES_z$, which is the sum of the means of $F_{i|z}$, $i = 1, \ldots, n$ and hence is independent of the choice of $X_Z \in A(H)$. It is easy to observe that $E(S_Z \mid Z) = \mu_Z$. We show that μ_Z is a simple lower bound with respect to convex order. Recall that $F_z = (F_{i|z})_{1 \leq i \leq n}$, $z \in D$, and let U be a $U(0, 1)$-distributed random variable that is independent of Z.

Proposition 9.10 (Lower bounds for convex risk measures)   *The following statements hold:*

a) For all $S \in \mathcal{S}(H)$ it holds that $\mu_Z \leq_{cx} S$.

b) For a law-invariant convex risk measure ϱ, we have $\varrho(\mu_Z) \leq \underline{\varrho}^f$.

c) $\mu_Z \in \mathcal{S}(H)$ if and only if F_z is jointly mixable for G-almost all $z \in D$.

d) For $n = 2$ and a law-invariant convex risk measure ϱ, we have $\underline{\varrho}^f = \varrho(S_Z^a)$, where $S_z^a = F_{1|z}^{-1}(U) + F_{2|z}^{-1}(1 - U)$, $z \in D$.

Proof

a) For any $S \in S(H)$, write $S = S_Z$, and the conditional Jensen's inequality implies that $E(S_Z \mid Z) = \mu_Z \leq_{cx} S_Z$.

b) From a), $\varrho(\mu_Z) \leq \varrho(S)$ for all $S \in S(H)$, which implies the result.

c) Suppose that F_z is jointly mixable for $z \in D_0$. By the definition of joint mixability and (9.4), there exist $U_z = (U_{1,z}, \ldots, U_{n,z})$, $z \in D_0$ such that $X_{i,z} = F_{i|z}^{-1}(U_{i,z})$, $1 \leq i \leq n$, and $S_z = \sum_{i=1}^n X_{i,z} = \mu_z$, for $z \in D_0$. This shows $\mu_Z = S_Z$, almost surely, and hence $\mu_Z \in S(H)$.

   For the other direction, take $X_Z \in A(H)$ such that $S_Z = \mu_Z$ almost surely. Then, since $\sum_{i=1}^n X_{i,Z} = \mu_Z$ almost surely, there exists D_0, $P(Z \in D_0) = 1$ such that $\sum_{i=1}^n X_{i,z} = S_z = \mu_z$ for each $z \in D_0$. That is, F^z is jointly mixable for each $z \in D_0$.

d) Note that for any $U(0,1)$ random variables U_1 and U_2, we have $F_{1|z}^{-1}(U)$ $+ F_{2|z}^{-1}(1-U) \leq_{cx} F_{1|z}^{-1}(U_1) + F_{2|z}^{-1}(U_2)$ since counter-monotonicity yields a sum that is minimum with respect to convex order. Thus, for any $S_Z \in S(H)$, we have $S_Z^a \leq_{cx} S_z$. By definition of convex order, for any convex function f, such that $E(f(S_Z))$ and $E(f(S_Z^a))$ are well defined, we have

$$E(f(S_Z^a)) = \int_D E(f(S_z^a)) \, dG(z) \leq \int_D E(f(S_z)) \, dG(z) = E(f(S_Z))$$

and hence $S_Z^a \leq_{cx} S_Z$, implying $\varrho(S_Z^a) \leq \varrho(S_Z)$.   □

As a consequence of Propositions 9.8 and 9.10, we obtain that, for any law-invariant convex risk measure ϱ and any $S \in S(H)$,

$$\varrho(\mu) \leq \varrho(\mu_Z) \leq \varrho(S) \leq \varrho(S_Z^c) \leq \varrho(S^c),$$

where $\mu = E(\mu_Z)$. In particular, Proposition 9.10 suggests that with respect to convex order, the best case risk $S \in S(H)$ is the one whose randomness derives entirely from the factor Z. However, to prove that $\mu_Z \in S(H)$, one needs to establish joint mixability, which is difficult and not valid in general. Some criteria for joint mixability are given in Chapter 1. An example of the sharp lower bound $\mu_Z \in S(H)$ in a location-scale family is provided next.

Example 9.11 (Convex order bounds in location-scale families)
Let $Z = (Z_1, Z_2)$ have an arbitrary distribution on $D = \mathbb{R}^2$. For some real numbers $a_i, b_i \in \mathbb{R}$, $i = 1, \ldots, n$, and positive numbers $\sigma_1, \ldots, \sigma_n$ satisfying $2 \max_{i=1,\ldots,n} \sigma_i \leq \sum_{i=1}^n \sigma_i$, let

$$X_i = a_i + Z_1 b_i + Z_2(\sigma_i \varepsilon_i), \quad i = 1, \ldots, n,$$

where the $\varepsilon_1, \ldots, \varepsilon_n$ are identically distributed and are independent of Z, but the joint distribution of $(\varepsilon_1, \ldots, \varepsilon_n)$ is not known. That is, Z_1 is a common location factor and Z_2 is a common scale factor for the n risks $X_1, \ldots, X_n$. Assume that the distribution F_0 of the ε_i has a unimodal and symmetric density; this includes the normal and t-distributions. Let ε be an F_0-distributed random variable independent of Z, and write

$$a = \sum_{i=1}^n a_i, \quad b = \sum_{i=1}^n b_i, \quad \sigma = \sum_{i=1}^n \sigma_i.$$

From Proposition 9.8, a largest element in $S(H)$ with respect to convex order is given by

$$S_Z^c = a + bZ_1 + \sigma\varepsilon Z_2.$$

From Rüschendorf and Uckelmann (2002), we know that, for $z \in \mathbb{R}^2$, the tuple of distributions of $X_{1,z}, \ldots, X_{n,z}$ is jointly mixable. This allows us to apply Proposition 9.10 to obtain a smallest element in $S(H)$ with respect to convex order, which is given by

$$\mu_Z = a + bZ_1,$$

where we used the fact that $E(\varepsilon) = 0$. This implies sharp bounds on a convex risk measure ϱ, i.e., for $S \in S(H)$,

$$a + \varrho b Z_1 \le \varrho(S) \le a + \varrho(bZ_1 + \sigma\varepsilon Z_2), \quad \alpha \in (0,1).$$

In the following example we illustrate the TVaR bounds for normally distributed risks.

Example 9.12 (TVaR bounds for normally distributed risks)
The set up is as in Example 9.4. Recall that, for a standard normally distributed risk X,

$$\text{TVaR}_\alpha(X) = \frac{\varphi(\Phi^{-1}(\alpha))}{1 - \alpha}, \quad \alpha \in (0,1),$$

where φ is the standard normal density. As for the *unconstrained bounds*, S^c has an $N(0,4)$ distribution, and $ES = 0$. We obtain that

$$\overline{\text{TVaR}}_\alpha = 2\frac{\varphi(\Phi^{-1}(\alpha))}{1 - \alpha} \quad \text{and} \quad \underline{\text{TVaR}}_\alpha = 0, \quad \alpha \in (0,1).$$

As for the *constrained bounds*, S_Z^c has an $N(0, \sigma_1^2)$ distribution with $\sigma_1^2 = 2(1 + r_1 r_2 + \sqrt{(1 - r_1^2)(1 - r_2^2)})$, and S_Z^a has an $N(0, \sigma_2^2)$ distribution with $\sigma_2^2 = 2(1 + r_1 r_2 - \sqrt{(1 - r_1^2)(1 - r_2^2)})$. Hence,

$$\overline{\text{TVaR}}_\alpha^f = \sigma_1\frac{\varphi(\Phi^{-1}(\alpha))}{1 - \alpha} \quad \text{and} \quad \underline{\text{TVaR}}_\alpha = \sigma_2\frac{\varphi(\Phi^{-1}(\alpha))}{1 - \alpha}, \quad \alpha \in (0,1).$$

Table 9.2, Panel A shows these TVaR bounds for different values of $r_1 = r_2$ and α. We observe that there is no difference between the unconstrained upper bound and the constrained one, whereas there is an improvement of the lower bound. In Table 9.2, Panel B we show the TVaR bounds for different values of $r_1 = -r_2$ and α. Note that in this case, $S_Z^a = \mu_Z = 0$. In this case, we find the opposite picture: The upper bounds improve significantly, whereas the lower bounds remain unchanged.

To assess the impact of heterogeneity, we fix $r_1 = -0.5$ where r_2 varies between -1 and 1. We represent the bounds and the improvements Δ_{TVaR} in Figure 9.1. When $|r_1| \ne |r_2|$, both the upper and the lower bounds improve.

From the following example, we obtain further insight into the influence of dependence information of a factor model on risk bounds.

Panel A	r_1	TVaR_α	$(\underline{\mathrm{TVaR}_\alpha}, \overline{\mathrm{TVaR}_\alpha})$	$(\underline{\mathrm{TVaR}_\alpha^f}, \overline{\mathrm{TVaR}_\alpha^f})$	Δ_{TVaR}
	0	2.917	(0.000, 4.125)	(0.000, 4.125)	0 %
$\alpha = 0.95$	0.5	3.261	(0.000, 4.125)	(2.063, 4.125)	50 %
	0.8	3.736	(0.000, 4.125)	(3.300, 4.125)	80 %
	1	4.125	(0.000, 4.125)	(4.125, 4.125)	100 %
	0	4.090	(0.000, 5.784)	(0.000, 5.784)	0 %
$\alpha = 0.995$	0.5	4.573	(0.000, 5.784)	(2.892, 5.784)	50 %
	0.8	5.238	(0.000, 5.784)	(4.627, 5.784)	80 %
	1	5.784	(0.000, 5.784)	(5.784, 5.784)	100 %

Panel B	r_1	TVaR_α	$(\underline{\mathrm{TVaR}_\alpha}, \overline{\mathrm{TVaR}_\alpha})$	$(\underline{\mathrm{TVaR}_\alpha^f}, \overline{\mathrm{TVaR}_\alpha^f})$	Δ_{TVaR}
	0	2.917	(0.000, 4.125)	(0.000, 4.125)	0 %
$\alpha = 0.95$	0.5	2.526	(0.000, 4.125)	(0.000, 3.573)	13.4 %
	0.8	1.750	(0.000, 4.125)	(0.000, 2.475)	40 %
	1	0.000	(0.000, 4.125)	(0.000, 0.000)	100 %
	0	4.090	(0.000, 5.784)	(0.000, 5.784)	0 %
$\alpha = 0.995$	0.5	5.009	(3.542, 5.784)	(0.000, 5.009)	13.4 %
	0.8	5.487	(2.454, 5.784)	(0.000, 3.470)	40 %
	1	0.000	(0.000, 5.784)	(0.000, 0.000)	100 %

Table 9.2 TVaR bounds for the normal case. Panel A: $r_1 = r_2$. Panel B: $r_1 = -r_2$. The column TVaR_α provides the TVaR in the case in which (X_1, X_2) is bivariate normally distributed with correlation r_1^2 (panel A) resp. $-r_1^2$ (panel B).

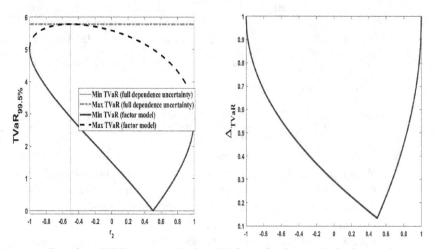

Figure 9.1 Bounds on $\mathrm{TVaR}_{99.5\%}$ are displayed (left panel), along with the degree of improvement (right panel). We consider $r_1 = -0.5$, and r_2 varies between -1 and 1.

Example 9.13 (Pareto risks)   We consider a risk factor model for the case $n = 2$ given by

$$X_1 = (1 - Z)^{-1/3} - 1 + \varepsilon_1,$$
$$X_2 = p((1 - Z)^{-1/3} - 1) + (1 - p)(Z^{-1/3} - 1) + \varepsilon_2,$$

where $Z \sim U(0, 1)$, ε_1 and ε_2 are Pareto(4) distributed and independent of Z, and $p \in (0, 1)$. We allow any dependence between the variables ε_1 and ε_2. In this example the common component $(1 - Z)^{-1/3} - 1$ is Pareto(3)-distributed and thus dominates the idiosyncratic risk components ε_i. Based on the risk bounds established in Section 9.3, Propositions 9.8 and 9.10, we obtain, for TVaR at level $\alpha = 0.95$, the dependence uncertainty spread as in Figure 9.2.

For $p \approx 0$ the common risk factor Z creates strong negative dependence between X_1 and X_2, and, as a consequence, we obtain a strong reduction in the upper risk bounds. For $p \approx 1$ the risk factor Z induces strong positive dependence between X_1 and X_2, and we obtain, as a consequence, a strong improvement in the lower bounds but not in the upper bounds. For all intermediate p we have a total reduction of a similar order.

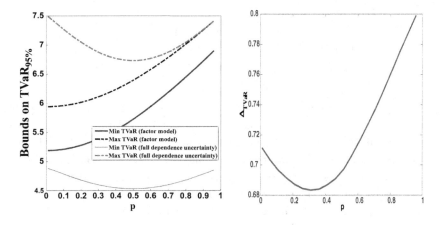

Figure 9.2 Bounds for TVaR at 95 % are displayed (left panel), along with the degree of improvement from Theorem 9.8 (right panel).

9.4   Relation between VaR and TVaR Bounds

In Example 9.12, we observed that adding factor information does not always yield improved bounds. In particular, the upper bound on TVaR does not improve when $F_{1|z} = F_{2|z}$, $z \in D$ (i.e., when $r_1 = r_2$), and the lower bound remains unchanged when $\mu_z = \mu$ for $z \in D$. The following proposition generalizes these observations and provides conditions under which upper and lower bounds on a law-invariant convex risk measure do not improve when using factor information.

Proposition 9.14 (No improvement for convex risk measures)

Let ϱ be a law-invariant convex risk measure.

a) *If $F_{1|z} = \cdots = F_{n|z}$ for all $z \in D$, then $\overline{\varrho}^f = \overline{\varrho} = \varrho(S^c)$.*
b) *If $\mu_z = \mu$ and F_z is jointly mixable for all $z \in D$, then $\underline{\varrho}^f = \underline{\varrho} = \varrho(\mu)$.*
c) *If the conditions of a) and b) both hold, then $\Delta_\varrho = 0$.*

Proof
a) Note that $\varrho(S_Z^c) = \overline{\varrho}^f \leq \overline{\varrho} = \varrho(S^c)$. It suffices to show that $\varrho(S_Z^c) = \varrho(S^c)$. Note that $F_1 = \cdots = F_n$, and we have $S_Z^c \stackrel{d}{=} nF_{1|z}^{-1}(U)$ and $S^c \stackrel{d}{=} nF_1^{-1}(U)$ for some $U \sim U(0,1)$, $z \in D$. We can check for $x \in \mathbb{R}$ that

$$P(S_Z^c \leq x) = \int P(nF_{1|z}^{-1}(U) \leq x)\,\mathrm{d}G(z)$$
$$= \int F_{1|z}\left(\frac{x}{n}\right)\,\mathrm{d}G(z)$$
$$= F_1\left(\frac{x}{n}\right)$$
$$= P(nF_1^{-1}(U) \leq x) = P(S^c \leq x).$$

Thus, $S_Z^c \stackrel{d}{=} S^c$, and $\overline{\varrho}^f = \varrho(S_Z^c) = \varrho(S^c) = \overline{\varrho}$.
b) Note that $\underline{\varrho}^f \geq \underline{\varrho} \geq \varrho(\mu)$. From Proposition 9.10, we have $\mu_Z \in S(H)$, and thus $\mu \in S(H)$ since $\mu_Z = \mu$.
   Therefore, $\underline{\varrho}^f \geq \underline{\varrho} \geq \varrho(\mu) \geq \underline{\varrho}^f$; that is, $\underline{\varrho}^f = \underline{\varrho} = \varrho(\mu)$.
c) This is a direct consequence of a) and b).   □

The above statements are applicable to TVaR and do not hold when VaR is used as a risk measure. Counterexamples can be easily constructed (see also Example 9.15 hereafter). In particular, there are situations in which using factor information yields improved bounds on VaR but not on TVaR. This observation raises the issue of whether the result of Embrechts et al. (2015) that, in the unconstrained setting, (asymptotically) large portfolios exhibit a larger dependence uncertainty spread for VaR than for TVaR carries over to the constrained setting. The answer to this question is negative, as the following example shows.

Example 9.15   (Dependence uncertainty spread of VaR is not necessarily larger than that of TVaR) Assume that the factor Z takes two values: $Z = 0$ with probability 0.95 and $Z = 1$ with probability 0.05. When $Z = 0$, we assume that all X_i ($i = 1, \ldots, n$) are degenerate and take value 0.5. When $Z = 1$, X_i ($i = 1, \ldots, n$) are Bernoulli distributed with parameter 0.5. Let n be even, and take $\alpha = 0.9$. As for the unconstrained bounds, observe that S^c satisfies $P(S^c = 0) = 0.025$, $P(S^c = 0.5n) = 0.95$, and $P(S^c = n) = 0.025$.

From results regarding VaR bounds (see, e.g., Bernard et al., 2014b), it is easy to see that $\overline{\mathrm{VaR}}_\alpha = \overline{\mathrm{TVaR}}_\alpha = \frac{n}{4} + \frac{3 \times 0.5n}{4} = 0.625n$ and that $\underline{\mathrm{VaR}}_\alpha < \underline{\mathrm{TVaR}}_\alpha = 0.5n$. Hence, the dependence uncertainty spread of VaR is larger than that of TVaR.

In the constrained setting, since for any $S_Z \in S(H)$, $P(S_Z = 0.5n) > 0.95$ and $\mathrm{VaR}_{0.9}(S_Z) = 0.5n$, there is no longer dependence uncertainty spread on $\mathrm{VaR}_{0.9}(S_Z)$.

However, the dependence uncertainty spread for TVaR remains unchanged. In particular, it is higher than in the case of VaR, and this holds also as $n \to \infty$.

Embrechts et al. (2015) show that in the unconstrained case, the VaR of large portfolios is more sensitive to misspecification of the model (i.e., has a higher dependence uncertainty spread) than the TVaR. Example 9.15 shows that whether TVaR can be seen as less sensitive in this regard (i.e., has less dependence uncertainty spread) depends merely on the available set of information. In particular, when structural factor information is available as a source of dependence information, TVaR has, in general, no such advantage over VaR.

9.4.1   Approximation of VaR Bounds Based on TVaR Bounds

Recall that formula (9.16) for $\overline{\mathrm{VaR}}_\alpha^f$ remains difficult to evaluate practically. However, the TVaR bounds that we developed in the previous section can be used to determine an easy-to-evaluate upper bound for $\overline{\mathrm{VaR}}_\alpha^f$.

To this aim, note that under the condition $Z = z \in D$, the conditional VaR is bounded above by the conditional comonotonic TVaR, i.e., for all conditional sums S_z and all $\beta \in (0,1)$, it holds that

$$q_z(\beta) = \mathrm{VaR}_\beta(S_z) \le \mathrm{TVaR}_\beta(S_z) \le t_z(\beta) := \mathrm{TVaR}_\beta(S_z^c).$$

The above inequality implies that $(\overline{q}_z)^{-1}(\gamma) \ge t_z^{-1}(\gamma)$, $\gamma \in \mathbb{R}$, and therefore,

$$\overline{\mathrm{VaR}}_\alpha^f = \overline{b}^*(\alpha) = \inf\left\{\gamma \in \mathbb{R}: \int (\overline{q}_z)^{-1}(\gamma)\, dG(z) \ge \alpha\right\}$$

$$\le \inf\left\{\gamma \in \mathbb{R}: \int t_z^{-1}(\gamma)\, dG(z) \ge \alpha\right\} =: b_t^*(\alpha). \qquad (9.21)$$

So, for each set $\{Z = z\}$, we replace the VaR upper bound with the TVaR upper bound.

Proposition 9.16 (TVaR-based bounds for $\overline{\mathrm{VaR}}^f$)   *For $\alpha \in (0,1)$ the sharp upper bound $\overline{\mathrm{VaR}}_\alpha^f$ in the partially specified risk factor model is bounded above by the TVaR-based bound $b_t^*(\alpha)$ in (9.21), i.e.,*

$$\overline{\mathrm{VaR}}_\alpha^f \le b_t^*(\alpha). \qquad (9.22)$$

Note that, conditional on $Z = z$, one has a simple expression for $t_z(\beta)$, i.e.,

$$t_z(\beta) = \mathrm{TVaR}_\beta(S_z^c) = \sum_{i=1}^{n} \mathrm{TVaR}_\beta(X_{i|z}), \qquad (9.23)$$

and thus $t_z(\beta)$ is easy to calculate. As a result, the calculation of the upper TVaR-based bound $b_t^*(\alpha)$ of $\overline{\mathrm{VaR}}_\alpha^f$ is much simpler than the calculation of $\overline{\mathrm{VaR}}_\alpha^f$. In particular, we avoid the iterated application of the RA algorithm.

There is also an alternative way to establish the simplified VaR bounds in (9.22). This method leads to a stochastic representation that is useful in evaluating the bounds by simulation. Define, for any $z \in D$, a random variable T_z^+ as

$$T_z^+ = \mathrm{TVaR}_V(S_z^c),$$

where $V \sim U(0, 1)$ is a random variable that is uniformly distributed on $(0, 1)$ and independent of Z and of $(S_z^c)_{z \in D}$. It is easy to simulate V and Z, and hence also the random variable T_Z^+, as well as to approximate its VaR. The following proposition therefore yields an interesting connection between T_Z^+ and the upper bound $b_t^*(\alpha)$ in Proposition 9.16.

Proposition 9.17 (Representation of TVaR-based bounds)
For $\alpha \in (0, 1)$, we have

$$\overline{\mathrm{VaR}}_\alpha^f \le \mathrm{VaR}_\alpha(T_Z^+) = b_t^*(\alpha). \tag{9.24}$$

Proof We only need to show that $\mathrm{VaR}_\alpha(T_Z^+) = b_t^*(\alpha)$. Using comonotone additivity of TVaR, we obtain

$$T_z^+ = \mathrm{TVaR}_V(S_z^c) = \sum_{i=1}^n \mathrm{TVaR}_V(X_{i|z}) =: \sum_{i=1}^n T_{i|z}^+.$$

Note that $\mathrm{VaR}_\alpha(T_{i|z}^+) = \mathrm{TVaR}_\alpha(X_{i|z})$. From the representation for VaR in (9.21), we obtain $\mathrm{VaR}_\alpha(T_Z^+) = b_t^*(\alpha)$.   □

Remark 9.18 In a similar way, we also obtain approximations of the lower sharp VaR-bound $\underline{\mathrm{VaR}}_\alpha^f$. Define $T_z^- := \mathrm{LTVaR}_V(S_z^c)$; then

$$\mathrm{VaR}_\alpha(T_Z^-) \le \underline{\mathrm{VaR}}_\alpha^f, \tag{9.25}$$

i.e., $\mathrm{VaR}_\alpha(T_Z^-)$ is a lower bound for $\underline{\mathrm{VaR}}_\alpha^f$. Here, $\mathrm{LTVaR}_\alpha(S_z^c)$ is the left TVaR; we have that $\mathrm{LTVaR}_\alpha(S_z^c) = \frac{1}{\alpha} \int_0^\alpha \mathrm{VaR}_u(S_z^c) \, du$.   ◇

9.4.2   Asymptotic Sharpness

In the unconstrained setting (Part I), we showed that $\mathrm{TVaR}_\alpha(S^c)$ is an asymptotically sharp bound on $\mathrm{VaR}_\alpha(S)$, i.e.,

$$\frac{\overline{\mathrm{VaR}}_\alpha}{\overline{\mathrm{TVaR}}_\alpha} \to 1, \text{ as } n \to \infty \tag{9.26}$$

holds under some moment conditions; more general conditions that also ensure asymptotic equivalence can be found in Puccetti and Rüschendorf (2014), Puccetti et al. (2013), Wang and Wang (2015), and Embrechts et al. (2015). In this section, we extend these results to the constrained setting by showing that the TVaR-based bounds $\mathrm{VaR}_\beta(T_Z^+)$ and $\mathrm{VaR}_\beta(T_Z^-)$ developed in Section 9.4.1 are asymptotically sharp bounds for VaR, which includes the equivalence (9.26) as a special case in which Z is independent of X.

In the constrained setting, it is clear that $\mathrm{TVaR}_\alpha(S_Z^c)$ is a bound on $\mathrm{VaR}_\alpha(S_Z)$, but an asymptotic equivalence for the ratio $\overline{\mathrm{VaR}}_\alpha^f / \overline{\mathrm{TVaR}}_\alpha^f$ fails to hold in general. A simple counterexample would be to choose $F_{i|z}$, $z \in D$, $i = 1, 2, \ldots$ to be all degenerate distributions. In this case, dependence uncertainty is no longer relevant, and all elements in $S(H)$ are distributed as μ_Z (which depends on n). There is no hope that $\mathrm{VaR}_\alpha(\mu_Z)/\mathrm{TVaR}_\alpha(\mu_Z) \to 1$ would hold generally. In fact, in order to ensure asymptotic equivalence for the ratio $\overline{\mathrm{VaR}}_\alpha^f / \overline{\mathrm{TVaR}}_\alpha^f$, one would need asymptotic mixability, which can intuitively be seen as a (strengthened) version of the law of large numbers; see the related discussions in Embrechts et al. (2014), Wang (2014) and Bernard et al. (2017c). Although an asymptotic equivalence fails to hold for $\overline{\mathrm{VaR}}_\alpha(S_Z)/\overline{\mathrm{TVaR}}_\alpha(S_Z)$, we have the following asymptotic equivalence theorem for risk aggregation with a risk factor.

Proposition 9.19 (Asymptotic sharpness of TVaR-based bounds) *Suppose that for any $z \in D$ and some $k > 1$,*

a) $E|X_{i,z} - EX_{i,z}|^k < M < \infty$ *for* $i = 1, \ldots, n;$
b) $\liminf_{n\to\infty} n^{-1/k} \sum_{i=1}^n E(X_{i,z}) = \infty.$

Then, for $\alpha \in (0,1)$, as $n \to \infty$,

$$\frac{\overline{\mathrm{VaR}}_\alpha^f}{\mathrm{VaR}_\alpha\left(T_Z^+\right)} \to 1 \text{ and } \frac{\underline{\mathrm{VaR}}_\alpha^f}{\mathrm{VaR}_\alpha\left(T_Z^-\right)} \to 1.$$

Proof From conditions a), b) for the sequence $\{F_{i,z}; \ i \in \mathbb{N}\}$, it follows that the conditions from Theorem 3.3 of Embrechts et al. (2015) are satisfied for all $\beta \in (0,1)$ and $z \in D$. Therefore, we have

$$\frac{\overline{\mathrm{VaR}}_\beta(S_z)}{\overline{\mathrm{TVaR}}_\beta(S_z)} = \frac{\bar{q}_z(\beta)}{t_z(\beta)} \to 1$$

as $n \to \infty$ for each $\beta \in (0,1)$ and $z \in D$. Thus, we have, for a $U(0,1)$-distributed random variable V independent of Z, $\bar{q}_Z(V)/t_Z(V) \to 1$ as $n \to \infty$, and this convergence holds for all $\omega \in \Omega$. Then, from (9.16), we conclude that

$$\frac{\overline{\mathrm{VaR}}_\alpha^f}{\mathrm{VaR}_\alpha\left(T_Z^+\right)} = \frac{\mathrm{VaR}_\alpha(\bar{q}_Z(V))}{\mathrm{VaR}_\alpha(t_Z(V))} \longrightarrow 1.$$

The case for lower bounds is similar. $\qquad\square$

In the following example we consider a low-dimensional portfolio, where $n = 2$, and compare the sharp VaR bounds with those that are asymptotically sharp.

Example 9.20 (Approximations using TVaR-based bounds for normal distributions) We consider the setting as in Example 9.4. In Table 9.3 we compare sharp VaR bounds with the TVaR-based bounds for VaR in Proposition 9.17 for various parameter values of the correlations r_1 and r_2.

The values of VaR_α, $\overline{\mathrm{VaR}}_\alpha^f$, and $\underline{\mathrm{VaR}}_\alpha^f$ are taken from Example 9.4. To compute $\mathrm{VaR}_\alpha\left(T_Z^+\right)$ and $\mathrm{VaR}_\alpha\left(T_Z^-\right)$, we simulate z from $\mathcal{N}(0,1)$ and v from $U(0,1)$.

Next, we compute, for $r_1 = r_2$, $T_z^+ = 2r_1 z + 2\sqrt{1 - r_1^2}\frac{\varphi(\Phi^{-1}(v))}{1-v}$ and, for $r_1 = -r_2$,

$T_z^+ = 2\sqrt{1 - r_1^2}\frac{\varphi(\Phi^{-1}(v))}{1-v}$. By generating many values for v and z, we can accurately approximate $\mathrm{VaR}_{95\%}$ of T_Z^+. We proceed similarly for T_Z^-.

$\alpha = 0.95$	VaR_α	$\overline{\text{VaR}}_\alpha^f$	$\text{VaR}_\alpha\left(T_Z^+\right)$	$\underline{\text{VaR}}_\alpha^f$	$\text{VaR}_\alpha\left(T_Z^-\right)$
$r_1 = r_2 = 0$	2.33	3.92	4.12	−0.12	−0.21
$r_1 = r_2 = 0.5$	2.60	3.92	4.11	0.82	0.68
$r_1 = r_2 = 0.8$	2.98	3.88	4.01	1.89	1.78
$r_1 = r_2 = 1$	3.29	3.28	3.28	3.28	3.28
$r_1 = -r_2 = 0$	2.33	3.92	4.13	−0.12	−0.21
$r_1 = -r_2 = -0.5$	2.01	3.39	3.57	−0.11	−0.18
$r_1 = -r_2 = -0.8$	1.40	2.35	2.47	−0.07	−0.13
$r_1 = -r_2 = -1$	0.00	0.00	0.00	0.00	0.00

Table 9.3 Comparison between sharp VaR bounds and their TVaR-based approximations.

For this small portfolio, $n = 2$, the asymptotic sharp bounds perform reasonably well. This is somewhat expected since for normally distributed risks, VaR and TVaR are not very different.

Example 9.21 (Approximations using TVaR-based bounds for Pareto distributions) We consider a Pareto risk model in which $n = 2$, $P(Z = 1) = P(Z = 2) = 1/2$, and X_1, X_2 are Pareto(ϑ, Z)-distributed conditional on $Z = 1, 2$. That is, $P(X_{1,z} > x) = z^\vartheta x^{-\vartheta}$, $x \geq z$, where $\vartheta > 1$. A smaller value of ϑ indicates a heavier-tailed distribution of X_1, X_2.

For this simple setting, using a result in Chapter 1 we can analytically determine the values $M_z(t), z = 1, 2, t \in \mathbb{R}$ and consequently obtain $M^f(t), t \in \mathbb{R}$ by Proposition 9.1. Taking an inverse of M^f, we calculate $\overline{\text{VaR}}_\alpha^f$. Omitting all simple intermediate steps, we have

$$\overline{\text{VaR}}_\alpha^f = (2^\vartheta + 4^\vartheta)^{1/\vartheta}(1 - \alpha)^{-1/\vartheta}, \quad \alpha \in (0, 1).$$

On the other hand, we can easily calculate $T_z^+ = \text{TVaR}_V(S_z^c), z = 1, 2$, and consequently we obtain the distribution of T_Z^+. Omitting all intermediate steps, we have

$$\text{VaR}_\alpha(T_Z^+) = 2^{-1/\vartheta}\frac{\vartheta}{\vartheta - 1}(2^\vartheta + 4^\vartheta)^{1/\vartheta}(1 - \alpha)^{-1/\vartheta}, \quad \alpha \in (0, 1).$$

Therefore, for this model in which $n = 2$, we have

$$\frac{\text{VaR}_\alpha(T_Z^+)}{\overline{\text{VaR}}_\alpha^f} = 2^{-1/\vartheta}\frac{\vartheta}{\vartheta - 1}.$$

It is clear that the above ratio is a decreasing function of ϑ and is independent of α. We report some numbers for different choices of (α, ϑ) in Table 9.4.

From Table 9.4, it is clear and not surprising that for lighter-tailed Pareto distributions, the TVaR-based approximation performs better than in the case of heavy-tailed distributions. Due to the fact that a reliable calculation of $\overline{\text{VaR}}_\alpha^f$ is not available in general (see discussions in Section 9.4.4), we only report numbers for $n = 2$, and the asymptotics of Proposition 9.19 have not fully kicked in. Puccetti and Rüschendorf (2014) studied the performance of TVaR-based bounds in the unconstrained case and

(α, ϑ)	$\overline{\mathrm{VaR}}_\alpha^f$	$\mathrm{VaR}_\alpha\left(T_Z^+\right)$	$\mathrm{VaR}_\alpha(T_Z^+)/\overline{\mathrm{VaR}}_\alpha^f$
(0.95,2)	20.000	28.284	1.414
(0.95,5)	7.327	7.973	1.088
(0.95,10)	5.398	5.596	1.037
(0.95,20)	4.646	4.724	1.017
(0.99,2)	44.721	63.246	1.414
(0.99,5)	10.110	11.001	1.088
(0.99,10)	6.340	6.573	1.037
(0.95,20)	5.036	5.120	1.017

Table 9.4 Comparison between sharp VaR bounds and their TVaR-based approximations.

used the RA to approximate $\overline{\mathrm{VaR}}_\alpha$. For a portfolio containing heavy-tailed Pareto distributions ($\vartheta = 2$), they concluded that TVaR-based bounds yield good approximations for portfolio size $n \geq 10$.

9.4.3   Adding Variance Information

In this section, we show that the bounds can be further sharpened if conditional variance information is also available. The idea of using the variance to sharpen the unconstrained bounds can be found in Chapter 6, where it is shown that doing so can have a significant impact on the unconstrained VaR bounds. Here, we consider conditional variance information in addition to factor information and discuss how this can be useful in improving the bounds on VaR and TVaR.

Consider $v := (v_z)_{z \in D}$, $v_z \geq 0$. We define the partially specified factor model with variance information as

$$A(H, v) = \{X \in A(H): \ \mathrm{Var}(S \mid Z = z) \leq v_z^2, \ z \in D\},$$

where we assume that it contains at least one element. Hence, v_z^2 provides a bound on the conditional variance of S_z, $z \in D$. We study bounds on VaR and TVaR, i.e., we consider the problems

$$\overline{\mathrm{VaR}}_\alpha^{f,v} = \sup\{\mathrm{VaR}_\alpha(S): X \in A(H, v)\}$$

and

$$\overline{\mathrm{TVaR}}_\alpha^{f,v} = \sup\{\mathrm{TVaR}_\alpha(S): X \in A(H, v)\}.$$

We can consider the lower bound problems in a similar way. We denote the corresponding infima by $\underline{\mathrm{VaR}}_\alpha^{f,v}$ and $\underline{\mathrm{TVaR}}_\alpha^{f,v}$.

Proposition 9.22   (VaR bounds in the factor model with variance information)
For $\alpha \in (0, 1)$, we have

$$\overline{\mathrm{VaR}}_\alpha^{f,v} \leq \mathrm{VaR}_\alpha\left(\min\left(\mathrm{TVaR}_U(S_Z^c), \mu_Z + v_Z\sqrt{\frac{U}{1-U}}\right)\right)$$

and

$$\underline{\mathrm{VaR}}_\alpha^{f,v} \geq \mathrm{VaR}_\alpha \left(\max \left(\mathrm{LTVaR}_U(S_Z^c), \mu_Z - v_Z \sqrt{\frac{1-U}{U}} \right) \right),$$

where $U \sim U(0,1)$ *is independent of* Z *and* $(S_z^c)_{z \in D}$.

Proof For any $X_Z \in A(H)$, it holds that

$$\mathrm{VaR}_\alpha(S_Z) = \mathrm{VaR}_\alpha(\mathrm{VaR}_U(S_Z))$$

$$\leq \mathrm{VaR}_\alpha \left(\min \left(\mathrm{TVaR}_U(S_Z^c), \mu_Z + v_Z \sqrt{\frac{U}{1-U}} \right) \right),$$

where have used by the Cantelli bound that for all $z \in D$, $u \in (0,1)$, $\mathrm{VaR}_u(S_z) \leq \mathrm{TVaR}_u(S_z^c)$, and $\mathrm{VaR}_u(S_z) \leq \mu_z + v_z \sqrt{\frac{u}{1-u}}$. This shows the desired result for $\overline{\mathrm{VaR}}_\alpha^f$. The case of $\underline{\mathrm{VaR}}_\alpha^f$ is similar. $\qquad\square$

Proposition 9.23 **(TVaR bounds in the factor model with variance information)** *For* $\alpha \in (0,1)$, *we have*

$$\overline{\mathrm{TVaR}}_\alpha^{f,v} \leq \mathrm{TVaR}_\alpha \left(\min \left(\mathrm{VaR}_U(S_Z^c), \mu_Z + v_Z \sqrt{\frac{U}{1-U}} \right) \right),$$

where $U \sim U(0,1)$ *independent of* Z *and* $(S_z^c)_{z \in D}$.

Proof For any $X_Z \in A(H)$, it holds that

$$\mathrm{TVaR}_\alpha(S_Z) \leq \mathrm{TVaR}_\alpha(S_Z^c) = \mathrm{TVaR}_\alpha(\mathrm{VaR}_U(S_Z^c)).$$

Furthermore, $\mathrm{TVaR}_\alpha(S_Z) = \mathrm{TVaR}_\alpha(\mathrm{VaR}_U(S_Z))$ and for all $z \in D$, $u \in (0,1)$, and $\mathrm{VaR}_u(S_z) \leq \mu_z + v_z \sqrt{\frac{u}{1-u}}$ by the Cantelli bound. Hence, by combining we obtain the desired result. $\qquad\square$

Remark 9.24 In Chapter 6, we study when, in addition to the marginal information, information on the (unconditional) variance of the sum is also provided; there we consider the problem

$$\overline{\mathrm{VaR}}_\alpha^v = \sup \{ \mathrm{VaR}_\alpha(S) \colon X \in A_1(F), \ \mathrm{Var}(S) \leq v^2 \},$$

where $v \geq 0$ is an admissible variance constraint. In a similar way, we consider $\underline{\mathrm{VaR}}_\alpha^v$ and obtain that

$$\overline{\mathrm{VaR}}_\alpha^v \leq \min \left(\mu + v \sqrt{\frac{\alpha}{1-\alpha}}, \ \mathrm{TVaR}_\alpha(S^c) \right) \qquad (9.27)$$

and

$$\underline{\mathrm{VaR}}_\alpha^v \geq \max \left(\mu - v \sqrt{\frac{1-\alpha}{\alpha}}, \ \mathrm{LTVaR}_\alpha(S^c) \right). \qquad (9.28)$$

$\diamond$

9.4.4   Numerical Methods for Calculating VaR Bounds

Evaluating the sharp bounds $\overline{\mathrm{VaR}}_\alpha^f$ and $\underline{\mathrm{VaR}}_\alpha^f$ is not straightforward. However, the results developed in Sections 9.1–9.4 make it possible to propose numerical methods to approximate the risk bounds. We explain these approximations for the case of the upper bounds.

1) Asymptotic Bounds:
We approximate $\overline{\mathrm{VaR}}_\alpha^f$ from above by $\mathrm{VaR}_\alpha(T_Z^+)$; see Proposition 9.17. Observe that $T_z^+ = \mathrm{TVaR}_V(S_z^c) = \sum_{i=1}^n \mathrm{TVaR}_V(X_{i|z})$, where $V \sim U(0,1)$ is a standard uniformly distributed random variable that is taken independent of Z. Hence, the computation of $\mathrm{VaR}_\alpha(T_Z^+)$ can be performed in a straightforward way using Monte Carlo simulations.

2) Repeated RA:
The second method is to use a discrete approximation for G. We consider the following steps:

a) Define $\beta_j := \frac{j-1/2}{m} \in [0,1]$, $j = 1, 2, \ldots, m$, where m is a large integer.
b) Use the RA to determine, for $z_i \in D$ and $\beta_j \in [0,1]$, the approximations for
$$\overline{q}_{z_i}(\beta_j) = \overline{\mathrm{VaR}}_{\beta_j}(S_{z_i}).$$
c) From the $\overline{q}_{z_i}(\beta)$ one obtains by Proposition 9.6 an approximation for $\mathrm{VaR}_\alpha(\overline{q}_Z(V)) = \overline{\mathrm{VaR}}_\alpha^f$, where V is a $U(0,1)$-distributed random variable that is taken independent of Z.

3) Analytical Method:
For some particular classes of distributions, M_z^f or $\overline{q}_z(\beta)$ may be available analytically as, for instance, in Example 9.21. In these cases, analytically calculating M^f and $\overline{\mathrm{VaR}}^f$ may be possible via Propositions 9.1 and 9.6. However, this procedure involves taking inverses and mixtures of functions repeatedly, and there is no guarantee that the resulting functions still have an explicit form, which slows down the procedure as a numerical inverse needs to be estimated.

Discussion of the Three Approaches Described Above:
Using the asymptotic TVaR-based upper bound allows us to compute bounds in a fast way and does not suffer from the curse of dimensionality. While this method overestimates the true sharp bound, the degree of overestimation is typically small, in particular for large portfolios (Proposition 9.19); see also the numerical evidence provided in Embrechts et al. (2014) and in Bernard et al. (2017c). Hence, we recommend it as a standard method.

The second method is essentially based on an application of the RA to conditional distributions. It is well known that the RA is a very suitable method for approximating numerically sharp bounds, and one can expect this method to provide excellent approximations for sharp bounds. However, a drawback of this method is that a repeated application of the RA is needed, which can make its application time consuming, especially when Z can take many values.

The third method relies heavily on the distributions in the model. In case $n = 2$, M_z^f can be calculated, which involves an inverse of the VaR bounds (see Example 9.21) for each $z \in D$.

9.5   Application to Credit Risk Portfolios

Financial crises have shown that credit portfolios require careful monitoring. In this regard, many financial institutions and regulatory frameworks, such as Basel III and Solvency II, rely on a Bernoulli mixture model to measure the risk. In the industry, this model is also known as the KMV model (Gordy, 2003), and we use this terminology hereafter. Specifically, the risks X_i $(i = 1, \dots, n)$ are modeled as

$$X_i = \begin{cases} 0, & \text{if } \sqrt{r_i}Z + \sqrt{1 - r_i}\varepsilon_i > \Phi^{-1}(q_i), \\ \frac{e^{a_i Z}}{1 + e^{a_i Z}}, & \text{otherwise,} \end{cases}$$

in which $q_i \in (0, 1)$ and Z, ε_i are standard normally distributed and independent. Note that $\text{Cor}(X_i, X_j) = r_i r_j$. Under the KMV specifications, it is further assumed that the idiosyncratic risks ε_i are mutually independent. Under these assumptions, risk measures of $S = \sum_{i=1}^n X_i$, such as VaR, can be computed using Monte Carlo simulations. We challenge the dependence assumptions among the X_i and compute the bounds on VaR correspondingly.

Assuming that only the marginal distributions are known, we assess the quality of the bounds $\underline{\text{VaR}}_\alpha$ and $\overline{\text{VaR}}_\alpha$ using their asymptotic versions $\text{VaR}_\alpha(\text{LTVaR}_U(S^c))$ and $\text{VaR}_\alpha(\text{TVaR}_U(S^c))$. Next, we add dependence information in various ways. First, using the structural factor information, we assess the bounds $\overline{\text{VaR}}_\alpha^f$ and $\underline{\text{VaR}}_\alpha^f$ using their asymptotic versions discussed in Chapter 3; see the expressions (9.24) and (9.25). Second, we add variance information and approximate $\underline{\text{VaR}}_\alpha^{f,v}$ and $\overline{\text{VaR}}_\alpha^{f,v}$ using the bounds established in Proposition 9.22. Finally, we approximate the variance bounds $\underline{\text{VaR}}_\alpha^v$ and $\overline{\text{VaR}}_\alpha^v$ using the expressions (9.27) and (9.28) (assuming that the variance constraint writes as $v_z = \beta\mu_z$).

We study four homogeneous cases (1)–(4) and two heterogeneous cases (5) and (6) and display the results in Table 9.5. Our findings can be summarized as follows. In the unconstrained case, the dependence uncertainty spread is very wide. In particular, the VaR numbers that one obtains by applying the standard KMV model (labeled as $\text{VaR}_\alpha^{\text{KMV}}$) lie far away from the unconstrained upper bounds. Adding factor information improves the lower bound significantly but not the upper bounds. This feature is to be expected, as the factor information induces positive dependence among the risks in that the risks are perfectly dependent, conditionally on the ε_i. By contrast, adding variance information improves the upper but not the lower bound. This feature is also to be expected, as the variance of a credit portfolio loss is driven mainly by high outcomes, and putting a constraint on variance thus implies that upper VaRs become reduced. All in all, using the factor information supplemented with conditional variance information reduces the unconstrained bounds tremendously.

α	$\mathrm{VaR}_\alpha^{\mathrm{KMV}}$	$(\underline{\mathrm{VaR}}_\alpha, \overline{\mathrm{VaR}}_\alpha)$	$(\underline{\mathrm{VaR}}_\alpha^f, \overline{\mathrm{VaR}}_\alpha^f)$	$(\underline{\mathrm{VaR}}_\alpha^v, \overline{\mathrm{VaR}}_\alpha^v)$	$(\underline{\mathrm{VaR}}_\alpha^{f,v}, \overline{\mathrm{VaR}}_\alpha^{f,v})$
(1) 95 %	16.5	(0, 125)	(0, 125)	(4.04, 47.76)	(6.18, 45.58)
99.5 %	29.5	(5.03, 250)	(8.18, 250)	(5.58, 140.6)	(13.89, 126.7)
(2) 95 %	29.5	(0, 125)	(0, 123.8)	(1.77, 91.27)	(7.03, 68.21)
99.5 %	83.5	(5.03, 250)	(29.35, 250)	(5.03, 250)	(32.88, 249.9)
(3) 95 %	32.96	(0, 189.8)	(0, 182.2)	(5.13, 92.25)	(11.94, 85.91)
99.5 %	59	(7.01, 499.7)	(16.74, 499.1)	(8.13, 277.2)	(28.39, 243.5)
(4) 95 %	58.89	(0, 235.2)	(0, 235.6)	(2.77, 182.4)	(13.99, 137.4)
99.5 %	168	(9.36, 500)	(62.6, 499.9)	(9.36, 500)	(68.46, 498)
(5) 95 %	66.92	(0.14, 275.3)	(0.42, 264.4)	(4.23, 197.6)	(17.12, 156.9)
99.5 %	175	(11.53, 484.9)	(75.42, 481.8)	(11.53, 484.9)	(75.77, 480)
(6) 95 %	56.88	(0, 228.9)	(0, 226.4)	(2.46, 182.2)	(13.24, 132.4)
99.5 %	175	(9.10, 499.9)	(68.02, 499.2)	(9.10, 499.9)	(71.4, 472.2)

Table 9.5 VaR bounds for credit risk portfolios.
Homogeneous cases: $(a, n, q, r) = (0, 500, 2.5\,\%, 10\,\%)$ in (1), $(0, 500, 2.5\,\%, 40\,\%)$ in (2), $(-4, 500, 2.5\,\%, 10\,\%)$ in (3), and $(-4, 500, 2.5\,\%, 40\,\%)$ in (4).
Heterogeneous cases: $(a, n, q_i, r_i) = (-4, 500, q_i \in \{0.1\,\%, \ldots, 9.9\,\%\}, r_i = r = 40\,\%)$ in (5), $(-4, 500, 2.5\,\%, r_i \in \{10.6\,\%, \ldots, 69.4\,\%\})$ in (6).

10 Models with a Specified Subgroup Structure

This chapter is concerned with models that are split into k subgroups. Reduction of unconstrained risk bounds and dependence uncertainty are induced by additional information on the dependence within the subgroups and/or on the structure and dependence between the subgroups. Subgroup models as tools for reduction of dependence uncertainty were introduced in Bignozzi et al. (2015). Puccetti et al. (2017) considered an effective class of models and risk reduction of this type with particular relevance to and consequences for hierarchical models, as typically used in insurance models. In Rüschendorf and Witting (2017) the various reduction options for risk estimates of these kinds of models were systematically investigated based on results on dependence orderings.

The basic assumption of this approach is the subgroup structure of the model. It is assumed that the risk set $\{1, \ldots, n\}$ is split into k disjoint subsets I_j, $j = 1, \ldots, k$, i.e.,

$$\{1, \ldots, n\} = \bigcup_{i=1}^{k} I_j, \tag{10.1}$$

and the risk vector X is split into k subgroup vectors $X_{I_j} = (X_\ell)_{\ell \in I_j}$ with corresponding subgroup sums $Z_j = \sum_{\ell \in I_j} X_\ell$. For comparison purposes it is assumed that a comparison vector $Y = (Y_1, \ldots, Y_n)$ is available with the same subgroup structure and with subvectors Y_{I_j} and subgroup sums $W_j = \sum_{\ell \in I_j} Y_l$.

Then in order to estimate the joint portfolio

$$S_n = S_n^X = \sum_{i=1}^{n} X_i = \sum_{j=1}^{k} Z_j = S_k^Z \tag{10.2}$$

by the portfolio

$$S_n^Y = \sum_{i=1}^{n} Y_i = \sum_{j=1}^{k} W_j = S_k^W \tag{10.3}$$

of the comparison vector Y, it will be of interest to compare the dependence of the risk vector X_{I_j} to that of Y_{I_j} within the groups I_j, $1 \le j \le k$ or to compare the dependence between the subgroup sums (Z_j) and (W_j) described by their copulas $C = C_Z$ and $D = C_W$, or to combine these points.

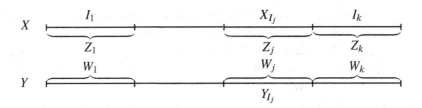

10.1   Subgroup Models with Positive and Negative Internal Dependence

Decomposition of a risk portfolio into independent subgroups is a natural assumption that can be statistically checked. Also, in many applications the subgroups exhibit some kind of strong positive dependence. A class of subgroup models as in the introduction with these properties was introduced in Bignozzi et al. (2015). For two random vectors X, Y, we define

- the **upper orthant order** $Y \leq_{\mathrm{uo}} X$ if $\overline{F}_y(x) \leq \overline{F}_Y(x)$ for all $x \in \mathbb{R}^n$;
- the **lower orthant order** $Y \leq_{\mathrm{lo}} X$ if $F_Y(x) \leq F_X(x)$ for all $x \in \mathbb{R}^n$;
- the **concordance order** $Y \leq_{\mathrm{co}} X$ if $Y \leq_{\mathrm{uo}} X$ and $Y \leq_{\mathrm{lo}} X$.

Bignozzi et al. (2015) considered the **positive internal dependence subgroup property (PISG)**. (X, Y) has the PISG if:

1) X has marginals F_i.

2)
$$Y \leq X, \tag{10.4}$$

where $\leq$ is one of the ordering $\leq_{\mathrm{lo}}$, $\leq_{\mathrm{uo}}$, or $\leq_{\mathrm{co}}$.

3)
$$F_Y(x) = \prod_{j=1}^{k} \min_{i \in I_j} G_j(x_i), \tag{10.5}$$

i.e., Y has comonotonic homogeneous subgroups with distribution functions G_j.

Equation (10.4) postulates a positive dependence restriction on X by a comparison model Y with independent subgroups; the subgroups themselves exhibit comonotonicity and thus a very strong form of positive dependence. In general, condition (10.5) does not imply that the marginals of Y are the same as the marginals of X. But in typical applications of this model, X and Y will have the same marginals, i.e.,

$$F_i = G_j \quad \text{if } i \in I_j. \tag{10.6}$$

Note that the dependence properties of the lower bound model Y get weaker with increasing number k of groups. For $k = 1$ they correspond to the strong comonotonicity condition; for $k = n$ they correspond to the weak positive upper resp. lower resp. concordance dependence (PUOD, PLOD, PCOD).

The assumption of comonotonicity within the subgroups of the comparison vector Y may be too strong for several applications, but it can serve to define worst case situations if one uses a convex risk measure to quantify the risk of the portfolio held. When using value-at-risk, this assumption rules out the worst case VaR distributions attained by negative dependence (see Chapter 1). It turns out that by the PISG assumption, the upper bounds on the VaR of the aggregate position will not be essentially reduced by these positive dependence assumptions. The lower VaR bounds may, however, be essentially improved.

The specific form of comonotonicity within the subgroups allows us to derive formulas for the VaR bounds for $S_n = \sum_{i=1}^{n} X_i$ under the PISG assumption.

Theorem 10.1 (Upper and lower VaR bounds in PISG models)
Let (X, Y) satisfy the PISG assumption. Then:

a) If $Y \leq_{\text{uo}} X$, then for any $\alpha \in (0, 1)$ it holds that

$$\text{VaR}_\alpha(S_n) \geq \sup_{u \in \mathcal{L}_\alpha} \sum_{j=1}^{k} n_j G_j^{-1}(u_j), \tag{10.7}$$

where $\mathcal{L}_\alpha := \{u \in [0, \alpha]^k; \ \prod_{j=1}^{k}(1 - u_j) = 1 - \alpha\}$.
b) If $Y \leq_{\text{lo}} X$, then for any $\alpha \in (0, 1)$ it holds that

$$\text{VaR}_\alpha(S_n) \leq \inf_{u \in \mathcal{U}_\alpha} \sum_{j=1}^{k} n_j G_j^{-1}(u_j), \tag{10.8}$$

where $\mathcal{U}_\alpha := \{u \in [\alpha, 1]^k; \ \prod_{j=1}^{k} u_j = \alpha\}$.
Proof   a) By the standard bounds (see Section 1.2) it holds that

$$P(S_n \geq s) \geq \sup_{x \in \mathcal{U}(s)} \overline{F}_Y(x) = \sup_{x \in \mathcal{U}(s)} \prod_{j=1}^{k} \overline{G}_j\left(\max_{i \in I_j} x_i\right) := B,$$

where $\mathcal{U}(s) = \{x \in \mathbb{R}^n : \sum x_i = s\}$. Defining $\overline{H}_j(x_j) := \overline{G}_j(\frac{x_j}{n_j})$, one finds by straightforward calculations (see Lemma 3.2 in Bignozzi et al. (2015)) that $B = \sup_{y_1 + \cdots + y_k = s} \prod_{j=1}^{k} \overline{H}_j(y_j)$.

This implies by the duality principle (see Theorem 5.10)

$$\text{VaR}_\alpha(S_n) \geq \sup_{\mathcal{L}_\alpha} \sum_{j=1}^{k} H_j^{-1}(u_j) = \sup_{u \in \mathcal{L}'_\alpha} \sum_{j=1}^{k} n_j G_j^{-1}(u_j),$$

where $\mathcal{L}'(\alpha) := \{u \in [0, 1]^k; \ \prod_{j=1}^{k}(1 - u_j) = 1 - \alpha\}$. The lower bound in (10.7) follows, noting that the constraints $\prod_{j=1}^{k}(1 - u_j) = 1 - \alpha$ and $u_j \in [0, 1]$ imply that $u_j \leq \alpha, 1 \leq j \leq k$.
b) The proof of b) is similar.   □

Under additional hypotheses on the marginals G_j, the bounds in Theorem 10.1 can be evaluated in analytical form.

Theorem 10.2 (Analytical VaR bounds under PISG)   *Let* (X, Y) *satisfy the PISG assumption, and assume that* $\Psi_j: [\ln(1-\alpha), 0] \to \mathbb{R}$ *defined by* $\Psi_j(x) = G_j^{-1}(1-e^x)$ *is continuous and convex,* $1 \le j \le k$. *Then*

$$\sup_{u \in \mathcal{L}_\alpha} \sum_{j=1}^{k} n_j G_j^{-1}(u_j) = \max_{1 \le j \le k} \left\{ n_j G_j^{-1}(\alpha) + \sum_{i \ne j} n_i G_i^{-1}(0) \right\}. \tag{10.9}$$

Proof   With the transformation $u_j = 1 - e^{v_j}$ it holds that

$$\sup_{u \in \mathcal{L}_\alpha} \sum_{j=1}^{k} n_j G_j^{-1}(u_j) = \sup_{v \in \mathcal{V}_\alpha} \sum_{j=1}^{k} n_j G_j^{-1}(1 - e^{v_j}),$$

where $\mathcal{V}_\alpha := \{v \in [\ln(1-\alpha), 0]^k; \sum_{j=1}^{k} v_j = \ln(1-\alpha)\}$.

By the assumed convexity of Ψ_j, we have that $\sum_{j=1}^{k} \Psi_j(v_j)$ is a convex function of v in the bounded convex domain $\mathcal{V}_\alpha$ and thus attains its maximum at an extreme point of $\mathcal{V}_\alpha$: thus at an point v with $v_j = \ln(1-\alpha)$ for some j and $v_i = 0$ else. This implies the result in (10.9).   □

Example 10.3 (Pareto marginal distributions)   Assume that all G_j are Pareto distributed with parameter $\vartheta_j > 0$, that is,

$$G_j(x) = 1 - (1+x)^{\vartheta_j}, \quad x > 0, 1 \le j \le k.$$

Then the functions $\Psi_j(x) = G_j^{-1}(1 - e^x) = \exp(-\frac{x}{\vartheta_j}) - 1$ satisfy the assumption of Theorem 10.2, and thus for a PISG model (X, Y) under the assumption $Y \le_{\text{uo}} X$, we get the lower VaR bound

$$\text{VaR}_\alpha(S_n) \ge \max_{1 \le j \le k} \left\{ n_j \left((1-\alpha)^{-\frac{1}{\vartheta_j}} - 1 \right) \right\}. \tag{10.10}$$

Under the assumption that X, Y have the same marginal distributions, i.e., $F_i = G_j$, $i \in I_j$, it is natural to compare the lower bounds in (10.10) called VaR_α^{lb} with the unconstrained lower VaR bound $\underline{\text{VaR}}_\alpha$ based only on the marginal information. In Table 10.1 we provide the comparison in the case of a Pareto(2) model, whereas in Table 10.2 an exponential model is considered. The lower bound VaR_α^{lb} from (10.7)

		VaR_α^{lb}			
α	$\underline{\text{VaR}}_\alpha$	$k = 1$	$k = 2$	$k = 4$	$k = 8$
0.990	9.00	72.00	36.00	18.00	9.00
0.995	13.14	105.14	52.57	26.28	13.14
0.999	30.62	244.98	122.49	61.25	30.62

Table 10.1 Lower VaR bounds in Pareto(2) model with $n = 8$.

and (10.9) improves the unconstrained bound for all $k < n$. In the case $k = n = 8$, i.e., under the weak assumption of PUOD of X, no improvement is attained. The PUOD condition alone is too weak to increase the lower risk bound. This behavior is confirmed in further examples.

Example 10.4 (Exponential marginal distribution)   Let $G_j(x) = 1 - e^{-\vartheta_j x}$, $x > 0$, $1 \le j \le k$ be exponential distributions. Since $G_j^{-1}(1 - e^x) = -\frac{x}{\vartheta_j}$ is linear, Theorem 10.2 implies

$$\text{VaR}_\alpha(S_n) \ge \max_{1 \le j \le k} \left\{ -\frac{n_j}{\vartheta_j} \ln(1 - \alpha) \right\}. \tag{10.11}$$

		VaR_α^{lb}			
α	VaR_α	$k = 1$	$k = 2$	$k = 4$	$k = 8$
0.990	2.30	13.82	9.21	4.61	2.30
0.995	2.65	15.89	10.6	5.30	2.65
0.999	3.45	20.72	13.82	6.91	3.45

Table 10.2 Lower VaR bounds in exponential model with 4 Exp(2) and 4 Exp(4) risks.

Again the lower VaR bound VaR_α^{lb} under the positive dependence condition PISG considerably improves the unconstrained bound for $k < n = 8$, while again PUOD alone, i.e., the case $k = n = 8$, does not improve the lower bounds (see Table 10.2).

Remark 10.5

a) **General solutions.** In the case that the condition of Theorem 10.2 is not satisfied, the supremum in (10.7) might be attained in an inner point of $\mathcal{L}_\alpha$. Assuming positive densities g_i of G_i, first-order conditions for a solution vector u^* of (10.7) then lead to the equations

$$\frac{n_i(1 - u_i^*)}{g_i(G_i^{-1}(u_i^*))} = \frac{n_k(1 - \alpha)}{\prod(1 - u_j^*)g_k(G_k^{-1}(1 - \frac{(1-\alpha)}{\prod(1-u_j^*)}))} \quad \text{if } u_i^* > 0. \tag{10.12}$$

In the simplified case $G_j = G$ and $n_j = \frac{n}{k}$, $1 \le j \le k$, the set of equations becomes

$$\frac{1 - u_i^*}{g(G^{-1}(u_i^*))} = \frac{1 - \alpha}{\prod(1 - u_j^*)g(G^{-1}(1 - \frac{1-\alpha}{\prod(1-u_j^*)}))} \quad \text{if } u_i^* > 0. \tag{10.13}$$

The equations in (10.13) are solved by

$$u_i^* = 1 - (1 - \alpha)^{\frac{1}{c}}, \quad \text{where } c = |\{i : u_i^* > 0\}|.$$

In the case that the density of G is monotone in $[G^{-1}(0), G^{-1}(\alpha)]$, the solution is unique. For this homogeneous framework, the lower bound for VaR becomes

$$\text{VaR}_\alpha(S_n) \ge \max_{0 \le j \le k-1} \left\{ \frac{n(k - j)}{k} G^{-1}\left(1 - (1 - \alpha)\frac{1}{k - j}\right) + \frac{nj}{k} G^{-1}(0) \right\}. \tag{10.14}$$

To obtain this solution, one considers the case of possible values of c and takes the corresponding maximum.

b) **Upper bound with positive dependence constraints.** Similar first order conditions as in a) can also be given to determine the upper bounds in (10.8) (see Bignozzi et al., 2015 Section 3.2) in the simplified homogeneous framework as in (10.14).

Assuming that the density g of G is monotone in $[G^{-1}(\alpha), G^{-1}(1)]$, one obtains the upper VaR bounds Var_α^{ub}:

$$\text{VaR}_\alpha(S_n) \leq \text{VaR}_\alpha^{ub} := \min_{1 \leq j \leq k} \left\{ \frac{nj}{k} G^{-1}\left(\alpha^{\frac{1}{j}}\right) + \frac{n(k-j)}{k} G^{-1}(1) \right\}. \qquad (10.15)$$

The bounds VaR_α^{ub} in (10.8) resp. (10.15) hold under the assumption that $Y \leq_{\text{lo}} X$. As expected the examples in Bignozzi et al. (2015) show that the positive dependence information PISG only improves the upper bound in the cases $k = 1, 2$.

c) **Simplified lower and upper bounds.** If Y has a set of marginal distributions that do not satisfy the assumptions of Theorem 10.2, we can use simplified lower bounds under the assumption $Y \leq_{\text{uo}} X$ as a consequence of (10.9):

$$\text{VaR}_\alpha(S_n) \geq \text{VaR}_\alpha^{ub} := \max_{1 \leq j \leq k} \left\{ n_j G_j^{-1}(\alpha) + \sum_{i \neq j} n_k G_i^{-1}(0) \right\}; \qquad (10.16)$$

and also under the assumption $Y \leq_{\text{lo}} X$, we can use

$$\text{VaR}_\alpha(S_n) \leq \sum_{j=1}^{k} n_j G_j^{-1}(\alpha^{\frac{1}{k}}) := \text{VaR}_\alpha^{ub}, \qquad (10.17)$$

since $(\alpha^{\frac{1}{k}}, \ldots, \alpha^{\frac{1}{k}})$ is an admissible vector in (10.8). Both bounds hold if $Y \leq_{\text{co}} X$ is assumed, implying also that X and Y have the same marginals. Table 10.3 shows the results for the case $k = 2$. For larger values of k, the upper bounds do not

α	VaR_α	VaR_α^{lb}	VaR_α^{ub}	$\overline{\text{VaR}}_\alpha$
0.990	9.00	36.00	73.68	89.05
0.995	13.14	52.57	99.91	120.58
0.999	30.62	122.49	205.27	248.24

Table 10.3 Simplified VaR bounds for $n = 8$ portfolio with $k = 2$ subgroups of 4 Pareto (2) and 4 Exp(1) risks.

improve any more. Table 10.4 shows improvements of the lower bounds for an inhomogeneous portfolio with $n = 8$, $k = 4$, and subgroups of two LN(0, 1), LN(1, 2), Pareto(2) and Pareto(3) risks.   ◊

α	VaR_α	VaR_α^{lb}	$\overline{\text{VaR}}_\alpha$
0.990	285.1	570.1	197.9
0.995	469.5	939.0	3 338.2
0.999	1 313.5	2 627.0	11 119.2

Table 10.4 Improved lower bounds for inhomogeneous portfolio $n = 8$, $k = 4$.

Reducing Dependency Spreads of Convex Risk Measures

As shown in the previous part of this section, the upper resp. lower positive orthant comparison of the risk vector X with a comparison vector Y is not strong enough to imply a reduction of the upper resp. lower risk bounds. Therefore, we consider a stronger positive dependence comparison.

For two random vectors X, Y, define the **weakly conditional in sequence order** **(WCS)** $Y \leq_{\text{wcs}} X$ if for all $i \leq n - 1$ and $x \in \mathbb{R}$,

$$\text{Cov}(1_{(Y_i > x)}, f(Y_{i+1}, \ldots, Y_n)) \leq (1_{(X_i > x)}, f(X_{i+1}, \ldots, X_n)) \tag{10.18}$$

for all increasing functions f.

In the case that Y has independent components and the same marginals as X, condition (10.18) is equivalent to

$$F_{X_{(i+1)}} \leq_{\text{st}} F_{X_{(i+1)}|X_i > x} \quad \text{for all } i \leq n - 1, x \in \mathbb{R}, \tag{10.19}$$

where $X_{(i+1)} = (X_{i+1}, \ldots, X_n)$. In this case, X is called **weakly associated in sequence (WAS)** and is very intuitive to understand. It is known that $\leq_{\text{wcs}}$ implies the supermodular ordering $\leq_{\text{sm}}$ in the case of identical marginals. The WCS-ordering can be easily checked in several functional models (see Rüschendorf, 2004).

We also make use of the **convex order** $\leq_{\text{cx}}$ between random variables U, V defined by

$$U \leq_{\text{cx}} V \quad \text{if } Ef(U) \leq Ef(V) \tag{10.20}$$

for all convex functions f such that the expectations exist.

Next, let X be a risk vector with marginals $F_1, \ldots, F_n$, and let Y be a comparison vector with the same marginals F_i as X such that

$$Y \leq_{\text{wcs}} X. \tag{10.21}$$

Equation (10.21) is a positive dependence condition on X. It implies the following convex ordering of the joint portfolio $S_n^Y = \sum_{i=1}^n Y_i$ and $S_n = S_n^X = \sum_{i=1}^n X_i$:

$$S_n^Y \leq_{\text{cx}} S_n \leq_{\text{cx}} S_n^c, \tag{10.22}$$

where $S_n^c = \sum_{i=1}^n X_i^c$ is the sum of the comonotonic vector $X^c = (X_1^c, \ldots, X_n^c)$ (see Theorem 2.1 in Rüschendorf (2004) and Meilijson and Nadas (1979)). Since convex law-invariant risk measures are consistent with respect to convex order, this implies the following result.

Theorem 10.6 (Lower bound for convex risk measures)   *Let ϱ be a law-invariant convex risk measure, and assume that X, Y are random vectors with the same marginals. Then $Y \leq_{\text{wcs}} X$ implies that*

$$\varrho(S_n^Y) \leq \varrho(S_n) \leq \varrho(S_n^c). \tag{10.23}$$

Note that the upper bound $\varrho(S_n^c)$ is valid without the positive dependence assumption $Y \leq_{\text{wcs}} X$; i.e., generally for a convex law-invariant risk measure ϱ, it holds that

$$\overline{\varrho}(S_n) := \sup\{\varrho(S_n^X); X_j \sim F_j, j \leq n\} = \varrho(S_n^c). \tag{10.24}$$

The ordering assumption $Y \leq_{\text{wcs}} X$ only influences the lower bound $\varrho(S_n^Y)$ of $\varrho(S_n)$.

In the following example we assume that (X, Y) is a PISG model with $F_i = G_j$ for $i \in I_j$. We consider Gamma-distributed risks, which allow us to calculate the bounds in (10.23) in explicit form. The unconstrained lower bounds $\underline{\varrho}(X)$ can be calculated by a version of the rearrangement algorithm, while $\varrho(S_n^Y)$ can be computed by Monte Carlo simulation.

Example 10.7 (TVaR bounds) We consider a PISG model with $G_j = \text{Gamma}(a_j, s) = F_{a_j, s}$, i.e., with density $f_{a_j, s}(x) = \frac{1}{s^{a_j}\Gamma(a_j)} x^{a_j - 1} e^{-\frac{x}{s}}$, $1 \leq j \leq k$, and $F_i = G_j$ for $i \in I_j$. Then with $n_j = |I_j| = \frac{n}{k}$, $W_j = \sum_{i \in I_j} Y_i$ are $F_{a_j, n_j s}$ distributed.

The sum $S_n^Y = S_k^W = \sum_{j=1}^{k} W_j$ has a Gamma distribution $F_{\sum_{j=1}^k a_j, \frac{ns}{k}}$. Its TVaR value has the explicit form

$$\text{TVaR}_\alpha(S_n^Y) = \frac{ns}{k(1 - \alpha)} \frac{\Gamma(1 + \sum_{j=1}^{k} a_j)}{\Gamma(\sum_{j=1}^{k} a_j)} \overline{F}_{1 + \sum_{j=1}^k a_j, \frac{ns}{k}} (F_{\sum_{j=1}^k a_j, \frac{ns}{k}}^{-1}(\alpha)). \quad (10.25)$$

The unconstrained lower bound $\underline{\text{TVaR}}_\alpha(S_n)$ is very well approximated by the mean $\mu = ES_n = ES_n^Y$,

$$\underline{\text{TVaR}}_\alpha(S_n) \approx \mu, \quad (10.26)$$

since it can be checked that the Gamma distributions are approximatively mixable on the intervals $[0, F_j^{-1}(1 - \alpha)]$ for $\alpha \leq 1 - 10^{-7}$.

The upper comonotonic bound from (10.23) is given by

$$\overline{\text{TVaR}}_\alpha(S_n^c) = \text{TVaR}_\alpha(S_n^c) = \sum_{i=1}^{n} \text{TVaR}_\alpha(Y_i)$$

$$= \frac{ns}{k(1 - \alpha)} \sum_{j=1}^{k} \frac{\Gamma(a_j + 1)}{\Gamma(a_j)} \overline{F}_{a_j + 1, s}(F_{a_j, s}^{-1}(\alpha)). \quad (10.27)$$

Table 10.5 gives the bounds $\underline{\text{TVaR}}_\alpha(S_n)$, $\text{TVaR}_\alpha^{lb} = \text{TVaR}_\alpha(S_n^Y)$, and $\overline{\text{TVaR}}_\alpha(S_n) = \text{TVaR}_\alpha(S_n^c)$ for an inhomogeneous portfolio with $n = 8$, k subgroups, and with four $\text{Gamma}(2, \frac{1}{2})$, four $\text{Gamma}(4, \frac{1}{2})$, and $n_j = \frac{n}{k}$. In this case $\mu = 12$.

α	$\underline{\text{TVaR}}_\alpha$	$\overline{\text{TVaR}}_\alpha$	TVaR_α^{lb}		
			$k = 2$	$k = 4$	$k = 8$
0.990	12.00	38.27	29.15	23.29	19.56
0.995	12.00	41.64	31.15	24.52	20.33
0.999	12.00	49.27	35.63	27.21	22.02

Table 10.5 TVaR bounds with subgroup structure.

As a result we find that, as expected, the improvement of the lower TVaR bounds TVaR_α^{lb} decreases with increasing number of subgroups k. Note that the upper TVaR bound $\overline{\text{TVaR}}_\alpha$ coincides with TVaR_α^{lb} in the case $k = 1$. But due to the stronger dependence ordering assumption by $\leq_{\mathrm{wcs}}$ in comparison to the related results on VaR using the weaker orthant ordering assumption, even in the case $k = n$, the improvement of the dependence uncertainty is approximatively 28 %. $\diamond$

To reduce the upper bounds of a convex risk measure $\varrho(X)$, one can consider – similarly to in Theorem 10.6 – a negative dependence assumption of the form

$$X \leq_{\mathrm{wcs}} Y. \tag{10.28}$$

A model (X, Y) with Y satisfying (10.5) and the negative dependence condition (10.28) we call a **negative internal dependence subgroup model** (NISG). In this kind of model, (10.28) does not pose a restriction on the dependence structure within the subgroups I_j but affects the joint dependence between the subgroup sums to be more negative dependent compared to the independent groups. The following comparison result is analogous to Theorem 10.6 and does not need this special NISG structure.

Theorem 10.8   (Risk bound with a negative dependence assumption) *Let ϱ be a law-invariant convex risk measure, and let X, Y be random vectors with the same marginals. Then $X \leq_{\mathrm{wcs}} Y$ implies that*

$$\varrho(S_n) \leq \varrho(S_n^Y) \leq \varrho(S_n^c). \tag{10.29}$$

Theorem 10.8 implies an improvement of the upper risk bound $\varrho(S_n^c)$. The assumption $X \leq_{\mathrm{wcs}} Y$ may be realistic in hierarchical order insurance models, where some branches (groups) of insurance companies are approximatively independent or possibly negative dependent. For example, bad weather insurance (against hail, heavy rain, storm, etc.) could be supposed realistically to be negatively dependent to hot weather insurances (against drought, fire, thunderstorm, etc.).

As a corollary, Theorem 10.8 also implies upper VaR bounds.

Corollary 10.9   *Let X, Y be random vectors with the same marginals such that $X \leq_{\mathrm{wcs}} Y$. Then for $\alpha \in (0, 1)$ it holds that*

$$\mathrm{VaR}_\alpha(S_n) \leq \mathrm{TVaR}_\alpha(S_n^Y). \tag{10.30}$$

A more detailed study of the application of dependence orderings in subgroup models will be given in Section 10.3.

10.2   Models with Independent Subgroups

A practically relevant and effective use of subgroup models concerns the case that the risk vector itself is split into k independent subgroups I_i. Let $\bigcup_{i=1}^{k} I_i = \{1, \ldots, n\}$ be a partition with $|I_i| = n_i$; then the basic assumption of this section is the following:

Assumption I) (Independent subgroup hypothesis) The risk vector $X = (X_1, \ldots, X_n)$ has marginals $F_1, \ldots, F_n$, and the risk subvectors X_{I_i}, $1 \leq i \leq k$ are independent.

Under Assumption I), the subgroup sums $Z_i = \sum_{j \in I_i} X_j$ are independent, and the aggregate risk position is thus given by the independent sum of the Z_i, i.e.,

$$S_n = \sum_{i=1}^{k} \sum_{j \in I_i} X_j = \sum_{i=1}^{k} Z_i. \tag{10.31}$$

Our aim is to investigate how much the unconstrained VaR bounds $\underline{\text{VaR}}_\alpha$, $\overline{\text{VaR}}_\alpha$ are improved by the independent subgroup Assumption I).

Note that in this section we do not assume a comonotonic structure within the subgroups, as in Section 10.1, but allow any dependence within the subgroups I_i.

We define $\underline{\text{VaR}}_\alpha^I$ and $\overline{\text{VaR}}_\alpha^I$ to be the minimal and maximal value for $\text{VaR}_\alpha(S_n)$ under Assumption I). Some simple bounds for these quantities are obtainable as in Section 10.1 (Theorems 10.6 and 10.8; Corollary 10.9) from a connection with convex ordering.

A classical result in Levy and Kroll (1978) implies that the convex order is equivalent to the ordering of TVaR_α, $\alpha \in [0, 1]$, i.e., for real random variables X, Y it holds that

$$X \leq_{\text{cx}} Y \quad \text{if and only if } \text{TVaR}_\alpha(X) \leq \text{TVaR}_\alpha(Y) \quad \text{for all } \alpha \in (0, 1). \tag{10.32}$$

Since $X \leq_{\text{cx}} Y$ implies that X and Y have the same mean, this equivalence implies further that

$$X \leq_{\text{cx}} Y \quad \text{if and only if } \text{LTVaR}_\alpha(X) \leq \text{LTVaR}_\alpha(Y) \quad \text{for all } \alpha \in (0, 1). \tag{10.33}$$

This connection leads to the following simple bounds.

Theorem 10.10 (VaR bounds with independent subgroups)
Let a risk vector X satisfy Assumption I). Let $U_1, \ldots, U_k$ be k independent $U(0, 1)$-distributed random variables, and define

$$S^{c,k} := \sum_{i=1}^{k} Z_i^c \quad \text{with} \quad Z_i^c := \sum_{j \in I_i} F_j^{-1}(U_i). \tag{10.34}$$

Then for any $\alpha \in (0, 1)$ it holds that

$$a^I := \text{LTVaR}_\alpha(S^{c,k}) \leq \text{VaR}_\alpha(S_n) \leq \text{TVaR}_\alpha(S^{c,k}) =: b^I. \tag{10.35}$$

Proof The comonotonic dependence is, by Meilijson and Nadas (1979), the worst case dependence in convex order, and thus

$$Z_i = \sum_{j \in I_i} X_j \leq_{\text{cx}} Z_i^c = \sum_{j \in I_i} F_j^{-1}(U_i).$$

Thus by (10.32), $\text{TVaR}_\alpha(Z_i) \leq \text{TVaR}_\alpha(Z_i^c)$. Since convolution preserves convex order, we obtain by the independent subgroup Assumption I) that

$$\sum_{i=1}^{k} Z_i \leq_{\text{cx}} \sum_{i=1}^{k} Z_i^c = S^{c,k},$$

and thus

$$\mathrm{VaR}_\alpha(S_n) \le \mathrm{TVaR}_\alpha(\sum_{i=1}^n Z_i) \le \mathrm{TVaR}_\alpha(S^{c,k}) = b^I.$$

The lower inequality $a^I \le \mathrm{VaR}_\alpha(S_n)$ follows similarly with (10.33) or from the relation

$$(1 - \alpha)\,\mathrm{TVaR}_\alpha(Y) + \alpha\,\mathrm{LTVaR}_\alpha(Y) = EY \tag{10.36}$$

being valid for any integrable random variable Y.   □

Remark 10.11

a) The bounds $a^I = \mathrm{LTVaR}_\alpha(\sum_{i=1}^k Z_i^c)$ and $b^I = \mathrm{TVaR}_\alpha(\sum_{i=1}^k Z_i^c)$ are given by the LTVaR resp. TVaR of independent sums and thus can be obtained from Monte Carlo simulation. In some specific cases they also can be obtained in closed form.

b) Under Assumption I), the bounds a^I, b^I are not the best possible bounds for $\mathrm{VaR}_\alpha(S_n)$ in general. However, they are the best bounds under this independence hypothesis, when TVaR is used as the reference risk measure.

c) Without dependence information, the TVaR bounds for $\mathrm{VaR}_\alpha(S_n)$ were given in Section 1.2. Let $S_n^c = \sum_{j=1}^n F_j^{-1}(U)$ with $U \sim U(0,1)$ be the comonotonic sum; then under Assumption I) we have

$$A = \mathrm{LTVaR}_\alpha(S_n^c) \le a^I = \mathrm{LTVaR}_\alpha(S_n^{c,k}) \le \underline{\mathrm{VaR}}_\alpha$$

$$\le \underline{\mathrm{VaR}}_\alpha^I \le \overline{\mathrm{VaR}}_\alpha^I \le \overline{\mathrm{VaR}}_\alpha \le b^I = \mathrm{TVaR}_\alpha(S_n^{c,k}) \tag{10.37}$$

$$\le B = \mathrm{TVaR}_\alpha(S_n^c).$$

The bounds A, B without independence assumption are easy to calculate:

$$A = \sum_{j=1}^n \mathrm{LTVaR}_\alpha(X_j), \quad B = \sum_{j=1}^n \mathrm{TVaR}_\alpha(X_j).$$

Note that under weak assumptions as entailed in Chapter 4, the unconstrained bounds are asymptotically sharp for the unconstrained VaRs. For comparison purposes, in the following examples we also list the comonotonic VaR,

$$\mathrm{VaR}_\alpha(S_n^c) = \sum_{i=1}^n \mathrm{VaR}_\alpha(X_i), \tag{10.38}$$

lying within the range $[\underline{\mathrm{VaR}}_\alpha, \overline{\mathrm{VaR}}_\alpha]$ but typically being strictly smaller than the upper VaR bound $\overline{\mathrm{VaR}}_\alpha$.   ◊

Example 10.12 (Gamma model)   Take as in Example 10.7 a portfolio of Gamma-distributed risks with $F_j = G_{a_i,s}$, $j \in I_i$, $1 \le i \le k$ for some positive a_i. Then the comonotonic sum $Z_i^c = \sum_{j \in I_i} F_j^{-1}(U_i)$ has distribution $G_{a_i,n_i s}$, $1 \le i \le k$, with U_i independent $U(0,1)$ variables. Under the independence Assumption I) and assuming subgroups with $n_i = \frac{n}{k}$, $1 \le i \le k$, the independent sum variable $S^{c,k}$ has a Gamma distribution $G_{\sum_{i=1}^k a_i, \frac{ns}{k}}$. Then $b^I = \mathrm{TVaR}_\alpha(S^{c,k})$ is given in formula (10.25). In Table 10.6 the case $n = 8$ with four Gamma$(2, \frac{1}{2})$ and four Gamma$(4, \frac{1}{2})$ risks is considered for k independent subgroups, $k = 1, 2, 4$.

α	$\mathrm{VaR}_\alpha(S_n^c)$	$\overline{\mathrm{VaR}}_\alpha$	b^I $k = 1$	$k = 2$	$k = 4$
0.990	33.37	38.26	38.27	29.15	23.29
0.995	36.82	41.63	41.63	31.15	24.52
0.999	44.59	49.27	49.27	35.63	27.21

Table 10.6 Gamma-distributed risks with independent subgroups, independence bounds b^I for $n = 8$, four Gamma$(2, \frac{1}{2})$, four Gamma$(4, \frac{1}{2})$ risks.

As a result one finds improvements of the worst VaR estimate $\overline{\mathrm{VaR}}_\alpha$ in the range of 23 % to 56 %. The improvement increases with the number of subgroups k. For $k = 1$, there is no essential improvement since by the asymptotic equivalence result in Chapter 4,

$$\overline{\mathrm{VaR}}_\alpha \underset{n\to\infty}{\simeq} b^I = \mathrm{TVaR}_\alpha(S_n^c). \tag{10.39}$$

The improvements of the lower bound by the independence hypothesis are consistent with the results in Section 10.1 of minor order and are not reported in the table.

Example 10.13 (Normal marginal distributions)   We consider k independent risk subgroups with $F_j = N(\mu_i, \sigma_i^2)$, $\mu_i \in \mathbb{R}$, $\sigma_i^2 > 0$, $j \in I_i$, $1 \leq i \leq k$. For a normal variable $Y \sim N(\mu, \sigma^2)$ it holds that

$$\mathrm{TVaR}_\alpha(Y) = \mu + \sigma \frac{\varphi(\Phi^{-1}(\alpha))}{1 - \alpha}, \quad \mathrm{LTVaR}_\alpha(Y) = \mu - \sigma \frac{\varphi(\Phi^{-1}(\alpha))}{1 - \alpha}, \tag{10.40}$$

where φ, Φ are the standard normal density and distribution functions. Under the independence hypothesis I) it holds that

$$S^{c,k} \sim N\left(\sum_{i=1}^n n_i \mu_i, \sum_{i=1}^k n_i^2 \sigma_i^2 \right), \tag{10.41}$$

and hence the independence bounds a^I, b^I are given in closed form:

$$
\begin{aligned}
a^I &= \sum_{i=1}^k n_i \mu_i - \sqrt{\sum_{i=1}^k n_i^2 \sigma_i^2 \, \frac{\varphi(\Phi^{-1}(\alpha))}{\alpha}}, \\[2mm]
b^I &= \sum_{i=1}^k n_i \mu_i + \sqrt{\sum_{i=1}^k n_i^2 \sigma_i^2 \, \frac{\varphi(\Phi^{-1}(\alpha))}{1 - \alpha}}.
\end{aligned}
\tag{10.42}
$$

Table 10.7 considers the case of homogeneous subgroups with $\mu_i = 0$, $\sigma_i^2 = 1$ and $n_i = \frac{n}{k}$, $1 \leq i \leq k$ to study the impact of the number of independent subgroups k on a^I, b^I.

The results show a considerable reduction of the DU-spread in the range from about 0.3 for $k = 2$ to 0.8 for $k = 25$. The magnitude of the reduction is mainly due to the

		(a^I, b^I)			
α	$(\underline{VaR}_\alpha, \overline{VaR}_\alpha)$	$k = 1$	$k = 2$	$k = 10$	$k = 25$
0.950	$(-5.41; 103.13)$	$(-5.43; 103.14)$	$(-3.84;\ 72.93)$	$(-1.72; 32.61)$	$(-1.09; 20.63)$
0.990	$(-1.33; 133.25)$	$(-1.35; 133.26)$	$(-0.95;\ 94.23)$	$(-0.43; 42.14)$	$(-0.27; 26.65)$
0.999	$(-0.73; 144.59)$	$(-0.73; 144.60)$	$(-0.51; 102.25)$	$(-0.23; 45.73)$	$(-0.15; 28.92)$

Table 10.7 Normally distributed risks with independent subgroups.

improvement of the upper bound b^I. For $k = 1$, the bounds a^I, b^I only slightly depart from the best possible unconstrained VaR bounds $\underline{VaR}_\alpha$, $\overline{VaR}_\alpha$. ◇

Example 10.14 (Pareto-distributed risks with independent subgroups) Let F_j be Pareto(ϑ_i) for $j \in I_i$ with tail index $\vartheta_i > 0$, i.e., $F_j(x) = 1 - \frac{1}{(1+x)^{\vartheta_i}}$, $x > 0$.

In this case the independence bounds a^I, b^I are computed by Monte Carlo simulation. For a portfolio with homogeneous subgroups with $n = 50$, $\vartheta_i = \vartheta = 3$, $n_i = \frac{n}{k}$. Table 10.8 shows considerable reductions of the DU spread, which are mainly achieved by the upper bound b^I.

(a^I, b^I)	$k = 1$	$k = 2$	$k = 10$	$k = 25$	$(\underline{VaR}_\alpha, \overline{VaR}_\alpha)$
0.950	$(18.23; 153.56)$	$(20.21; 116.10)$	$(22.95;\ 63.88)$	$(23.76; 48.56)$	$(18.24; 153.30)$
0.990	$(22.24; 298.17)$	$(23.14; 208.76)$	$(24.28;\ 95.97)$	$(24.59; 65.91)$	$(22.25; 297.64)$
0.999	$(23.18; 388.69)$	$(23.79; 265.83)$	$(24.55; 115.34)$	$(24.74; 76.23)$	$(23.19; 388.00)$

Table 10.8 Pareto-distributed risks with independent subgroups, independent homogeneous Pareto subgroups, $n = 50$, $\vartheta_i = \vartheta = 3$.

Table 10.9 displays an inhomogeneous portfolio with $\vartheta_i = i + 1$, $n_i = \frac{n}{k}$. Compared to the homogeneous case, the total reductions are smaller but still considerable. This is due to the fact that the total aggregate loss is dominated by the groups with lower tail index, and therefore, the influence of the independent subgroup is reduced.

(a^I, b^I)	$k = 1$	$k = 2$	$k = 10$	$k = 25$	$k = 50$
$d = 10$	$(6.34;\ 79.47)$	$(5.33;\ 44.95)$			
$d = 50$	$(31.72;\ 297.46)$	$(27.65;\ 224.73)$	$(12.62;\ 53.02)$	$(6.82;\ 23.02)$	
$d = 100$	$(63.45;\ 793.87)$	$(55.29;\ 449.45)$	$(25.25; 106.08)$	$(13.64;\ 46.03)$	$(8.19; 24.39)$
$d = 250$	$(158.63; 1985.38)$	$(138.24; 1123.47)$	$(63.12; 265.15)$	$(34.11; 115.12)$	$(20.47; 60.98)$

$(\underline{VaR}_\alpha; \overline{VaR}_\alpha)$	$k = 1$	$k = 2$	$k = 10$	$k = 25$	$k = 50$
$d = 10$	$(6.35;\ 74.83)$	$(5.00;\ 51.60)$			
$d = 50$	$(31.73;\ 392.17)$	$(24.99;\ 271.65)$	$(10.67;\ 87.85)$	$(5.74\ 41.68)$	
$d = 100$	$(63.47;\ 787.88)$	$(49.97;\ 546.35)$	$(21.22; 178.01)$	$(11.47;\ 85.28)$	$(6.88;\ 47.47)$
$d = 250$	$(158.67; 1969.79)$	$(124.93; 1368.16)$	$(53.34; 448.42)$	$(28.68; 216.22)$	$(17.19; 121.34)$

Table 10.9 Inhomogeneous Pareto-distributed risks.

It is possible to weaken the independence Assumption I) to partial independent subgroups. Let $\bigcup_{j=1}^{k} I_j = \{1, \ldots, n\}$ be a partition with $|I_j| = n_j$, and let $H \subset \{1, \ldots, k\}$, $|H| = r \leq k$, be a class of r subgroups.

Assumption I^p) (Partial independent subgroups) The risk vector $X = (X_1, \ldots, X_n)$ is assumed to have marginals F_j, and the subgroup risk vectors $\{X_{I_i}, i \in H\}$ are assumed to be independent and to be independent of $(X_{I_i})_{i \in H^c}$. For illustration, see Figure 10.1.

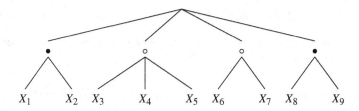

Figure 10.1 Partial independent subgroups with $n = 9$, $k = 4$, and $H = \{1, 4\}$ with subgroup risk vectors $I_1 = \{1, 2\}$, $I_4 = \{8, 9\}$ resp. $I_2 = \{3, 4, 5\}$ and $I_3 = \{6, 7\}$.

This partial independent subgroup assumption might be more realistic in various models, such as hierarchical insurance models. The VaR bounds in Theorem 10.10 with independence information can be extended to the case of partial dependence information and results by similar arguments in the following theorem.

Theorem 10.15 (VaR bounds for partial independent substructures) *Assume that a risk vector X satisfies the partial independence Assumption I^p). Let U_i, $i \in H$ denote r independent $U(0, 1)$-distributed random variables, and define $Z_i^c := \sum_{j \in I_i} F_j^{-1}(U_i)$, $i \in H$.*

Then for $\alpha \in (0, 1)$ it holds that

$$a^p := \sum_{i \notin H} \sum_{j \in I_i} \mathrm{LTVaR}_\alpha(X_j) + \mathrm{LTVaR}_\alpha\left(\sum_{i \in H} Z_i^c\right) \leq \mathrm{VaR}_\alpha(S_n)$$

$$\leq b^p := \mathrm{TVaR}_\alpha\left(\sum_{i \in H} Z_i^c\right) + \sum_{i \notin H} \sum_{j \in I_i} \mathrm{TVaR}_\alpha(X_j).$$

(10.43)

In (10.43), $\sum_{i \in H} Z_i^c$ is an independent sum, and its (L)TVaR can be computed by Monte Carlo simulation. The following example shows the increasing reduction of the bounds in (10.32) with increasing size of the independence set H.

Example 10.16 (Homogeneous partial independent subgroups) Let a risk vector X satisfy the partial independence Assumption I^p) with marginal distributions $F_j = G_i$, $j \in I_i$, $1 \leq i \leq k$. We consider the case of $n = 50$ risk variables split into $k = 5$ subgroups I_j. Table 10.10 shows the partial independence bounds a^p, b^p obtained for varying H and the unconstrained bounds $\underline{\mathrm{VaR}}_\alpha$, $\overline{\mathrm{VaR}}_\alpha$.

The partial independent subgroup can also be combined with the assumption of a bounded variance of S_n as discussed in Chapter 6.

	$\alpha = 0.95$	$\alpha = 0.995$	$\alpha = 0.995$
	$G_i = \Gamma(\alpha_i; 1)$	$G_i = N(\mu_i; 1)$	$G_i = N(0; 1)$
$H = [1, 2, 3, 4, 5]$	(27.58; 76.02)	(149.67; 214.67)	(−0.33; 64.66)
$H = \quad [2, 3, 4, 5]$	(26.83; 90.40)	(149.57; 236.76)	(−0.44; 86.76)
$H = \qquad [3, 4, 5]$	(25.85; 108.70)	(149.47; 257.93)	(−0.55; 107.93)
$H = \qquad\quad [4, 5]$	(24.80; 128.81)	(149.36; 277.66)	(−0.64; 127.66)
$H = \qquad\qquad [5]$	(23.75; 148.66)	(149.28; 294.60)	(−0.73; 144.60)
$(\underline{\text{VaR}}_\alpha; \overline{\text{VaR}}_\alpha)$	(23.76; 148.63)	(149.29; 294.59)	(−0.71; 144.59)

Table 10.10 VaR bounds a^p, b^p for different homogeneous partial independent substructures. VaR bounds, $n = 50$, $k = 5$ for Gamma and normal portfolios, $a_i = \frac{i}{k}$, $\mu_i = i$.

Assumption V) (Bounded variance assumption) Assume that the risk vector $X = (X_1, \ldots, X_n)$ has marginals F_i, $1 \le i \le n$ and, for some $s > 0$, satisfies

$$\text{Var}(S_n) \le s^2. \tag{10.44}$$

Then the following combined VaR bounds can be given.

Theorem 10.17   (VaR bounds with partial independent subgroups and variance bound) *Assume that a risk portfolio X satisfies both assumptions I^p) and V). Then for $\alpha \in (0, 1)$ it holds that*

$$a^{p,V} := \max\left(\mu - s\sqrt{\frac{1-\alpha}{\alpha}}, a^p\right)$$

$$\le \text{VaR}_\alpha(S_n) \le \min\left(\mu + s\sqrt{\frac{\alpha}{1-\alpha}}, a^p\right) =: b^{p,V}. \tag{10.45}$$

The following example shows the typical behavior of the bounds $a^{p,V}$, $b^{p,V}$ in dependence on the variance bounds s^2.

Example 10.18 (Dependence of VaR bounds on s^2)   In this example a risk portfolio of $n = 100$ normal $N(0, 1)$-distributed risk variables is split into $k = 10$ independent subgroups. Table 10.11 shows variance bounds a^V, b^V (see Chapter 6) and compares them with $(a^{p,V}, b^{p,V})$ in (10.45) for various levels of the parameters s^2 and α.

s^2 $\diagdown$ α	20	50	100	200	500
$(a^{p,V}; b^{p,V})$ 0.950	(−1.03; 19.49)	(−1.62; 30.82)	(−2.29; 43.59)	(−3.24; 61.64)	(−3.43; 65.23)
0.990	(−0.45; 44.50)	(−0.71; 70.36)	(−0.85; 84.28)	(−0.85; 84.28)	(−0.86; 84.28)
0.995	(−0.32; 63.09)	(−0.46; 91.45)	(−0.46; 91.45)	(−0.45; 91.45)	(−0.46; 91.45)
$(a^V; b^V)$ 0.950	(−1.03; 19.49)	(−1.62; 30.82)	(−2.29; 43.59)	(−3.24; 61.64)	(−5.13; 97.47)
0.990	(−0.45; 44.50)	(−0.71; 70.36)	(−1.01; 99.50)	(−1.42; 140.71)	(−2.25; 222.49)
0.995	(−0.32; 63.09)	(−0.5; 99.75)	(−0.71; 141.07)	(−1.00; 199.50)	(−1.45; 289.20)

Table 10.11 VaR bounds (a^V, b^V) and $(a^{p,V}, b^{p,V})$, $n = 100$, $k = 10$.

Figure 10.2 plots the bounds in the case $\alpha = 0.990$ as a function of the variance constraint s^2. There are two critical values: $s^*_{p,V}$, s^*_V. Below $s^*_{p,V}$ the variance information dominates; between $s^*_{p,V}$ and s^*_V the independence information dominates. Above s^*_V the variance bounds are improved by the unconstrained marginal bounds.

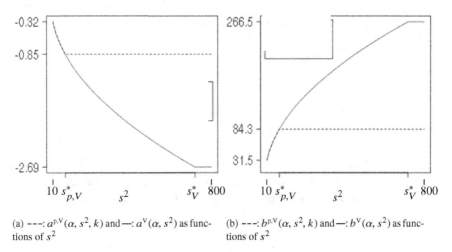

(a) ---: $a^{p,V}(\alpha, s^2, k)$ and —: $a^V(\alpha, s^2)$ as functions of s^2

(b) ---: $b^{p,V}(\alpha, s^2, k)$ and —: $b^V(\alpha, s^2)$ as functions of s^2

Figure 10.2 Variance-constrained versus independence + variance-constrained bounds a^V, $a^{p,V}$ resp. b^V, $b^{p,V}$.

The final example studies a risk portfolio resembling that held by two large European (re-)insurance groups.

Example 10.19 (Application to an insurance portfolio)   First, we consider the portfolio of a large European insurance group who uses it to perform its own risk and solvency assessment (ORSA), as required under Pillar II of the Solvency II regulatory framework. The total portfolio loss can be modeled as a sum of $n = 11$ marginal losses, representing different loss categories that are already aggregated across the different subsidiary companies; see Figure 10.3 for a graphical representation. Marginal losses are divided into four subgroups, which are considered by the insurance company to be independent. These independent subgroups, relying on the different nature of some of the risk types involved, is supported by statistical evidence. Formally, the portfolio satisfies Assumption I) with $d = 11$, $I_1 = \{1, \ldots, 8\}$, $I_2 = \{9\}$, $I_3 = \{10\}$, and $I_4 = \{11\}$. The eight loss distributions F_j, $1 \leq j \leq 8$, in the first group are $N(3, 4)$ distributed. The three remaining loss types have much more pronounced tails. Precisely, we find that $F_9 = LN(1, 1)$, $F_{10} = LN(2, 1)$, $F_{11} = Pareto(3)$. The VaR bounds a^I and b^I defined in (10.35) are given by

$$a^I = \text{LTVaR}_\alpha \left(\sum_{i=1}^k \sum_{j \in I_i} F_j^{-1}(U_i) \right) \quad \text{and} \quad b^I = \text{TVaR}_\alpha \left(\sum_{i=1}^k \sum_{j \in I_i} F_j^{-1}(U_i) \right),$$

where $U_1, \ldots, U_k$ are independent $U(0, 1)$ random variables. Considering the Gaussian nature of the first eight random variables, the bounds simplify to

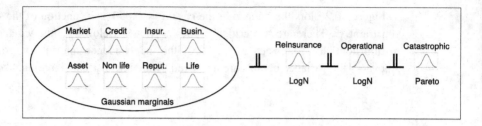

Figure 10.3 Insurance risk portfolio.

$$a^I = \mathrm{LTVaR}_\alpha\left(\sum_{i=1}^{4} G_i^{-1}(U_i)\right) \quad \text{and} \quad b^I = \mathrm{TVaR}_\alpha\left(\sum_{i=1}^{4} G_i^{-1}(U_i)\right), \qquad (10.46)$$

where $G_1 := N(24, 256)$, $G_i = F_{i+7}$, $2 \le i \le 4$. In Table 10.12 we compare the bound b^I, evaluated via Monte Carlo simulation, with the worst VaR estimate $\overline{\mathrm{VaR}}_\alpha$ based on marginals information only. The bound b^I provides a reduction of the bound $\overline{\mathrm{VaR}}_\alpha$ of roughly 30 %.

	b^I	$\mathrm{VaR}_\alpha(S_n^c)$	$\overline{\mathrm{VaR}}_\alpha$	$\bar{e}_\alpha$
$\alpha = 0.990$	147.34 – 148.46 – 149.66	168.37	209.59	29.2 %
$\alpha = 0.995$	173.37 – 175.18 – 176.96	202.89	249.55	29.8 %
$\alpha = 0.999$	250.41 – 256.04 – 262.47	304.63	367.70	30.4 %

Table 10.12 VaR bounds b^I (see (10.46)), $\mathrm{VaR}_\alpha(S_n^c)$, and $\overline{\mathrm{VaR}}_\alpha$ for the 11-dimensional risk portfolio in Figure 10.3. For the VaR bound b^I we give the median estimate and the end points of a 95 % empirical confidence interval that is obtained using 103 repetitions of 106 simulations. The improvement $\bar{e}_\alpha$ is computed with respect to the median estimate.

10.3   VaR Bounds in Subgroup Models with Dependence Information within and between the Subgroups

In this section, subgroup models with k subgroups are considered for the risk vector X as described in the introduction, which is compared to a comparison vector $Y = (Y_1, \ldots, Y_n)$. The subgroup sums are denoted by $Z_i = \sum_{j \in I_i} X_j$ and $W_i = \sum_{j \in I_i} Y_j$. Then $S = S_n = \sum_{i=1}^{k} Z_i$ has to be compared to $T = T_n = \sum_{i=1}^{k} W_i$. We denote the distribution function of Z_i by G_i and that of W_i by H_i.

In comparison to Sections 10.1 and 10.2, it is no longer assumed that the different subgroups are independent or that the subgroup vector Y_{I_i} is comonotonic. The basic comparison results between S and T use dependence conditions inside the groups I_i in order to infer that $G_i \leq H_i$ or $G_i \geq H_i$ with respect to suitable orderings. Furthermore, they use dependence ordering conditions between the groups – more precisely between the copulas $C = C_Z$ for the subgroup sums $Z = (Z_1, \ldots, Z_k)$ and $D = C_W$ for the subgroup sums $W = (W_1, \ldots, W_k)$ of the comparison vector – to conclude the comparison between the aggregated risks S and T. Thus the risk of T serves as a bound for the risk of S, the comparison being based on the assumed dependence information. The results in this section make essential use of Rüschendorf and Witting (2017) and Ansari and Rüschendorf (2021).

10.3.1 Some Results from Stochastic Ordering

In this subsection we recall some results from stochastic ordering that are useful for the ordering of subgroup structure models. Let $\leq_{cx}$, $\leq_{dcx}$, and $\leq_{sm}$ denote the convex, the directionally convex, and the supermodular ordering on the class of random vectors resp. probability measures on $\mathbb{R}^m$, i.e.,

$$X \leq_{cx} Y \quad \text{iff} \quad Ef(X) \leq Ef(Y) \tag{10.47}$$

for all convex functions $f: \mathbb{R}^m \to \mathbb{R}^1$ such that the integrals exist. Similarly, $\leq_{dcx}$ and $\leq_{sm}$ are defined via the class of all directionally convex resp. supermodular functions f. The orderings $\leq_{dcx}$ and $\leq_{sm}$ are positive dependence orders, while $\leq_{cx}$ is an order of diffusiveness. For general properties, we refer the reader to Müller and Stoyan (2002) and Shaked and Shantikumar (2007).

The related WCS-ordering, which induces the positive dependence notion weakly associated in sequence (WAS) in (10.19), was introduced in (10.18). Two further positive dependence notions are given in the following definition.

Definition 10.20 (Conditional increasing in sequence)

a) X is called **conditionally increasing in sequence** (CIS) if $X_i \uparrow_{st} (X_1, \ldots, X_{i-1})$, $2 \leq i \leq m$, i.e., $E(f(X_i) \mid (X_1, \ldots, X_{i-1}) = (x_1, \ldots, x_{i-1}))$ increases in $(x_1, \ldots, x_{i-1})$ for increasing functions f.

b) X is called **conditionally increasing** (CI) if for all $X_i \uparrow_{st} X_J$ for all $J \subset \{1, \ldots, m\} \setminus \{i\}$.

Some basic connections between these dependence and variability orderings are the following (see Rüschendorf, 2004).

Theorem 10.21 (Relations between dependence orderings)

a)
$$CI \Rightarrow CIS \Rightarrow WA \Rightarrow WAS. \tag{10.48}$$

For random vectors X, Y in $\mathbb{R}^m$ it holds that

b)   *If for all i, $X_i \overset{d}{=} Y_i$ and $X \leq_{\mathrm{wcs}} Y$, then $X \leq_{\mathrm{sm}} Y$.* $\tag{10.49}$

c)   *If for all i, $X_i \leq_{\mathrm{cx}} Y_i$ and $X \leq_{\mathrm{wcs}} Y$, then $X \leq_{\mathrm{dcx}} Y$.* $\tag{10.50}$

d) If X and Y have the same CI copula and if $X_i \leq_{\mathrm{cx}} Y_i$, then

$$X \leq_{\mathrm{wcs}} Y. \tag{10.51}$$

e) If $C_X \leq_{\mathrm{sm}} C_Y$ and if C_Y is CI, then

$$X_i \leq_{\mathrm{cx}} Y \text{ for all } i \text{ implies } X \leq_{\mathrm{wcs}} Y. \tag{10.52}$$

Remark 10.22   Müller and Scarsini (2001) established that for X and Y having the same CI copula, the condition

$$X_i \leq_{\mathrm{cx}} Y_i, \quad 1 \leq i \leq m \text{ implies that } X \leq_{\mathrm{dcx}} Y. \tag{10.53}$$

This conclusion also follows from a combination of (10.50) and (10.51).   ◊

In the following examples, various criteria for these orderings are given for the classes of elliptical and Archimedean copulas.

A) Elliptical distributions

Elliptical distributions are generalizations of the multivariate normal and the multivariate t-distributions. They also include multivariate Cauchy, logistic, and generalized multivariate hyperbolic distributions. A random vector $X \in \mathbb{R}^n$ is **elliptically distributed** $X \sim E_n(\mu, \Sigma, \Phi)$ if the characteristic function φ_X of X has a representation of the form

$$\varphi_X(t) = e^{it^\top \mu} \, \Phi(t^\top \Sigma t) \tag{10.54}$$

for some $\Phi \colon \mathbb{R}_+ \to \mathbb{R}$, the characteristic generator. Elliptical distributions are characterized by a stochastic representation of the form

$$X \overset{d}{=} \mu + RAU, \tag{10.55}$$

where $\Sigma = AA^\top$, U is uniformly distributed on the unit sphere S_{n-1}, and the radial part R is independent of U. Σ is called the correlation matrix of X. In the case when $\mu = 0$ and R is absolutely continuous distributed, the density of X is of the form

$$f(x) = |\Sigma|^{-\frac{1}{2}} \Phi(x^\top \Sigma^{-1} x),$$

where Φ is a scale function uniquely determined by the distribution of the radial part R.

The lower orthant comparison between two elliptical distributions with the same generator is characterized by Das Gupta et al. (1972, Theorem 5.1) by the ordering of the correlations, i.e., for $X \sim E_n(\mu, \sigma, \Phi)$, $Y \sim E_n(\mu, \Sigma', \Phi)$ it holds that

$$X \leq_{lo} Y \iff \sigma_{ij} \leq \sigma'_{ij}, \text{ for all } i, j. \qquad (10.56)$$

The comparison result in (10.56) is extendable to the stronger supermodular ordering (independently established in Ansari and Rüschendorf (2021) and Yin (2021)).

Theorem 10.23 ($\leq_{sm}$ ordering of elliptical distributions) *Let $X \sim E_n(\mu, \Sigma, \Phi)$, $Y \sim E_n(\mu', \Sigma', \Phi')$; then the following are equivalent:*
1) $X \leq_{sm} Y$;
2) $\mu = \mu'$, $\Phi = \Phi'$, $\sigma_{ii} = \sigma'_{ii}$ for all i and $\sigma_{ij} \leq \sigma'_{ij}$, for all $i \neq j$;
3) X, Y have the same marginals and $\sigma_{ij} \leq \sigma'_{ij}$, for all i, j.

The result for the $\leq_{sm}$ ordering for fixed generators in 10.23 also holds in similar form for the directionally convex order $\leq_{dcx}$.

Theorem 10.24 ($\leq_{dcx}$ ordering of elliptical distributions with a fixed generator) *Let $X \sim E_n(\mu, \Sigma, \Phi)$, $Y \sim E_n(\mu', \Sigma', \Phi)$ be integrable, then*
1) $\mu = \mu'$ and $\sigma_{ij} \leq \sigma'_{ij}$ for all i, j implies $X \leq_{dcx} Y$.
2) If $X \leq_{dcx} Y$, then $\mu = \mu'$. If additionally X, Y are square-integrable, then $\sigma_{ij} \leq \sigma'_{ij}$ for all i, j.

The comparison in $\leq_{dcx}$ ordering is extended in Ansari and Rüschendorf (2021) to the case when the radial part of Y is stochastically larger than the radial part of X.

To transfer a $\leq_{dcx}$ ordering result as in Theorem 10.24 for vectors X, Y with the same copula to the case of convexly increasing marginals, i.e., $X_i \leq_{cx} Y_i$, it is of interest by Remark 10.22 to establish conditions that imply the CI-property of the joint copula.

In the normal case this question is connected with the notion of an M-matrix.

Recall that a matrix $A = (a_{ij})_{i,j \leq n}$ is called an M-matrix if $a_{ij} \leq 0$, for all $i \neq j$ and all principal minors are positive. For a positive definite matrix A it holds that

$$A \text{ is an } M\text{-matrix} \iff \begin{array}{l} \text{There exists a lower triangular } M\text{-matrix } L \\ \text{such that } A = LL^{\top}. \end{array}$$

The following result due to Rüschendorf (1981a) says that for the multivariate normal distribution, the CI property is equivalent to the inverse of the covariance matrix being an M-matrix and is further equivalent to the MTP_2 property.

Proposition 10.25 *Let $X \sim N(0, \Sigma)$ be normally distributed with positive definite correlation matrix Σ. Then it holds that*

$$\begin{aligned} X \text{ is CI} &\iff \Sigma^{-1} \text{ is an } M\text{-matrix} \\ &\iff X \text{ has the } MTP_2 \text{ property.} \end{aligned} \qquad (10.57)$$

The corresponding characterization of the CI-property of elliptical distributions by the M-matrix property is, however, not true.

A random vector X with distribution P^X having a density f with respect to a sigma-finite product measure $\mu = \mu_1 \otimes \cdots \otimes \mu_n$ is said to be MTP_2 if

$$f(x)f(y) \leq f(x \wedge y)f(x \vee y) \text{ for all } x, y. \tag{10.58}$$

For twice-differentiable f, it is equivalent to

$$\frac{\partial^2}{\partial x_i \partial x_j} \ln f(x) \geq 0 \text{ for all } x \text{ and } i < j. \tag{10.59}$$

This condition can be checked in several examples. The MTP_2 property is a strong positive dependence property and implies the CI property.

In the two-dimensional case, the following criterion for positive likelihood ratio dependence (MTP_2) is known (see Abdous et al., 2005, Proposition 1.2), thus implying a sufficient condition for CI.

Proposition 10.26　(CI-criterion for bivariate elliptical distributions) *Let* $X \sim E_2(\mu, \Sigma_\varrho, \Phi)$ *with* $\Sigma_\varrho = \left(\begin{smallmatrix} 1 & \varrho \\ \varrho & 1 \end{smallmatrix}\right)$ *and with scale function* Φ *such that* $\beta(t) = \log(\Phi(t))$ *is twice-differentiable.*

1) X is positive likelihood ratio-dependent if and only if $\beta''(t) = 0$ whenever $\beta'(t) = 0$ and

$$-\frac{\varrho}{1+\varrho} \leq \inf_{t \in T} \frac{t\beta'(t)}{\beta'(t)} \leq \sup_{t \in T} \frac{t\beta''(t)}{\beta'(t)} \leq \frac{\varrho}{1-\varrho}, \tag{10.60}$$

where $T = \{t \in \mathbb{R}_+; \beta(t) < 0\}$.

2) Under condition (10.60), X is conditionally increasing.

Several classes of examples fulfilling condition (10.60) are given in Abdous et al. (2005).

Example 10.27 For $\Phi(t) \sim e^{-at^\alpha}$, $a, \alpha \in \mathbb{R}_+$, it holds that $\frac{t\beta''(t)}{\beta'(t)} = \alpha - 1$. Thus the CI condition (10.60) can hold only when $\varrho \geq \max\{1 - \frac{1}{\alpha}, \frac{1}{\alpha} - 1\}$ and $\varrho \neq 0$. This can hold only when $\alpha > \frac{1}{2}$. In particular for $\Phi(t) \sim t^{\gamma-1}e^{-at^\alpha}$, $a, \alpha, \gamma \in \mathbb{R}_+$, $\gamma \neq 1$, the corresponding Kotz-type elliptically contoured distribution does not satisfy (10.60) for any $\varrho \in [0, 1)$ since

$$\lim_{t \to 0} \frac{t\beta''(t)}{\beta'(t)} = -1 < -\frac{\varphi}{1+\varphi}.$$

In the normal case where $\varphi(t) \sim e^{-at}$, $a \in R_+$, the following supplement to Proposition 10.26 holds:

For $\varrho = 0$ an elliptical distribution is positive quadrant dependent (PQD) only in the Gaussian case.

The CI result is applied in Ansari and Rüschendorf (2021) to establish a dcx-ordering result for PSFMs with elliptical specifications of the distributions of X_i and the risk factor Z (see Chapter 9) under the assumption that the marginals increase convexly. Finally, the following increasing convex ordering result is given in Pan et al. (2016).

Theorem 10.28 (icx-ordering of elliptical distributions) *Let $X \sim E_n(\mu_1, \Sigma_1, \Phi)$, $Y \sim E_n(\mu_2, \Sigma_2, \Phi)$ be elliptically distributed with common generator Φ and correlation matrices $\Sigma_1 = (\sigma_{ij}^{(1)})$, $\Sigma_2 = (\sigma_{ij}^{(2)})$. Then it holds that if $\mu_1 \leq \mu_2$ and $\Sigma_1 \leq_{\mathrm{psd}} \Sigma_2$, then $X \leq_{\mathrm{icx}} Y$.*

B) Archimedean copulas

A copula is called Archimedean if it has the form

$$C(x_1, \ldots, x_n) = \Psi\left(\sum_{i=1}^{n} \Psi^{-1}(x_i)\right) = C_\Psi(x), \tag{10.61}$$

where $\Psi: \mathbb{R}_+ \to [0, 1]$. $\Psi \in C^n(\mathbb{R}_+)$ is called n-alternating if $(-1)^k \Psi^{(k)} \geq 0$ for $k \in \{1, \ldots, n\}$. If $\Psi \in C^n(\mathbb{R}_+)$ is n-alternating and further $\Psi(0) = 1$ and $\lim_{x \to \infty} \Psi(x) = 0$, then Ψ is the generator of an Archimedean copula. In the following we restrict to this subclass of Archimedean copulas.

The following characterization of positive dependence properties is due to Müller and Scarsini (2001, 2005).

Theorem 10.29 (Positive dependence properties of Archimedean copulas)
a) *An Archimedean copula C_Ψ is positive lower orthant dependent (PLOD) if and only if*

$$\Psi \circ \exp \text{ is superadditive.} \tag{10.62}$$

b) *For an Archimedean copula C_Ψ, the following conditions are equivalent:*
   1) *C_Ψ is CIS.*
   2) *C_Ψ is CI.*
   3) *$(-1)^k \Psi^{(k)}$ is log-convex for $1 \leq k \leq n - 1$.*
   4) *$(-1)^{n-1} \Psi^{(n-1)}$ is log-convex.* $\tag{10.63}$

c) *If Ψ is completely monotone, i.e., n-alternating for all $n \geq 1$, then C_Ψ is CI.*
d) *C_Ψ is MTP_2, i.e., C_Ψ has a density which is log-supermodular if and only if*

$$(-1)^n \Psi^{(n)} \text{ is log-convex.} \tag{10.64}$$

The condition for MTP_2 is strictly stronger than the condition for CI.

Let C_∞^* denote the class of completely monotone generators, i.e., those which are n-alternating for all $n \geq 1$. The following criterion for the $\leq_{\mathrm{sm}}$ ordering of Archimedean copulas is due to Wei and Hu (2002).

Theorem 10.30 (Supermodular ordering of Archimedean copulas) *Let $C_i = C_{\phi_i}$ be Archimedean copulas with generators $\phi_i \in C_\infty^*$, $i = 1, 2$. Then the convolution condition*

$$\phi_1^{-1} \circ \phi_2 \in C_\infty^* \text{ implies } C_1 \leq_{\mathrm{sm}} C_2. \tag{10.65}$$

Example 10.31

a) *Gumbel copula:*

The Gumbel copula is an Archimedean copula with generator $\phi_\vartheta(s) = \exp(-s^{1/\vartheta})$, $\vartheta \geq 1$, and thus

$$C_\vartheta^{\mathrm{Gu}}(u) = \exp\left(-\left(\sum_{i=1}^{n}(-\ln u_i)^\vartheta\right)^{1/\vartheta}\right). \tag{10.66}$$

For $\vartheta = 1$ it is identical to the independence copula; for $\vartheta \to \infty$ it approaches the comonotonic copula. ϕ_ϑ is completely monotone (see Nelsen, 2006, Example 4.23). For $\vartheta_1 < \vartheta_2$ it holds that $\phi_{\vartheta_1}^{-1} \circ \phi_{\vartheta_2} \in C_\infty^*$ (see Joe, 1997, pg. 375). Therefore, by Theorem 10.30 it holds that

$$C_{\vartheta_1}^{\mathrm{Gu}} \leq_{\mathrm{sm}} C_{\vartheta_2}^{\mathrm{Gu}}. \tag{10.67}$$

b) *Clayton copula:*

The Clayton copula has a completely monotone generator $\phi_\vartheta(s) = (1 + s\vartheta)^{-1/\vartheta}$, $\vartheta \geq 0$ and is given by

$$C_\vartheta^{\mathrm{Cl}}(u) = \left(\sum_{i=1}^{n} u_i^{-\vartheta} - n + 1\right)^{-1/\vartheta}. \tag{10.68}$$

For $\vartheta \to 0$ it approaches the independence copula; for $\vartheta \to \infty$ it approaches the comonotonic copula. For $\vartheta_1 < \vartheta_2$, holds that $\phi_{\vartheta_1}^{-1} \circ \phi_{\vartheta_2} \in C_\infty^*$ (see Joe, 1997, pg. 275) and, therefore, it holds that

$$\vartheta_1 < \vartheta_2 \quad \text{implies} \quad C_{\vartheta_1}^{\mathrm{Cl}} \leq_{\mathrm{sm}} C_{\vartheta_2}^{\mathrm{Cl}}. \tag{10.69}$$

10.3.2   Dependence Structures within the Subgroups

In this subsection we consider various partial dependence assumptions within the subgroups I_i while keeping the copula $C = C_Z$ of the subgroup sums $Z_1, \ldots, Z_k$ fixed. We assume in the following proposition that C is CI, which holds in particular in the case that the subgroups X_{I_i} are independent.

Proposition 10.32   *Under the assumption that C is CI it holds that*

a) *If $Z_i \leq_{\mathrm{cx}} W_i$, $1 \leq i \leq k$, and the comparison vector $W = (W_1, \ldots, W_k)$ has the same copula as Z, i.e., $C_W = C = C_Z$, then*

$$S = \sum_{i=1}^{k} Z_i \leq_{\mathrm{cx}} T = \sum_{i=1}^{k} W_i. \tag{10.70}$$

In particular,

$$\mathrm{LTVaR}_\alpha(T) \leq \mathrm{VaR}_\alpha(S) \leq \mathrm{TVaR}_\alpha(T). \tag{10.71}$$

b) *If $W_i \leq_{\mathrm{cx}} Z_i$, $1 \leq i \leq k$, and $C_W = C = C_Z$, then*

$$T \leq_{\mathrm{cx}} S \quad \text{and} \quad \mathrm{TVaR}_\alpha(T) \leq \mathrm{TVaR}_\alpha(S). \tag{10.72}$$

Proof

a) The assumption that C is CI implies the ordering condition $C^\perp \le_{\text{wcs}} C$, where $C^\perp$ is the independence copula.

From the assumption $Z_i \le_{\text{cx}} W_i$, we therefore conclude by Theorem 10.21 c) that $Z = (Z_1, \ldots, Z_k) \le_{\text{dcx}} W = (W_1, \ldots, W_k)$.

This implies that $S = \sum_{i=1}^{k} Z_i \le_{\text{cx}} T = \sum_{i=1}^{k} W_i$.

b) Part b) is a consequence of Part a), where Z_i and W_i change their roles. $\square$

Remark 10.33 a) If for the comparison vector Y it holds that

$$X_{I_i} \le_{\text{wcs}} Y_{I_i}, \quad 1 \le i \le k, \quad \text{and} \quad X_j \le_{\text{cx}} Y_j \text{ for } 1 \le j \le n, \qquad (10.73)$$

then by Theorem 10.21 c), $X_{I_i} \le_{\text{dcx}} Y_{I_i}$, and therefore

$$Z_i \le_{\text{cx}} W_i, \quad 1 \le i \le k. \qquad (10.74)$$

Similarly,

$$\text{if } X_{I_i} \le_{\text{sm}} Y_{I_i}, \text{ then } Z_i \le_{\text{cx}} W_i, \qquad (10.75)$$

and the converse conclusion holds if switching the role of X and Y to get lower bounds for the portfolio risk vector X. Consequently, under these ordering conditions Proposition 10.32 applies and delivers risk bounds.

If, for example, the subgroup vectors are modeled by elliptical copulas where only upper (or lower) estimates for the covariances are available $\sigma_{ij} \le \sigma_{ij}^{(1)}$ (or $\sigma_{ij} \ge \sigma_{ij}^{(2)}$), then with elliptical vectors Y_{I_i} with the same generator ϕ_i and with the covariance matrix $\Sigma^{(1)} = (\sigma_{ij}^{(1)})$ (resp. $\Sigma^{(2)} = (\sigma_{ij}^{(2)})$), we obtain from Theorem 10.21 the $\le_{\text{dcx}}$ resp. $\le_{\text{sm}}$ ordering estimates and, therefore, the convex ordering condition $Z_i \le_{\text{cx}} W_i$ (resp. $W_i \le_{\text{cx}} Z_i$).

b) **Completely unknown dependence structure within subgroups:**

If besides the marginal distributions F_j of X_j no dependence information is available on the ith subgroups X_{I_i}, then with $Y_j = F_j^{-1}(U_i)$, $j \in I_i$, $U_i \sim U(0,1)$, we have by the well-known ordering property of the comonotonic vector that

$$X_{I_i} \le_{\text{sm}} Y_{I_i} = (F_j^{-1}(U_i))_{j \in I_i}, \qquad (10.76)$$

and therefore

$$Z_i \le_{\text{cx}} W_i = \sum_{j \in I_i} F_j^{-1}(U_i). \qquad (10.77)$$

If $(U_1, \ldots, U_k) \sim C$, then we obtain

$$X \le_{\text{sm}} Y \quad \text{and} \quad S \le_{\text{cx}} T. \qquad (10.78)$$

c) **Partially specified risk factor model**

Let the ith subgroup be modeled by a partially specified risk factor model, i.e.,

$$X_j = f_j(Y_i^f, \varepsilon_j), \quad j \in I_i, \qquad (10.79)$$

where Y_i^f are systemic risk factors, and ε_j are individual risk factors. It is assumed that the joint distributions of (X_j, Y_i^f), $j \in I_i$, are known, but the joint distribution of (ε_j) and of (Y_i^f) is not specified. It was shown in Chapter 9 that

$$Z_i = \sum_{j \in I_i} X_j \leq_{cx} W_i = \sum_{j \in I_i} X_{j|Y_i^f}^c, \qquad (10.80)$$

where $X_{j|Y_i^f}^c := F_{j|Y_i^f}^{-1}(U_i)$, $j \in I_i$ is the conditionally comonotonic vector given Y_i^f, and $F_{j|y_i}$ are the conditional distribution functions of X_j given $Y_i^f = y_i$.

If $(W_1, \ldots, W_k)$ have the copula C, specifying dependence between the groups, then with

$$R_y^+ := \text{TVaR}_V(S_y^c) = \text{TVaR}_V \left(\sum_{i=1}^{k} \sum_{j \in I_i} F_{j|y_i}^{-1}(U_i) \right)$$

$$\text{and} \quad R_y^- := \text{LTVaR}_V(S_y^c),$$

where $V \sim U(0, 1)$ is independent of $(U_i) \sim C$, the following bounds were shown in Chapter 9 for $S = \sum_{j=1}^{n} X_j$:

$$\text{VaR}_\alpha(R_{Yf}^-) \leq \text{VaR}_\alpha(S) \leq \text{VaR}_\alpha(R_{Yf}^+). \qquad (10.82)$$

Equation (10.82) gives upper resp. lower estimates of the VaR in the partially specified risk factor models based on upper resp. lower estimates by TVaR in the conditional models. ◊

In the following example, we consider the case where the different subgroups are independent, i.e., $C = C^\perp$ is the independence copula. We compare various risk bounds and demonstrate the effects of various dependence restrictions within the subgroups. In particular we compare the case of marginal information only, the case with independent subgroups and no dependence information within the subgroups, and the case of independent subgroups with additional partial factor information within the subgroups.

Example 10.34 We consider a portfolio X of n risks with k independent subgroups I_i of the same size m, i.e., $n = km$. Within the subgroups we consider partially specified risk factor models. One half of the elements in the ith subgroup are of the form $X_j = (1 - U_i)^{-1/3} - 1 + \varepsilon_j$, while the other half are of the form $X_j = p((1 - U_i)^{-1/3} - 1) + (1 - p)(U_i^{-1/3} - 1) + \varepsilon_j$, where $U_i \sim U(0, 1)$, $\varepsilon_j \sim \text{Pareto}(4)$, and $p \in (0, 1)$ is a parameter describing the dependence of the systemic risk factors within the subgroups; $p = 0 \sim$ antimonotonic, $p = 1 \sim$ comonotonic behavior. The variables ε_j and U_i are independent for $j \in I_i$, while the $(\varepsilon_j)_{j \in I_i}$ may have any dependence. An example of this kind of factor model without different subgroups (i.e., for $k = 1$) was considered in Bernard et al. (2017b). If we use only the marginal information for this case, we get for $d = 100$ and $\alpha = 0.95$ by the rearrangement algorithm (RA) the sharp VaR bounds $\underline{\text{VaR}}_\alpha$ resp. $\overline{\text{VaR}}_\alpha$ shown in Table 10.13.

$p = 0.0$	$p = 0.2$	$p = 0.5$	$p = 0.8$	$p = 1.0$
(68; 392)	(69; 367)	(70; 349)	(69; 368)	(68; 391)

Table 10.13 Sharp VaR bounds with marginal information only; $d = 100$, $\alpha = 0.95$.

Table 10.14 displays the TVaR bounds

$$b(\text{TVaR}_\alpha) = (\text{LTVaR}_\alpha(S^c), \text{TVaR}_\alpha(S^c))$$

with marginal information only. It also shows the TVaR bounds with partial factor

		$p = 0.0$	$p = 0.2$	$p = 0.5$	$p = 0.8$	$p = 1.0$
$k = 1$	$b(\text{TVaR}_\alpha)$	(68; 474)	(69; 376)	(70; 372)	(69; 384)	(68; 402)
	$b(\text{TVaR}_\alpha^f)$	(72; 297)	(72; 301)	(71; 320)	(69; 351)	(68; 376)
	$b(\text{VaR}_\alpha^f)$	(132; 263)	(134; 265)	(145; 273)	(164; 286)	(182; 296)
$k = 2$	$b(\text{TVaR}_\alpha)$	(72; 385)	(74; 295)	(74; 295)	(74; 301)	(73; 313)
	$b(\text{TVaR}_\alpha^f)$	(76; 231)	(75; 234)	(75; 247)	(74; 269)	(73; 287)
	$b(\text{VaR}_\alpha^f)$	(121; 209)	(122; 210)	(130; 216)	(146; 227)	(158; 237)
$k = 5$	$b(\text{TVaR}_\alpha)$	(77; 305)	(77; 222)	(77; 226)	(77; 229)	(77; 234)
	$b(\text{TVaR}_\alpha^f)$	(79; 173)	(79; 174)	(78; 183)	(77; 197)	(77; 208)
	$b(\text{VaR}_\alpha^f)$	(110; 161)	(110; 162)	(116; 167)	(125; 174)	(133; 180)
$k = 10$	$b(\text{TVaR}_\alpha)$	(79; 266)	(79; 186)	(79; 193)	(79; 193)	(79; 195)
	$b(\text{TVaR}_\alpha^f)$	(80; 144)	(80; 145)	(80; 151)	(79; 161)	(79; 169)
	$b(\text{VaR}_\alpha^f)$	(101; 137)	(102; 138)	(107; 141)	(113; 146)	(119; 151)

Table 10.14 VaR bounds with and without factor model information for various group sizes, $d = 100$, $\alpha = 0.95$, $k = 1, 2, 5, 10$.

information within the subgroups from (10.82),

$$b(\text{TVaR}_\alpha^f) = (\text{VaR}_\alpha(R_{Yf}^-), \text{VaR}_\alpha(R_{Yf}^+))$$

and the sharp VaR bounds with factor information

$$b(\text{VaR}_\alpha^f) = (\underline{\text{VaR}_\alpha^f}, \overline{\text{VaR}}_\alpha^f).$$

By the results of Chapter 9, the determination of the bounds $\underline{\text{VaR}_\alpha^f}$, $\overline{\text{VaR}}_\alpha^f$ can be reduced to VaR upper and lower bounds for the conditional models given the values of the risk factors.

As a result one finds, as expected, a strong improvement of the upper bounds with increasing number k of independent subgroups. Also, the assumption of partially specified factor models within the subgroups strongly reduces the risk bounds compared to the case of arbitrary dependence within the subgroups. The improvements of the bounds are shown in Table 10.15.

For $p \approx 0$ the chosen structure produces strong negative dependence; for $p \approx 1$ it produces strong positive dependence between the two parts in the subgroups. In the

		$p = 0.0$	$p = 0.2$	$p = 0.5$	$p = 0.8$	$p = 1.0$
$k = 1$	$\Delta\,\mathrm{TVaR}_\alpha^f$	(4; 177)	(3; 75)	(1; 52)	(0; 33)	(0; 26)
	$\Delta\,\mathrm{VaR}_\alpha^f$	(64; 211)	(65; 111)	(75; 99)	(95; 98)	(114; 106)
$k = 2$	$\Delta\,\mathrm{TVaR}_\alpha^f$	(4; 154)	(1; 61)	(1; 48)	(0; 32)	(0; 26)
	$\Delta\,\mathrm{VaR}_\alpha^f$	(49; 176)	(48; 85)	(56; 79)	(72; 74)	(85; 76)
$k = 5$	$\Delta\,\mathrm{TVaR}_\alpha^f$	(2; 132)	(2; 48)	(1; 53)	(0; 32)	(0; 26)
	$\Delta\,\mathrm{VaR}_\alpha^f$	(33; 144)	(33; 60)	(39; 59)	(48; 55)	(85; 76)
$k = 10$	$\Delta\,\mathrm{TVaR}_\alpha^f$	(1; 122)	(1; 41)	(1; 42)	(0; 32)	(0; 26)
	$\Delta\,\mathrm{VaR}_\alpha^f$	(22; 129)	(22; 48)	(28; 52)	(34; 47)	(40; 44)

Table 10.15 Improvement of bounds in Table 10.14,
$\Delta\,\mathrm{TVaR}_\alpha^f = b(\mathrm{TVaR}_\alpha) - b(\mathrm{TVaR}_\alpha^f)$, $\Delta\,\mathrm{VaR}_\alpha^f = b(\mathrm{TVaR}_\alpha) - b(\mathrm{VaR}_\alpha^f)$.

first case this leads to a strong improvement of the upper bounds, while in the second case the improvement of the lower bound is more pronounced.

Example 10.35 (Partially specified Gauss factor submodels)
a) **Gauss factor model:** Let $X = (X_i)$ be a one-factor Gauss model of the form

$$X_i = r_i Y^f + \sqrt{1 - r_i^2}\,\varepsilon_i, \qquad 1 \le i \le n, \tag{10.83}$$

where $\{\varepsilon_i\}$ are independent $N(0, 1)$-distributed, and Y^f, $\{\varepsilon_i\}$ are independent. If Y^f is $N(0, 1)$-distributed, and $r_i \in [-1, 1]$, then $X \sim N(0, \Sigma)$ is multivariate normal with $\sigma_{ij} = r_i r_j$. If Y^f is t-distributed, then X is multivariate t-distributed.
b) **Partially specified Gauss factor submodels:** Let X be a portfolio of n risks with k independent subgroups I_i of the same size m. For the subgroups we assume homogeneous Gauss factor models as in (10.83), i.e.,

$$X_j = r_i Y_i^f + \sqrt{1 - r_i^2}\,\varepsilon_j, \qquad j \in I_i, \tag{10.84}$$

where $r_i = r \in [-1, 1]$, $Y_i^f \sim N(0, 1)$ are independent of ε_j for $j \in I_i$, and (Y_i^f) and (ε_{I_i}) are independent.

If the ε_i within the subgroups are independent, then with the assumption of independence between the subgroups, the distribution of X is uniquely determined, and $\mathrm{VaR}_\alpha(\sum_{i=1}^n X_i)$ can be easily calculated (see Table 10.16).

VaR_α	$k = 1$	$k = 2$	$k = 5$	$k = 10$
$r = 0.0$	12	12	12	12
$r = 0.2$	20	16	14	13
$r = 0.5$	43	31	21	16
$r = 0.8$	66	47	30	21
$r = 1.0$	83	58	27	26

Table 10.16 Simulation of $\mathrm{VaR}_\alpha(S)$ for $n = 50$, $\alpha = 0.95$; fully specified factor model.

With increasing value of r, the dependence within the subgroups increases, and the VaR of the sum increases. With increasing number of independent subgroups, it decreases.

If the factor models within the groups are only partially specified, i.e., the $(\varepsilon_j)_{j \in I_i}$ are possibly dependent, we have that

$$X_{j|Y_i^f = y_i} \sim N(ry_i, 1 - r^2);$$

we obtain from (10.80) the convex bounds

$$S = \sum_{i=1}^{k} Z_i \leq_{cx} \sum_{i=1}^{k} W_i = \sum_{i=1}^{k} \sum_{j \in I_i} F_{j|Y_i^f}^{-1}(U_i) \sim N(0, km^2), \qquad (10.85)$$

since the distributions of $F_{j|Y_i^f}^{-1}(U_i) \sim N(0, 1)$ are the same for all j and thus $\{\sum_{j \in I_i} F_{j|Y_i^f}^{-1}\}$ are independent $N(0, m^2)$ random variables.

Generally, for $X \sim N(\mu, \sigma^2)$ it holds that $\mathrm{TVaR}_\alpha(X) = \mu - \sigma^2 \frac{\varphi(\Phi^{-1}(\alpha))}{\alpha}$.

Here, φ and Φ are the pdf and the cdf of $N(0, 1)$. Therefore, we obtain from (10.85),

$$-\sqrt{k}\, m \frac{\varphi(\Phi^{-1}(\alpha))}{\alpha} \leq \mathrm{VaR}_\alpha(S) \leq \sqrt{k}\, m \frac{\varphi(\Phi^{-1}(\alpha))}{1 - \alpha}. \qquad (10.86)$$

The conditional distributions $F_{j|Y_i^f}$ are independent of j. Therefore, the convex upper bounds in (10.85), i.e., the conditional comonotonic sums for the partially specified risk factor models in the subgroups, coincide with the comonotonic sums $\sum_{j \in I_i} F_j^{-1}(U_i)$. Here $F_j = F_{j|Y_i^f} \sim N(0, 1)$ are the marginal distributions for the unconstrained model within the subgroups. These bounds are given in Table 10.17 for the case $n = 50$ and $\alpha = 0.95$.

$k = 1$	$k = 2$	$k = 5$	$k = 10$
$(-5; 103)$	$(-4; 73)$	$(-3; 46)$	$(-2; 33)$

Table 10.17 TVaR bounds for $n = 50$, $\alpha = 0.95$.

In comparison, the VaR bounds in (10.82) are for $r > 0$ improvements over the TVaR bounds in Table 10.17, in particular of the lower bounds.

10.3.3 Dependence Structure between the Subgroups

In contrast to Section 10.3.2, we now consider the case where the dependence structure between the subgroups is estimated from above or below. Let C again denote the copula of the vector Z of subgroup sums, and let D denote the copula of the comparison vector $W = (W_1, \ldots, W_k)$ of subgroup sums of Y.

Proposition 10.36

a) Assume that $C \leq_{wcs} D$ and that $Z_i \leq_{cx} W_i$, $1 \leq i \leq k$; then

$$S = \sum_{i=1}^{k} Z_i \leq_{cx} T = \sum_{r=1}^{k} W_i.$$

In particular

$$\text{LTVaR}_\alpha(T) \leq \text{VaR}_\alpha(S) \leq \text{TVaR}_\alpha(S) \leq \text{TVaR}_\alpha(T). \tag{10.87}$$

b) If $W_i \leq_{cx} Z_i$, $1 \leq i \leq k$ and $Z \leq_{wcs} C$, then $T \leq_{cx} S$ and $\text{TVaR}_\alpha(T) \leq \text{TVaR}_\alpha(S)$.

Proof The assumption that $C \leq_{wcs} D$ and that $Z_i \leq_{cx} W_i$ implies by Theorem 10.21 c) that $Z = (Z_1, \ldots, Z_k) \leq_{dcx} W = (W_1, \ldots, W_k)$.

This implies that $S = \sum_{i=1}^k Z_i \leq_{cx} T = \sum_{i=1}^k W_i$. Thus part a) follows.

Part b) follows by an analogous argument. □

Also, the following related criterion in terms of the supermodular ordering can be given.

Proposition 10.37 *Assume that C or D is CI.*

a) If $C \leq_{sm} D$ and $Z_i \leq_{cx} W_i$, $1 \leq i \leq k$, then

$$S = \sum_{i=1}^k Z_i \leq_{cx} T = \sum_{i=1}^k W_i \tag{10.88}$$

and $\text{LTVaR}_\alpha(T) \leq \text{VaR}_\alpha(S) \leq \text{TVaR}_\alpha(S) \leq \text{TVaR}_\alpha(T)$.

b) If $D \leq_{sm} C$ and $W_i \leq_{cx} Z_i$, $1 \leq i \leq k$, then

$$T \leq_{cx} S \quad and \quad \text{TVaR}_\alpha(T) \leq \text{TVaR}_\alpha(S). \tag{10.89}$$

Proof Let D be CI, and let V be a random vector with $C_V = D = C_W$ and $V_i \stackrel{d}{=} Z_i$. Then by (10.53) it holds $V \leq_{dcx} W$, which implies that

$$\sum_{i=1}^k V_i \leq_{cx} \sum_{i=1}^k W_i.$$

Since $Z_i \stackrel{d}{=} V_i$ the assumption $C \leq_{sm} D$ implies that $Z \leq_{dcx} V$. Thus

$$S = \sum_{i=1}^k Z_i \leq_{cx} \sum_{i=1}^k V_i \leq_{cx} \sum_{i=1}^k W_i,$$

and a) follows from transitivity of the convex ordering $\leq_{cx}$.

The proof under the assumption that C is CI is similar. The proof of b) is analogous. □

In Section 10.3.1 we recollected stochastic ordering results for a variety of concrete copulas: the independence copula, the Gauss copula, the t-copula, the Clayton copula, and the Gumbel copula. We use these kinds of copulas in the following examples as bounds for the dependence structure between the groups. Within the subgroups we allow any dependence structure.

Example 10.38 (Bounds for the copulas between subgroups) Let X be a risk portfolio of $n = 50$ risk variables. We assume that $X_i \sim \text{Pareto}(3, 1)$, $1 \leq i \leq n$, i.e.,

$$F_i(x) = 1 - \frac{1}{(1 + x)^3}, \qquad 1 \leq x. \tag{10.90}$$

The portfolio is split into k subgroups of equal size. The copula C between the subgroups can be bounded by one of the abovementioned copulas D in the sense of supermodular order, as in Proposition 10.37. For the application of Proposition 10.37 we need to verify the CI condition of D.

If D is a Gauss copula or a t-copula and all elements σ_{ij} of the correlation matrix Σ are positive, then Σ^{-1} is an M-matrix, and therefore by Proposition 10.25, D is CI. If D is a Clayton or a Gumbel copula, then D is an Archimedean copula with completely monotone generator, and thus by Theorem 10.29, D is CI. Thus by application of Proposition 10.37, we obtain by means of the comparison vector W estimates for the value-at-risk of S. Since there are no usable convolution properties for the Pareto distribution, we calculate all bounds in the following by means of Monte Carlo simulations.

In Table 10.18 we determine upper and lower VaR bounds using no subgroup structure but only the information on the marginals. The calculations of VaR estimates are based on the rearrangement algorithm (RA); $(a; b)$ denote the TVaR estimates.

	$(\underline{\text{VaR}}_\alpha; \overline{\text{VaR}}_\alpha)$	$(a; b)$
$\alpha = 0.950$	(18; 153)	(18; 154)
$\alpha = 0.990$	(22; 298)	(22; 298)
$\alpha = 0.995$	(23; 388)	(22; 389)

Table 10.18 Unconstrained VaR bounds $(\underline{\text{VaR}}_\alpha, \overline{\text{VaR}}_\alpha)$ calculated by RA, case without subgroups.

In Table 10.19 we consider the case that the copula C (negative dependence assumption) of the vector Z of subgroup sums is bounded above by the independence copula $D = C^\perp$. Within the subgroups we allow arbitrary dependence. From (10.87) we obtain the improved TVaR bounds with this subgroup information.

	$k = 2$	$k = 5$	$k = 10$	$k = 25$
$\alpha = 0.950$	(20; 116)	(22; 82)	(23; 64)	(24; 49)
$\alpha = 0.990$	(23; 209)	(24; 132)	(24; 96)	(25; 66)
$\alpha = 0.995$	(24; 266)	(24; 163)	(25; 115)	(25; 76)

Table 10.19 VaR bounds in subgroup model with independence copula between subgroups.

We observe strong improvements of the upper bounds and minor improvements of the lower bounds. This is to be expected when posing the independence copula as upper bound, as this only restricts the positive dependence from above.

In Table 10.20 we consider as examples for the comparison copula D the case of Gaussian copulas and t-copulas with various values of the correlation parameter, $\varrho = \varrho_{ij}, i \neq j$.

	$k = 2$	$k = 5$	$k = 10$	$k = 25$	$\overline{\Delta}$
Table A: Cor = 0.1					
$\alpha = 0.950$	(20; 119)	(22; 88)	(22; 73)	(23; 71)	58
$\alpha = 0.990$	(23; 214)	(24; 142)	(24; 116)	(24; 110)	130
$\alpha = 0.995$	(24; 271)	(24; 174)	(24; 135)	(24; 131)	174
Table B: Cor = 0.25					
$\alpha = 0.950$	(20; 124)	(21; 98)	(22; 86)	(22; 78)	58
$\alpha = 0.990$	(23; 222)	(24; 161)	(24; 134)	(24; 115)	107
$\alpha = 0.995$	(24; 283)	(24; 197)	(24; 160)	(25; 135)	135
Table C: Cor = 0.5					
$\alpha = 0.950$	(19; 132)	(20; 116)	(21; 109)	(21; 105)	27
$\alpha = 0.990$	(23; 242)	(24; 200)	(23; 183)	(24; 172)	70
$\alpha = 0.995$	(24; 308)	(24; 248)	(24; 225)	(25; 210)	98
Table D: $\nu = 50$, Cor = 0.1					
$\alpha = 0.950$	(20; 119)	(22; 89)	(22; 74)	(23; 63)	56
$\alpha = 0.990$	(23; 215)	(24; 146)	(24; 114)	(24; 90)	125
$\alpha = 0.995$	(24; 274)	(24; 179)	(24; 137)	(25; 105)	169
Table E: $\nu = 50$, Cor = 0.25					
$\alpha = 0.950$	(20; 124)	(21; 99)	(22; 88)	(23; 80)	44
$\alpha = 0.990$	(23; 224)	(24; 164)	(24; 139)	(24; 122)	102
$\alpha = 0.995$	(24; 285)	(24; 202)	(24; 168)	(24; 144)	143
Table F: $\nu = 10$, Cor = 0.25					
$\alpha = 0.950$	(20; 125)	(21; 102)	(21; 93)	(23; 87)	38
$\alpha = 0.990$	(23; 230)	(23; 177)	(24; 157)	(24; 144)	86
$\alpha = 0.995$	(24; 294)	(24; 223)	(24; 196)	(24; 177)	117

Table 10.20 VaR bounds in subgroup model with Gauss copula in A, B, and C and with t-copula in D, E, and F. $\overline{\Delta}$ denotes the difference between upper bounds for $k = 2$ and $k = 25$.

Again we see a strong improvement of the upper bounds. The improvements increase with decreasing correlation. For the case of t-copulas, the improvements increase with increasing degrees of freedom ν. The independence copula gives the strongest improvement of upper bounds resulting from CI copulas as considered in Table 10.20.

Table 10.21 is concerned with the case of a Clayton copula and a Gumbel copula. Again, the upper bounds improve considerably with decreasing value of the dependence parameter.

	$k = 2$	$k = 5$	$k = 10$	$k = 25$	$\overline{\Delta}$
Table A: $\vartheta = 1$					
$\alpha = 0.950$	(20; 122)	(22; 94)	(22; 81)	(23; 71)	51
$\alpha = 0.990$	(23; 216)	(24; 147)	(24; 116)	(24; 92)	124
$\alpha = 0.995$	(24; 274)	(24; 179)	(24; 135)	(25; 103)	171
Table B: $\vartheta = 3$					
$\alpha = 0.950$	(20; 130)	(21; 108)	(21; 98)	(22; 90)	40
$\alpha = 0.990$	(23; 227)	(24; 166)	(24; 138)	(24; 119)	108
$\alpha = 0.995$	(24; 285)	(24; 198)	(24; 160)	(25; 132)	153
Table C: $\vartheta = 10$					
$\alpha = 0.950$	(19; 140)	(20; 128)	(20; 122)	(20; 118)	22
$\alpha = 0.990$	(23; 244)	(23; 196)	(23; 176)	(24; 162)	82
$\alpha = 0.995$	(24; 304)	(24; 232)	(24; 202)	(24; 180)	124
Table D: $\vartheta = 1.5$					
$\alpha = 0.950$	(19; 140)	(19; 132)	(20; 129)	(20; 127)	13
$\alpha = 0.990$	(23; 272)	(23; 258)	(23; 254)	(23; 250)	22
$\alpha = 0.995$	(23; 353)	(23; 338)	(23; 329)	(23; 327)	26
Table E: $\vartheta = 3$					
$\alpha = 0.950$	(18; 151)	(18; 150)	(18; 149)	(18; 148)	3
$\alpha = 0.990$	(22; 294)	(22; 290)	(22; 290)	(22; 289)	5
$\alpha = 0.995$	(23; 383)	(23; 379)	(23; 379)	(23; 375)	8

Table 10.21 VaR bounds in subgroup model with Clayton copula in A, B, and C and Gumbel copula in D and E.

Remark 10.39

a) Proposition 10.36 also allows us to derive lower bounds for $\text{TVaR}_\alpha(S)$ assuming that $D \leq_{\text{wcs}} C$ and $W_i \leq_{\text{cx}} Z_i$.

b) Lower bounds for $\text{VaR}_\alpha(S)$ are also obtainable by the conclusion:

$$S \leq_{\text{cx}} T \text{ implies } \text{LTVaR}_\alpha(T) \leq \text{VaR}_\alpha(S). \qquad (10.91)$$

c) If the copula C between the subgroups is modeled by an elliptical copula with negative correlations $\sigma_{ij} < 0$ between subgroups $i \neq j$, then by Proposition 10.37 an upper bound for $\text{VaR}_\alpha(S)$ is obtained by the independence copula $D = C^\perp$. Table 10.19 shows some results for the VaR bound resulting from the independence copula D, for Pareto(3,1)-marginals. ◊

The results in this section show that the assumption of a CI upper bound D for the copula C between the subgroups leads to considerable reduction of the upper VaR and TVaR bounds and of the DU spread of the portfolio. The largest reduction is obtained by this method in the case that D is the independence copula.

We consider in this section the case of no dependence information within the subgroups. However, by the results in Sections 10.3.2 and 10.3.3, both kinds of dependence information, that between and that within subgroups, can be combined, leading to accumulated reduction effects. The magnitudes of the single reduction effects are considerable and can be well estimated from the examples given here. Altogether, this approach gives quite flexible tools with promising potential, which seems to be of interest for various real applications.

Part IV

Risk Bounds under Moment Information

In Parts II and III of this book, we have studied risk bounds when in addition to the information on the marginal distributions, dependence information or structural information is also available. In Chapter 6, we showed that adding a variance constraint on the portfolio loss $S = \sum X_i$ (equivalently, a constraint on the average correlation among the risks X_i) may lead to a significantly improved upper bound for the VaR of S and moreover that in this case the improved VaR upper bound is closely related to the classic Cantelli moment bound (involving mean and variance) for distributions.

This observation suggests an interesting parallel between, on the one hand, the problem of finding risk bounds when the marginal distributions and higher-order moments of the portfolio sum are given and, on the other hand, the problem of finding risk bounds when the mean of the sum (but not necessarily the marginal distributions) and some higher-order moments of S are given. In fact, by making use of the information on higher-order moments, one can effectively exploit the available dependence information while retaining some aspects of marginal distributions (as higher-order moments depend both on the marginal distributions and on the dependence structure). Moreover, in practice, it might happen that loss statistics are only available at an aggregate level. In an insurance context, for instance, the observed aggregated losses are available on a monthly basis at best. Hence, the number of data points that can be used in the risk analysis is limited, and one cannot use the observations to accurately estimate one-year VaRs of a portfolio sum S at a high probability level (e.g., 99.5 %) directly. By contrast, there could be enough data points available to estimate some lower moments of S with a sufficient degree of precision. Therefore, using moment information only to determine the risk bounds of a portfolio sum S provides an interesting perspective for studying this problem, which we consider in this final part of the book.

In Chapter 11, we study moment bounds for the risk measures VaR, TVaR, and RVaR. For the important case of VaR, we propose, under an additional domain restriction, a method that can in principle deal with any number of known moments. As a main tool to derive the bounds, we use convex ordering and its generalization, the s-convex order.

In Chapter 12, we study bounds for distortion risk measures. In Section 12.1, we build on results in Cornilly et al. (2018) to derive moment bounds under an additional domain restriction. In Section 12.2 we dispense with domain restrictions and derive bounds using the tool of isotonic projections combined with the use of the Cauchy–Schwarz inequality.

In Chapters 13 and 14, we study the influence of additional structural information on the moment bounds. First, in Chapter 13, we study bounds when in addition to moment information, it is also known that the distribution is unimodal. Second, in Chapter 14, we study bounds when the distribution is assumed to be in the neighborhood of a reference distribution.

11 Bounds on VaR, TVaR, and RVaR under Moment Information

In this chapter, we derive bounds for VaR, TVaR, and their generalization range value at risk (RVaR) under the sole knowledge of moment information on the risk at hand. This setting is, for instance, well adapted to the situation in which one has sufficient data to estimate lower moments of the portfolio loss, but the data are not sufficient to estimate VaR, TVaR, and RVaR, respectively, at high confidence level. Note, however, that the higher the order of the moments, the more vulnerable their estimators become to outliers. We deal with this issue by also considering the statistical uncertainty with respect to moment estimation and the impact this has on risk bounds.

11.1 VaR Bounds: Bounded Domain

In this section, we study VaR bounds when in addition to moment information, the loss distribution has a known bounded domain. The feature of bounded losses may appear as somewhat awkward in that most distributions used in practice for modeling losses do not have this property. However, the upper limit to the financial loss for which an insurance company underwrites is generally fixed by the contract or determined through reinsurance techniques. Moreover, insurers have limited liability (up to their capital) and finite resources to meet claims.

Bounds for VaR under moment information have been studied in the insurance literature by several authors, including Kaas and Goovaerts (1986), Hürlimann (2002), De Schepper and Heijnen (2010), and de Vylder (1982). In particular, Hürlimann (2002) finds analytical VaR bounds based on the mean, variance, skewness, and kurtosis. The approaches available in the literature for deriving VaR bounds are based on linear programming or special polynomials. The following results are mainly based on Bernard et al. (2018a), who exploit the stochastic ordering concept of s-convex order to derive VaR bounds in a straightforward fashion.

11.1.1 Moment Spaces

Description

Let $s \geq 2$ be a positive integer and $a, b \in \mathbb{R}, a < b$. We denote by $\mathcal{F}([a, b], \mu_1, \mu_2, \ldots, \mu_{s-1})$ the set of distribution functions (dfs) on $[a, b]$ for which the first $s - 1$ moments $\mu_1, \mu_2, \ldots, \mu_{s-1}$ are given, i.e.,

$$\mathcal{F}([a, b], \mu_1, \mu_2, \ldots, \mu_{s-1}) \tag{11.1}$$

$$= \left\{ F \text{ is a df on [a,b]} \,\Big|\, \int_a^b x^i \, dF(x) = \mu_i, \ i = 1, \ldots, s - 1 \right\}.$$

In what follows, we often need to separately discuss two cases according to the parity of s:

- When s is even, then $s = 2m$ for some positive integer m, and an odd number of moments $\mu_1, \ldots, \mu_{2m-1}$ are known.
- When s is odd, then $s = 2m + 1$ for some positive integer m, and an even number of moments $\mu_1, \mu_2, \ldots, \mu_{2m}$ are known.

It is clear that not every vector $(\mu_1, \mu_2, \ldots, \mu_{s-1}) \in \mathbb{R}^{s-1}$ corresponds to a sequence of moments, i.e., we may have that $\mathcal{F}([a, b], \mu_1, \mu_2, \ldots, \mu_{s-1}) = \phi$. Furthermore, other sequences of moments may correspond to a unique probability distribution. For example, if $\xi \in [a, b]$ then the moment sequence $(\xi, \xi^2, \ldots, \xi^{s-1})$ corresponds to the degenerate distribution function F_X having a single mass point ξ for any $s \geq 3$. These situations are clearly without further interest, and we impose conditions on moment sequences to ensure that the moment space at hand contains strictly more than one element (and thus many more elements as the moment space is convex). It is possible to describe these conditions precisely in terms of some moment matrices.[1] In what follows, we tacitly assume that these conditions are satisfied.

Distribution Functions with a Fixed Number of Mass Points

De Vylder (1982) provides a general result for finding a discrete distribution F_Y in a given moment space $\mathcal{F}([a, b], \mu_1, \mu_2, \ldots, \mu_{s-1})$ when some mass points are fixed. This construction is of key importance in our derivation of moment bounds for VaR. Specifically, if $\eta_1, \ldots, \eta_i$ are fixed mass points in $[a, b]$ and $\vartheta_1, \ldots, \vartheta_j$ are unknown mass points in (a, b), de Vylder (1996) explains in Section 8.2.2, Theorem 4, how to obtain the unique distribution function F_Y in $\mathcal{F}([a, b], \mu_1, \mu_2, \ldots, \mu_{i+2j-1})$ having $\eta_1, \ldots, \eta_i$ and $\vartheta_1, \ldots, \vartheta_j$ as mass points. This result can be obtained using the following two steps that are straightforward to implement.

Step 1: The unknown mass points $\vartheta_1, \ldots, \vartheta_j$ are the roots of the following equation based on a Vandermonde determinant:

$$\begin{Vmatrix} 1 & x & x^2 & \cdots & x^j \\ \mu_{i;\eta_1,\ldots,\eta_i} & \mu_{i+1;\eta_1,\ldots,\eta_i} & \mu_{i+2;\eta_1,\ldots,\eta_i} & \cdots & \mu_{i+j;\eta_1,\ldots,\eta_i} \\ \mu_{i+1;\eta_1,\ldots,\eta_i} & \mu_{i+2;\eta_1,\ldots,\eta_i} & \mu_{i+3;\eta_1,\ldots,\eta_i} & \cdots & \mu_{i+j+1;\eta_1,\ldots,\eta_i} \\ \vdots & \vdots & \vdots & \ddots & \vdots \\ \mu_{i+j-1;\eta_1,\ldots,\eta_i} & \mu_{i+j;\eta_1,\ldots,\eta_i} & \mu_{i+j+1;\eta_1,\ldots,\eta_i} & \cdots & \mu_{i+2j-1;\eta_1,\ldots,\eta_i} \end{Vmatrix} = 0, \tag{11.2}$$

---

[1] There are numerous papers and books where conditions on sequences $(\mu_1, \mu_2, \ldots, \mu_{s-1})$ are stated to ensure that the associated moment space is not void; Karlin and Studden (1966) is a classic textbook.

where $\| \cdot \|$ denotes the matrix determinant, and the elements of the matrix are defined recursively as

$$\mu_{k;\eta_1,\ldots,\eta_\ell} = \mu_{k;\eta_1,\ldots,\eta_{\ell-1}} - \eta_\ell \mu_{k-1;\eta_1,\ldots,\eta_{\ell-1}}$$

starting from $\mu_{k;\eta_1} = \mu_k - \eta_1 \mu_{k-1} (\mu_1 = 1, \mu_{-1} = 0)$.[2]

Step 2: The probability masses $P(Y = \eta_\ell)$, $\ell = 1, 2, \ldots, i$ and $P(Y = \vartheta_\ell)$, $\ell = 1, 2, \ldots, j$ can now easily be obtained from solving the linear system of equations

$$\sum_{l=1}^{i} P(Y = \eta_\ell)\eta_\ell^k + \sum_{l=1}^{j} P(Y = \vartheta_\ell)\vartheta_\ell^k = \mu_k, \quad k = 0, 1, \ldots, i + j - 1.$$

(11.3)

s-Convex Extrema

In a given moment space $\mathcal{F}([a, b], \mu_1, \mu_2, \ldots, \mu_{s-1})$, the two distribution functions with a minimum number of mass points play a special role. These distribution functions correspond to the s-convex minimum resp. maximum, and their properties are useful in the construction of risk bounds. We first recall the concept of s-convex order. Next, we use the method of de Vylder to construct the distributions of the s-convex extrema explicitly (see also Denuit et al., 1998).

Definition 11.1 (s-convex order) Let F_X and F_Y be two distribution functions in $\mathcal{F}([a, b], \mu_1, \mu_2, \ldots, \mu_{s-1})$ with corresponding random variables X and Y. If the inequality

$$E(g(X)) \leq E(g(Y))$$

holds for every function g with a non-negative sth order derivative such that the expectations exist, then F_X is said to be smaller than F_Y in the s-convex order sense, and we write $F_X \leq_{s-cx} F_Y$.

Note that the 1-convex order is the usual first-degree stochastic dominance, and 2-convex order is the usual convex order or mean-preserving increase in risk. Many properties of the s-convex stochastic order can be found in Denuit et al. (1998). As the sign of the sth derivative of the test function g is controlled, the s-convex order is closely related to the higher-degree increases in risk introduced by Ekern (1980).[3] It is clear that when F_X and F_Y belong to $\mathcal{F}([a, b], \mu_1)$, then, as they share the same mean, they must cross at least once. It is also well known that if they cross exactly once then they are ordered in the 2-convex order sense. Interestingly, these crossing properties can be generalized to any value of s. To this end, we recall that for a real-valued function g defined on a subset I of the real line, the number of changes of sign of g on this subset I is

$$S^-(g) = \sup S^-[g(\xi_1), g(\xi_2), \ldots, g(\xi_n)],$$

(11.4)

---

[2] This result can be proved directly based on a useful representation of the matrix whose determinant appears in (11.2).

[3] Specifically, the two relationships coincide for odd s, but the inequality has to be reversed for even s.

where the supremum is extended over all sets $\xi_1 < \xi_2 < \cdots < \xi_n \in I$, n is arbitrary but finite, and $S^-[\vartheta_1, \vartheta_2, \ldots, \vartheta_n]$ is the number of sign changes of the indicated sequence $(\vartheta_1, \vartheta_2, \ldots, \vartheta_n)$, zero terms being discarded (see, for example, Karlin, 1968). The functions g_1 and g_2 are said to cross each other k times if $S^-(g_1 - g_2) = k$. We can now formulate the following proposition; see also Theorems 4.2 and 4.3 in Denuit et al. (1998) and Proposition 3.2 in Denuit (1999).

Proposition 11.2 (Crossing conditions) *Let F_X and F_Y be in the set $\mathcal{F}([a, b], \mu_1, \mu_2, \ldots, \mu_{s-1})$. Then, $S^-(F_X - F_Y) \geq s - 1$. Moreover, $S^-(F_X - F_Y) = s - 1$ implies that $F_X \preceq_{s-cx} F_Y$ provided the last sign of $F_X - F_Y$ is a "+". By symmetry of the argument, $S^-(F_X - F_Y) = s - 1$ implies that $F_Y \preceq_{s-cx} F_X$ provided the last sign of $F_X - F_Y$ is a "−".*

Using equations (11.2) and (11.3), we can easily obtain the s-convex smallest and s-convex largest element in $\mathcal{F}([a, b], \mu_1, \mu_2, \ldots, \mu_{s-1})$, which we denote by $F_{X_{\min}^{(s)}}$ and $F_{X_{\max}^{(s)}}$, respectively.

Proposition 11.3 (Construction of s-convex extrema) *According to the parity of the number of moments, the s-convex extrema can be obtained as follows:*

- *odd number of moments fixed ($s = 2m$):*

  In $\mathcal{F}([a, b], \mu_1, \mu_2, \ldots, \mu_{2m-1})$, the m mass points $x_1, x_2, \ldots, x_m$ of $F_{X_{\min}^{(2m)}}$, $a < x_1 < x_2 < \cdots < x_m < b$, are the m distinct roots of equation (11.2) with $i = 0$ and $j = m$. The positive masses $p_1, p_2, \ldots, p_m$ on the mass points $x_1, x_2, \ldots, x_m$ follow from (11.3).

  The $m + 1$ mass points $a = y_1, y_2, \ldots, y_m, y_{m+1} = b$ of $F_{X_{\max}^{(2m)}}$, $a = y_1 < y_2 < \cdots < y_m < y_{m+1} = b$, are such that $y_2, \ldots, y_m$ are the $m - 1$ distinct roots of (11.2) with $i = 1$, $\eta_1 = a$, $\eta_2 = b$ and $j = m - 1$. The positive masses $q_1, q_2, \ldots, q_{m+1}$ on the mass points $a, y_2, \ldots, y_m, b$ follow from (11.3).

- *even number of moments fixed ($s = 2m + 1$):*

  In $\mathcal{F}(\mu_1, \mu_2, \ldots, \mu_{2m})$, the $m + 1$ mass points $a = x_1, x_2, \ldots, x_{m+1}$ of $F_{X_{\min}^{(2m+1)}}$, $a = x_1 < x_2 < \cdots < x_{m+1} < b$, are such that $x_2, \ldots, x_{m+1}$ are the m distinct roots of equation (11.2) with $i = 1$, $\eta_1 = a$ and $j = m$. The positive masses $p_1, p_2, \ldots, p_{m+1}$ on the mass points $a, x_2, \ldots, x_{m+1}$ follow from (11.3).

  The $m + 1$ mass points $y_1, y_2, \ldots, y_{m+1} = b$ of $F_{X_{\max}^{(2m+1)}}$, $a < y_1 < y_2 < \cdots < y_m < y_{m+1} = b$, are such that $y_1, y_2, \ldots, y_m$ are the m distinct roots of (11.2) with $i = 1$, $\eta_1 = b$ and $j = m$. The positive masses $q_1, q_2, \ldots, q_{m+1}$ on the mass points $y_1, y_2, \ldots, y_m, b$ follow from (11.3).

We refer to the proof in Section 11.4.1 for some more details of this construction.

We end this section with the following proposition that appears to be useful in the remainder, as it constrains the position of the graph of any distribution function in a given moment space. Its proof follows immediately from Proposition 11.2 and the structure of $F_{X_{\min}^{(s)}}$ and $F_{X_{\max}^{(s)}}$.

Proposition 11.4 (Number of crossing points) *For any $F_X \in \mathcal{F}([a, b], \mu_1, \mu_2, \ldots, \mu_{s-1})$ it holds that*

$$S^-(F_X - F_{X_{\min}^{(s)}}) = S^-(F_X - F_{X_{\max}^{(s)}}) = s - 1. \tag{11.5}$$

In particular, as $F_{X_{\min}^{(s)}}$ and $F_{X_{\max}^{(s)}}$ both belong to $\mathcal{F}([a, b]; \mu_1, \mu_2, \ldots, \mu_{s-1})$, we must have

$$S^-(F_{X_{\min}^{(s)}} - F_{X_{\max}^{(s)}}) = s - 1, \tag{11.6}$$

so that the mass points of $F_{X_{\min}^{(s)}}$ and $F_{X_{\max}^{(s)}}$ alternate in $[a, b]$ (see also (11.10) and (11.11) below).

11.1.2   Bounds on Distribution Functions

For any given $t \in (a, b)$ we aim to determine the maximum and minimum possible value that a distribution function can take in the moment space $\mathcal{F}([a, b], \mu_1, \mu_2, \ldots, \mu_{s-1})$. Formally, we are interested in

$$M(t) := \sup \{F_X(t) \mid F_X \in \mathcal{F}([a, b], \mu_1, \mu_2, \ldots, \mu_{s-1})\}, \tag{11.7}$$

$$m(t) := \inf \{F_X(t-) \mid F_X \in \mathcal{F}([a, b], \mu_1, \mu_2, \ldots, \mu_{s-1})\}, \tag{11.8}$$

where $F_X(t-) := P(X < t)$. In what follows, we use an elementary probabilistic reasoning to show that for every t there exists a unique distribution function $F_{Z^*} \in \mathcal{F}([a, b], \mu_1, \mu_2, \ldots, \mu_{s-1})$ such that

$$M(t) = F_{Z^*}(t) \quad \text{and} \quad m(t) = F_{Z^*}(t-). \tag{11.9}$$

Furthermore, using de Vylder's method for the construction of distributions in a moment space, we are able to build F_{Z^*} explicitly (see Algorithm 1 and Algorithm 2 below).

Odd Number of Moments Fixed ($s = 2m$)

Let $s = 2m$ for some positive integer m. Let F_Z be some discrete distribution in the moment space $\mathcal{F}([a, b], \mu_1, \mu_2, \ldots, \mu_{s-1})$ having mass points $z_1, \ldots, z_{m+1} \in [a, b]$, $z_1 < \cdots < z_{m+1}$. Then, for any other distribution function F_X in $\mathcal{F}([a, b], \mu_1, \mu_2, \ldots, \mu_{s-1})$, $F_X \neq F_{X_{\min}^{(s)}}$, $F_X \neq F_{X_{\max}^{(s)}}$, we know from Proposition 11.2 that $S^-(F_X - F_Z) \geq s - 1$. A simple reasoning shows then that it is possible to select the appropriate locations for the mass points of F_Z so that $S^-(F_X - F_Z) = s$ always holds. Next, we show that for a given $t \in (a, b)$, there is exactly one such F_Z, denoted by F_{Z^*}, that can be used to bound $F_X(t)$ and $F_X(t-)$. We explain this further as follows:

First, for $S^-(F_X - F_Z) = s$ to hold, we must have that either $z_1 = a$ or $z_{m+1} = b$, and the other m mass points must be taken in (a, b). Specifically, if one has $z_1 = a < z_2 < \cdots < z_{m+1} < b$ then we have a total of $2m = s$ crossings, namely
- one crossing over each flat part of F_Z on the intervals (z_j, z_{j+1}), $j = 1, \ldots, m$, i.e., a total of m crossings;
- one crossing at each mass point $z_2, \ldots, z_{m+1}$, i.e., a total of m crossings.

Next, if $a < z_1 < \cdots < z_m < z_{m+1} = b$ then we also have $2m = s$ crossings, which can be described as follows:
- one crossing over each flat part of F_Z on the intervals (z_j, z_{j+1}), $j = 1, \ldots, m$, i.e., a total of m crossings;

- one crossing at each mass point $z_1, \ldots, z_m$, i.e., a total of m crossings.

Second, for all F_Z such that $S^-(F_X - F_Z) = s$, it must hold that for each $z_j \in (a, b)$, $F_Z(z_j-) < F_X(z_j) < F_Z(z_j)$.

Hence, since one mass point of F_Z is given (a or b), we have $2m+1 = s+1$ unknown mass points and probability masses. However, we face s constraints, namely $s-1$ given moments and the probabilities that sum up to 1, and there remains thus a single degree of freedom. We are therefore free to locate one mass point of F_Z at t in order to bound $F_X(t)$. Hence, we only have to build a distribution function F_{Z^*} having the structure described above in which we take t as a particular mass point, to get (11.9).

However, at this point, it appears that there are still two possibilities for choosing F_Z. Fortunately, depending on the location of t with respect to the mass points of the extrema $F_{X_{\min}^{(s)}}$ and $F_{X_{\max}^{(s)}}$, one can determine whether either a or b is to be considered as a mass point of F_Z. Specifically, when $s = 2m$, the mass points of $F_{X_{\min}^{(s)}}$ are $a < x_1 < x_2 < \cdots < x_m < b$ and the mass points of $F_{X_{\max}^{(s)}}$ are $a = y_1 < y_2 < \cdots < y_m < y_{m+1} = b$. The x_j and y_j satisfy

$$a = y_1 < x_1 < y_2 < x_2 < \cdots < y_m < x_m < y_{m+1} = b, \tag{11.10}$$

see also Proposition 11.4, equation (11.6). At each mass point z_j of F_Z, the distribution function F_Z must cross both $F_{X_{\min}^{(s)}}$ and $F_{X_{\max}^{(s)}}$ (see Proposition 11.4). This implies that
- if $t \in (y_j, x_j)$ then each mass point of F_Z must belong to an interval (y_ℓ, x_ℓ), $\ell = 1, \ldots, m$, and b must belong to the mass points of F_Z. The mass points of F_Z thus consist of

$$a < z_1 < z_2 < \cdots < z_m < z_{m+1} = b,$$

with each $z_k \in (y_k, x_k)$ and $z_j = t$.
- if $t \in (x_j, y_{j+1})$ then each mass point of F_Z must belong to an interval $(x_\ell, y_{\ell+1})$, $\ell = 1, \ldots, m$ and a must belong to the mass points of F_Z. The mass points of F_Z thus consist of

$$a = z_1 < z_2 < \cdots < z_{m+1} < b,$$

with each $z_k \in (x_k, y_{k+1})$ and $z_j = t$.

Hence, the location of t with respect to the partition of $[a, b]$ created by the union of the domains of $F_{X_{\min}^{(s)}}$ and $F_{X_{\max}^{(s)}}$ satisfying (11.10) determines whether a or b belongs to the domain of the optimal F_Z.

This line of reasoning shows that one can apply de Vylder's method (11.2)–(11.3) for the construction of distributions in a moment space to obtain the bounds $m(t)$ and $M(t)$. Note that these bounds are sharp and moreover attainable, as F_{Z^*} belongs to the moment space. We find the following algorithm:

Algorithm 1 for $s = 2m$ (Distributional bounds, odd number of moments)

The mass points of the optimal distribution function F_{Z^*} allowing us to bound $F_X(t-)$ and $F_X(t)$ in the moment space $\mathcal{F}([a, b], \mu_1, \mu_2, \ldots, \mu_{s-1})$ can be found as follows:

- Determine the mass points x_j and y_j of $F_{X^{(2m)}_{\min}}$ resp. $F_{X^{(2m)}_{\max}}$ (see Proposition 11.3).
- When $t \in (x_j, y_{j+1})$ for some $j = 1, \ldots, m + 1$, solve equations (11.2) and (11.3) in which $i = 2$, $\eta_1 = a$, and $\eta_2 = t$.
- When $t \in (y_j, x_j)$ for some $j = 1, \ldots, m + 1$, solve equations (11.2) and (11.3) in which $i = 2$, $\eta_1 = t$, and $\eta_2 = b$.

Even Number of Moments Fixed ($s = 2m + 1$)

When there is an even number of moments, i.e., when $s = 2m + 1$ for some positive integer m, the construction goes along the same lines as in the case of an odd number of known moments. We first characterize distribution functions F_Z with the property that $S^-(F_X - F_Z) = s$ for any other distribution function F_X in $\mathcal{F}([a, b], \mu_1, \mu_2, \ldots, \mu_{s-1})$, $F_X \neq F_{X^{(s)}_{\min}}$, $F_X \neq F_{X^{(s)}_{\max}}$, to hold. The mass points x_j and y_j of $F_{X^{(s)}_{\min}}$ and $F_{X^{(s)}_{\max}}$ satisfy

$$a = x_1 < y_1 < x_2 < y_2 < \cdots < y_m < x_{m+1} < y_{m+1} = b. \tag{11.11}$$

As before, depending on the location of t with respect to these support points, one can select one particular distribution function, denoted by F_{Z^*}, to bound $F_X(t)$. We refer to Section 11.4.2 for more details and here merely provide the final algorithm.

Algorithm 2 for $s = 2m + 1$ (Distributional bounds, even number of moments)

The mass points of the optimal distribution function F_{Z^*} (making it possible to bound $F_X(t-)$ and $F_X(t)$ in the moment space $\mathcal{F}([a, b], \mu_1, \mu_2, \ldots, \mu_{s-1})$) can be found as follows:

- Determine the mass points x_j and y_j of $F_{X^{(2m+1)}_{\min}}$ resp. $F_{X^{(2m+1)}_{\max}}$ (see Section 11.1.1).
- When $t \in (x_j, y_j)$ for some $j = 1, \ldots, m + 1$, solve equations (11.2) and (11.3) in which $i = 1$ and $\eta_1 = t$. Note that a and b are not included.
- When $t \in (y_j, x_{j+1})$ for some $j = 1, \ldots, m$, solve equations (11.2) and (11.3) in which $i = 3$, $\eta_1 = a$, $\eta_2 = t$, and $\eta_3 = b$. Note that a and b are included.

We provide two examples to illustrate the algorithms.

Example 11.5 (Known mean, $s = 2$) In $\mathcal{F}([a, b], \mu_1)$, $F_{X^{(2)}_{\min}}$ and $F_{X^{(2)}_{\max}}$ are distribution functions with corresponding variables $X^{(2)}_{\min} = \mu_1$ and

$$X^{(2)}_{\max} = \begin{cases} a & \text{with probability } \frac{b-\mu_1}{b-a}, \\ b & \text{with probability } \frac{\mu_1-a}{b-a}. \end{cases}$$

The construction of F_{Z^*} that makes it possible to bound $F_X(t)$ depends on the location of t with respect to $\text{supp}(X^{(2)}_{\min}) \cup \text{supp}(X^{(2)}_{\max}) = \{a, \mu_1, b\}$. Using Algorithm 1, we find that (see Figure 11.1)

- If $a < t < \mu_1$, then $0 = P(Z^* < t) \leq F_X(t) \leq P(Z^* \leq t) = P(Z^* = t) = \frac{b-\mu_1}{b-t}$.
- If $\mu_1 < t < b$, then $\frac{t-\mu_1}{t-a} = P(Z^* = a] = P(Z^* < t) \leq F_X(t) \leq P(Z^* \leq t) = 1$.

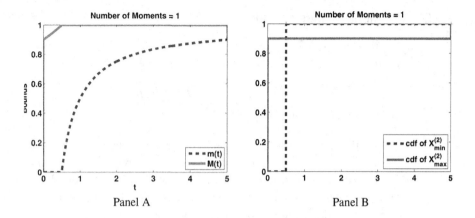

Figure 11.1 Case $s = 2$: Panel A displays the distribution functions of $X^{(2)}_{\min}$ and $X^{(2)}_{\max}$ when $\mu_1 = \frac{1}{2}$, $a = 0$, and $b = 5$. Panel B displays the lower and upper bounds $m(t)$ and $M(t)$, respectively, i.e., the lower and upper curves in Panel B represent bounds on $F_X(t)$ when only the mean is fixed.

Example 11.6 (Known mean and variance, $s = 3$) In $\mathcal{F}([a, b], \mu_1, \mu_2)$ we find that $F_{X^{(3)}_{\min}}$ and $F_{X^{(3)}_{\max}}$ are distribution functions with corresponding variables $X^{(3)}_{\min}$ and $X^{(3)}_{\max}$,

$$
X^{(3)}_{\min} = \begin{cases} a & \text{with probability } \frac{\mu_2 - \mu_1^2}{(a - \mu_1)^2 + \mu_2 - \mu_1^2}, \\ \mu_1 + \frac{\mu_2 - \mu_1^2}{\mu_1 - a} & \text{with probability } \frac{(a - \mu_1)^2}{(a - \mu_1)^2 + \mu_2 - \mu_1^2}, \end{cases}
\tag{11.12}
$$

and

$$
X^{(3)}_{\max} = \begin{cases} \mu_1 - \frac{\mu_2 - \mu_1^2}{b - \mu_1} & \text{with probability } \frac{(b - \mu_1)^2}{(b - \mu_1)^2 + \mu_2 - \mu_1^2}, \\ b & \text{with probability } \frac{\mu_2 - \mu_1^2}{(b - \mu_1)^2 + \mu_2 - \mu_1^2}. \end{cases}
\tag{11.13}
$$

In this case, the construction of F_{Z^*} that makes it possible to bound $F_X(t)$ depends on the position of t with respect to

$$
\text{supp}(X^{(3)}_{\min}) \cup \text{supp}(X^{(3)}_{\max}) = \left\{ a, \mu_1 - \frac{\mu_2 - \mu_1^2}{b - \mu_1}, \mu_1 + \frac{\mu_2 - \mu_1^2}{\mu_1 - a}, b \right\}.
$$

Using Algorithm 2, the final result we get is as follows:

- $0 = P(Z^* < t) \le F_X(t) \le P(Z^* \le t) = \frac{\mu_2 - \mu_1^2}{(\mu_1 - t)^2 + \mu_2 - \mu_1^2}$.

- If $\mu_1 - \frac{\mu_2 - \mu_1^2}{b - \mu_1} < t < \mu_1 + \frac{\mu_2 - \mu_1^2}{\mu_1 - a}$, then $P(Z^* = a) = P(Z^* < t) \le F_X(t) \le P(Z^* \le t) = P(Z^* = b]$ where $P(Z^* = a) = \frac{\mu_2 - (t + b)\mu_1 + bt}{(a - t)(a - b)}$ and $P(Z^* = b) = \frac{\mu_2 - (a + b)\mu_1 + ab}{(t - a)(t - b)}$.

- If $\mu_1 + \frac{\mu_2 - \mu_1^2}{\mu_1 - a} < t < b$, then $\frac{(x - \mu_1)^2}{(t - \mu_1)^2 + \mu_2 - \mu_1^2} = P\left(Z^* = \mu_1 + \frac{\mu_2 - \mu_1^2}{\mu_1 - t}\right) = P(Z^* < t) \le F_X(t) \le P(Z^* \le t) = 1$.

11.1.3   Moment Bounds on VaR

We now derive VaR bounds for random losses that have some prescribed moments $\mu_1, \mu_2, \ldots, \mu_{s-1}$. Hence, we consider the problems

$$\underline{\text{VaR}}_\alpha := \inf\{\text{VaR}_\alpha(X) \mid F_X \in \mathcal{F}([a,b], \mu_1, \mu_2, \ldots, \mu_{s-1})\}, \tag{11.14}$$

$$\overline{\text{VaR}}_\alpha := \sup\{\text{VaR}_\alpha^+(X) \mid F_X \in \mathcal{F}([a,b], \mu_1, \mu_2, \ldots, \mu_{s-1})\}. \tag{11.15}$$

Inverting the bounds on distribution functions obtained in the preceding section provides us with the corresponding VaR bounds provided some conditions are fulfilled, as demonstrated in Barrieu and Scandolo (2015). A more direct derivation is nevertheless possible, and we propose here an approach to obtaining bounds on VaR without resorting to bounds on distribution functions.

To obtain VaR bounds we proceed in a similar way as in Section 11.1.2 for finding bounds on distribution functions. Hence, to bound $F_X(t)$ in $\mathcal{F}([a,b], \mu_1, \mu_2, \ldots, \mu_{s-1})$ we use de Vylder's method to construct, among the distribution functions F_Z in $\mathcal{F}([a,b]; \mu_1, \mu_2, \ldots, \mu_{s-1})$ having the property that $S^-(F_Y - F_Z) = s$ for all other F_Y in the moment space, the element (i.e., F_{Z^*}) having t as a particular mass point.

It is then clear that in order to find VaR bounds on the moment space $\mathcal{F}([a,b], \mu_1, \mu_2, \ldots, \mu_{s-1})$, we need to consider as distribution function F_{Z^*} such that $F_{Z^*}(z_j) = \alpha$ for some mass point z_j. Indeed, any other distribution F_Y in the moment space must then cross F_{Z^*} on $(z_j, z_{j+1}))$ and cross $(\text{VaR}_\alpha(Z^*); \text{VaR}_\alpha^+(Z^*))$. We obtain that

$$\underline{\text{VaR}}_\alpha = \text{VaR}_\alpha(Z^*) \quad \text{and} \quad \overline{\text{VaR}}_\alpha = \text{VaR}_\alpha^+(Z^*]. \tag{11.16}$$

Unfortunately, one cannot use de Vylder's method to identify F_{Z^*} directly (because we need to fix a probability and not a mass point) but rather have to consider all possible F_Z (constructed with de Vylder's method) and take the one that satisfies $F_Z(z_j) = \alpha$. However, the position of the level α in the union of the ranges of the distribution functions of the s-convex extrema $F_{X_{\min}^{(s)}}$ and $F_{X_{\max}^{(s)}}$ indicates the mass point at which F_Z must reach level α (to ensure that we indeed have s crossings). We explain this further as follows:

Even number of moments fixed $(s = 2m + 1)$

The mass points of $F_{X_{\min}^{(s)}}$ are $a = x_1 < x_2 < \cdots < x_{m+1} < b$ and the mass points of $F_{X_{\max}^{(s)}}$ are $a < y_1 < y_2 < \cdots < y_m < y_{m+1} = b$. The x_j and y_j satisfy (11.11). We distinguish the following cases:

- If $\alpha \in \left(F_{X_{\min}^{(s)}}(x_j), F_{X_{\max}^{(s)}}(y_j)\right)$, i.e., $\alpha \in \left(\sum_{k=1}^j p_k, \sum_{k=1}^j q_k\right)$, then F_Z must take the value α from before $F_{X_{\max}^{(s)}}^{-1}(\alpha) = y_j$ to after $F_{X_{\min}^{(s)}}^{-1}(\alpha) = x_{j+1}$. This means that F_Z has no mass point in (y_j, x_{j+1}) so that each of the $m + 1$ mass points of F_Z belongs to an interval (x_ℓ, y_ℓ), $\ell = 1, \ldots, m + 1$, and a and b do not belong to the support of Z. Level α is reached by F_Z at the mass point located in (x_j, y_j), that is, $F_Z(z_j) = \alpha$, and

$$\underline{\text{VaR}}_\alpha = z_j \quad \text{and} \quad \overline{\text{VaR}}_\alpha = z_{j+1}. \tag{11.17}$$

- If $\alpha \in \left(F_{X_{\max}^{(s)}}(y_j), F_{X_{\min}^{(s)}}(x_{j+1})\right)$, i.e., $\alpha \in \left(\sum_{k=1}^j q_k, \sum_{k=1}^{j+1} p_k\right)$, then F_Z must take the value α from before $F_{X_{\min}^{(s)}}^{-1}(\alpha) = x_{j+1}$ to after $F_{X_{\min}^{(s)}}^{-1}(\alpha) = y_{j+1}$. This means that F_Z has no mass point in (x_{j+1}, y_{j+1}), and the mass points of F_Z must be a, b, and the (unique) elements in each interval $(y_\ell, x_{\ell+1})$, $\ell = 1, \dots, m$. Level α is reached by F_Z at the mass point located in (y_j, x_{j+1}), that is, $F_Z(z_{j+1}) = \alpha$, and

$$\underline{\text{VaR}}_\alpha = z_{j+1} \quad \text{and} \quad \overline{\text{VaR}}_\alpha = z_{j+2}. \tag{11.18}$$

- If $\alpha \in \left(0, F_{X_{\min}^{(s)}}(x_1)\right)$, i.e., $\alpha \le p_1$, then we consider F_Z having mass points $a = z_1 < z_2 < \cdots < z_{m+2} = b$ with probability $P(Z = a) = \alpha$, and

$$\underline{\text{VaR}}_\alpha = z_1 \quad \text{and} \quad \overline{\text{VaR}}_\alpha = z_2. \tag{11.19}$$

Odd number of moments fixed ($s = 2m$)

The mass points of $F_{X_{\min}^{(s)}}$ are $a < x_1 < x_2 < \cdots < x_m < b$, and the mass points of $F_{X_{\max}^{(s)}}$ are $a = y_1 < y_2 < \cdots < y_m < y_{m+1} = b$. The x_j and y_j satisfy (11.10). The following cases can be identified:

- If $\alpha \in (F_{X_{\max}^{(s)}}(y_j), F_{X_{\min}^{(s)}}(x_j))$, i.e., $\alpha \in \left(\sum_{k=1}^j q_k, \sum_{k=1}^j p_k\right)$, then F_Z must take the value α from before $F_{X_{\min}^{(s)}}^{-1}(\alpha) = x_j$ to after $F_{X_{\max}^{(s)}}^{-1}(\alpha) = y_{j+1}$. Hence, Z has no mass point in (x_j, y_{j+1}) so that the mass points of F_Z must include b and the (unique) elements in each interval (y_ℓ, x_ℓ), $\ell = 1, \dots, m$. Level α is reached by F_Z at the mass point located in (y_j, x_j), that is, $F_Z(z_j) = \alpha$, and

$$\underline{\text{VaR}}_\alpha = z_j \quad \text{and} \quad \overline{\text{VaR}}_\alpha = z_{j+1}.$$

- If $\alpha \in \left(F_{X_{\min}^{(s)}}(x_j), F_{X_{\max}^{(s)}}(y_{j+1})\right)$, i.e., $\alpha \in \left(\sum_{k=1}^j p_k, \sum_{k=1}^{j+1} q_k\right)$, then F_Z must take the value α from before $F_{X_{\max}^{(s)}}^{-1}(\alpha) = y_{j+1}$ to after $F_{X_{\min}^{(s)}}^{-1}(\alpha) = x_{j+1}$. This means that F_Z has no mass point in (y_{j+1}, x_{j+1}), and the mass points of F_Z must be a and the (unique) elements in each interval $(x_\ell, y_{\ell+1})$, $\ell = 1, \dots, m$. Level α is reached by F_Z at the mass point located in (x_j, y_{j+1}), that is, $F_Z(z_j) = \alpha$, and

$$\underline{\text{VaR}}_\alpha = z_j \quad \text{and} \quad \overline{\text{VaR}}_\alpha = z_{j+1}.$$

- If $\alpha \in \left(0, F_{X_{\min}^{(s)}}(x_j)\right)$, then we consider F_Z having mass points $a = z_1 < z_2 < \cdots < z_{m+1} < b$ with probability $P(Z = a) = \alpha$, and

$$\underline{\text{VaR}}_\alpha = z_1 = a \quad \text{and} \quad \overline{\text{VaR}}_\alpha = z_2.$$

Finally, we point out that the mass points of the s-convex extrema surrounding z_j provide easy-to-compute VaR bounds (that are, however, not sharp). An important application is the case of VaR bounds when mean and variance are given.

Theorem 11.7 (VaR bounds, bounded domain, known mean and variance) *The VaR bounds on $\mathcal{F}([a, b], \mu_1, \mu_2)$ are as follows:*

- For $\alpha < \frac{\mu_2 - \mu_1^2}{(a-\mu_1)^2 + \mu_2 - \mu}$,

$$\underline{\text{VaR}}_\alpha = a \quad and \quad \overline{\text{VaR}}_\alpha = \frac{\alpha a(a-b) + b\mu_1 - \mu_2}{b - \mu_1 + \alpha(a-b)}.$$

- For $\frac{\mu_2 - \mu_1^2}{(a-\mu_1)^2 + \mu_2 - \mu_1^2} < \alpha < \frac{(b-\mu_1)^2}{(b-\mu_1)^2 + \mu_2 - \mu_1^2}$,

$$\underline{\text{VaR}}_\alpha = \mu_1 - \sqrt{\frac{1-\alpha}{\alpha}(\mu_2 - \mu_1^2)}$$

$$and \quad \overline{\text{VaR}}_\alpha = \mu_1 + \sqrt{\frac{\alpha}{1-\alpha}(\mu_2 - \mu_1^2)}.$$

- For $\alpha > \frac{(b-\mu_1)^2}{(b-\mu_1)^2 + \mu_2 - \mu_1^2}$,

$$\underline{\text{VaR}}_\alpha = \frac{(1-\alpha)b(b-a) + a\mu_1 - \mu_2}{a - \mu_1 + (1-\alpha)(b-a)} \quad and \quad \overline{\text{VaR}}_\alpha = b.$$

11.1.4   Application: VaR Bounds for a Portfolio of Risks

The so-called Pareto–Clayton model is of interest in the field of risk modeling; we refer here to Albrecher et al. (2011) and Cuberos et al. (2015), as well as to some works cited below. Here, we consider an insurance portfolio $(X_1, X_2, \ldots, X_n)$ and assume that, conditionally on some factor $\Lambda = \lambda$, the claims X_i are independent and exponentially distributed with parameter λ. When Λ is degenerate, the claims X_i have a fixed failure rate, and the model can be seen as the least dangerous one in insurance applications. However, by letting Λ vary we obtain marginal risks that are more heavily tailed and that are interrelated. Specifically, if we take Λ as a Gamma(α_1, β_1)-distributed variable, we find that the different X_i are Pareto-distributed and exhibit a Clayton-type dependence (Cuberos et al., 2015).

Under the Pareto–Clayton model, the exact distribution of the portfolio sum $S = \sum_{i=1}^n X_i$ can be obtained. Indeed, conditionally on $\Lambda = \lambda$, we have that S is Gamma(n, λ)-distributed. Invoking results from equations (1.1) and (1.3) of Dubey (1970), we find that S/β_1 has a Beta distribution of the second kind (also called a Beta prime distribution) with parameters n and α_1.[4] The moments of S are thus given as

$$ES^k = \frac{\beta_1^k B(\alpha_1 - k, n + k)}{B(n, \alpha_1)}. \tag{11.20}$$

Furthermore, its quantile function $F_S^{-1}(p)$ can be expressed as $F_S^{-1}(p) = \beta_1 \frac{G^{-1}(p)}{1 - G^{-1}(p)}$, in which G denotes the distribution function of a Beta-distributed random variable with parameters n and α_1. While available data may indicate that using a Pareto–Clayton model is reasonable, such a finding cannot guarantee that this model is "true." We shed light on this model uncertainty by assuming that data are only "rich enough" to trust the first k moments μ_i, $i = 1, 2, \ldots, k$ of the portfolio sum S. Specifically, we

---

[4] S/β_1 has density $f_{S/\beta_1}(x) = \frac{x^{n-1}}{B(n,\alpha_1)(1+x)^{n+\alpha_1}}$, $x > 0$, in which $B(\cdot, \cdot)$ is the Beta function.

assume that these moments match those given in (11.20) and that S takes values in the interval $[0, 1000]$. In Table 11.1 we consider $n = 100$, $\alpha_1 = 10$, $\beta_1 = 1$ and provide VaR bounds[5] for various values of the confidence level α. It can be checked that the moment sequence (11.20) is admissible for the interval $[0, 1000]$.

# of moments k	$\alpha = 0.9$	$\alpha = 0.95$	$\alpha = 0.99$
1	(0; 111.11)	(0; 222.22)	(1.12; 1000)
2	(9.74; 23.41)	(10.17; 28.99)	(10.70; 51.92)
3	(10.13; 22.47)	(10.81; 26.19)	(11.68; 38.71)
4	(10.13; 22.46)	(11.46; 25.56)	(15.08; 32.94)
5	(10.15; 22.41)	(11.51; 25.53)	(15.09; 32.92)

Table 11.1 Minimum and maximum possible values of VaR when the first k moments μ_i, $i = 1, 2, \ldots, k$ of the loss are fixed ($a = 0, b = 1000$). The VaR bounds are reported as (;), in which the first value is the lower bound $\underline{\text{VaR}}_\alpha$ and the second value is the upper bound $\overline{\text{VaR}}_\alpha$ ($\alpha = 90\%$, $\alpha = 95\%$, or $\alpha = 99\%$). The VaR numbers are obtained by assuming that the loss is distributed according to a Pareto–Clayton model with parameters $\alpha_1 = 10$ and $\beta_1 = 1$ and are given by $\text{VaR}_{90\%} = 16.29$, $\text{VaR}_{95\%} = 18.75$, and $\text{VaR}_{99\%} = 24.79$.

Note that when only two moments are given, we can directly build on Example 11.7 to obtain closed-form expressions for the VaR bounds. Hence, for the stated parameter values $n = 100$, $a = 0$, and $b = 1000$, we find from (11.20) that $\mu_1 = 11.11$, $\mu_2 = 140.28$, and we obtain:

- If $\alpha < 0.1199$, then $\underline{\text{VaR}}_\alpha = 0$ and $\overline{\text{VaR}}_\alpha = \frac{10970.83}{988.89 - 1000\alpha}$.
- If $0.1199 \leq \alpha \leq 0.99998$, then $\underline{\text{VaR}}_\alpha = 11.11 - 4.10\sqrt{\frac{1-\alpha}{\alpha}}$ and $\overline{\text{VaR}}_\alpha = 11.11 + 4.10\sqrt{\frac{\alpha}{1-\alpha}}$.
- If $\alpha > 0.99998$, then $\underline{\text{VaR}}_\alpha = \frac{1000^2(1-\alpha) - 140.28}{(1-\alpha)988.89 - 11.11}$ and $\overline{\text{VaR}}_\alpha = 1000$.

From Table 11.1 we observe that there can still be a substantial amount of model risk on VaR assessments even when several moments are known. Note that the VaRs that result from the Pareto–Clayton model fall nicely within the central region of the VaR admissible range. Also, for lower probability levels ($\alpha = 0.9$, for instance), no significant improvement in the VaR bounds is achieved when more than three moments are given. For higher probability levels ($\alpha = 0.95$ and 0.99, for instance), the same comment applies when four moments are given.

11.1.5   The Impact of Statistical Uncertainty

The VaR bounds that we have derived so far assume that moments are known exactly. In practice, these moments have to be estimated from data and are not known with certainty. Thus, we extend our framework to account for statistical uncertainty with respect to moment estimates. It turns out that the setting we propose makes it possible to greatly simplify the analysis, and the VaR bounds are very easy to compute for any given number of moment constraints.

---

[5] Note that only the first nine moments of S are finite.

To account for statistical uncertainty and the impact it may have on the risk bounds, we propose to fix the first moment but to allow moments of order k to take values up to some finite maximum d_k, $k = 2, 3, \ldots, s - 1$, reflecting, for instance, the upper limit of a confidence interval around the best estimate. It is clear that doing so is also prudent in that the risk bounds will become wider, as compared to the situation we considered previously (fixed moments). Note that we could also consider the minimum values that moments have to take based on a two-sided confidence interval around the best estimate. In most cases, doing so would not affect the upper bound on VaR, which is the most interesting bound, as it quantifies the maximum risk exposure. Indeed, we will observe that the maximizing distribution that we obtain by the sole use of upper inequality constraints typically shows moments that are close to these upper constraints. In these instances, adding lower constraints on the moments would not affect the results.

Hence, we denote by $\mathcal{F}_{\leq}([a, b], \mu_1, d_2, \ldots, d_{s-1})$ the class of all distribution function F_X having domain $[a, b]$ and for which the first moment μ_1 is given as well as upper bounds d_k, $k = 2, 3, \ldots, s - 1$, for the $s - 2$ higher order moments; i.e., $F_X \in \mathcal{F}_{\leq}([a, b], \mu_1, d_2, \ldots, d_{s-1})$ if and only if F_X is a distribution function with domain $[a, b]$, $\int_a^b x \, dF(x) = \mu_1$, and $\int_a^b x^d \, dF(x) \leq d_k$, $k = 2, 3, \ldots, s - 1$.

To identify the maximum and minimum VaR in this class, we consider now the problems:

$$\underline{\text{VaR}}_\alpha^{\leq} := \inf\{\text{VaR}_\alpha(X) \mid F_X \in \mathcal{F}_{\leq}([a, b], \mu_1, d_2, \ldots, d_{s-1})\}, \tag{11.21}$$

$$\overline{\text{VaR}}_\alpha^{\leq} := \sup\{\text{VaR}_\alpha^+(X) \mid F_X \in \mathcal{F}_{\leq}([a, b], \mu_1, d_2, \ldots, d_{s-1})\}. \tag{11.22}$$

In what follows, we restrict our analysis to distributions with non-negative domains, i.e., we take $0 \leq a < b$. Note, however, that this is not a real restriction because one can always transform a distribution into another one that has positive domain (as well as adjust the corresponding moment (bounds) $m_1, d_2, \ldots, d_{s-1}$) and adjust the VaR bounds obtained after the transformation to obtain the VaR bounds for the original distribution functions. Without loss of generality, we also assume $d_k \geq \mu_1^k$ so that $\mathcal{F}_{\leq}([a, b], \mu_1, d_2, \ldots, d_{s-1})$ contains at least one element, namely the degenerate distribution with masspoint μ_1. It turns out that in the construction of VaR bounds, the following random variables Z_η are of interest:

$$Z_\eta = \begin{cases} \mu_1 - \eta \frac{1-\alpha}{\alpha} & \text{with probability } \alpha, \\ \mu_1 + \eta & \text{with probability } 1 - \alpha, \end{cases} \tag{11.23}$$

where $\eta \geq 0$ and $0 < \alpha < 1$. The random variables Z_η have two-point distributions[6] with

$$\text{VaR}_\alpha(Z_\eta) = \mu_1 - \eta \frac{1-\alpha}{\alpha} \quad \text{and} \quad \text{VaR}_\alpha^+(Z_\eta) = \mu_1 + \eta.$$

We formulate the following theorem:

---

[6] Recall that when the moments are known (no uncertainty), the analysis in Section 11.1.2 involves variables Z with more than two support points, making it harder to determine explicit expressions for VaR bounds.

Theorem 11.8 (VaR bound under statistical uncertainty)
The bounds $\underline{\text{VaR}}_\alpha$ *and* $\overline{\text{VaR}}_\alpha$ *in* $\mathcal{F}_\le([a, b], \mu_1, d_2, \dots, d_{s-1})$ *are explicitly given as*

$$\underline{\text{VaR}}_\alpha = \text{VaR}_\alpha(Z_{\eta^*}) = \mu_1 - \eta^* \frac{1-\alpha}{\alpha}$$

$$and \quad \overline{\text{VaR}}_\alpha = \text{VaR}_\alpha^+(Z_{\eta^*}) = \mu_1 + \eta^*, \tag{11.24}$$

where η^* *is the maximum value in* $\left[0, \min\left\{b - \mu_1, (\mu_1 - a)\frac{\alpha}{1-\alpha}\right\}\right]$ *such that the inequalities*

$$(\mu_1 + \eta^*)^k (1 - \alpha) + \left(\mu_1 - \eta^* \frac{1-\alpha}{\alpha}\right)^k \alpha \le d_k \tag{11.25}$$

hold for $k = 1, \dots, s - 1$.

The proof of this result can be found in Section 11.4.4. Note that the VaR bounds are given in explicit form in that no inversion of a distribution function is required. They are also attainable as $F_{Z_{\eta^*}} \in \mathcal{F}_\le([a, b], \mu_1, d_2, \dots, d_{s-1})$.

We now compare, for the portfolio of dependent risks, VaR bounds from Theorem 11.8 with uncertainty on the moments with VaR bounds obtained assuming fixed moments. To do so, we use the same Pareto–Clayton-type portfolio of policies as in Section 11.1.4. Specifically, we consider upper bound constraints d_k given as $d_k = \frac{\beta_1^k B(\alpha_1 - k, n+k)}{B(n, \alpha_1)}$; see (11.20).

# of moments k	$\alpha = 0.9$	$\alpha = 0.95$	$\alpha = 0.99$
1	(0; 111.11)	(0; 222.22)	(1.12; 1000)
2	(9.74; 23.42)	(10.17; 28.99)	(10.70; 51.91)
3	(9.80; 22.87)	(10.26; 27.31)	(10.77; 43.10)
4	(9.82; 22.74)	(10.30; 26.42)	(10.82; 38.26)
5	(9.82; 22.74)	(10.31; 26.21)	(10.86; 34.72)
$6 \le k \le 10$	(9.82; 22.74)	(10.31; 26.21)	(10.87; 34.72)

Table 11.2 VaR bounds when $\mu_1 = 11.11$ and $\mu_i \le \frac{\beta_1^i B(\alpha_1 - i, n+i)}{B(n, \alpha_1)}$ for $i = 2, 3, \dots, 10$, $a = 0$, $b = 1000$, $\alpha_1 = 10$, $\beta_1 = 1$. The VaR bounds are reported as (;), in which the first value is the lower bound $\underline{\text{VaR}}_\alpha$ and the second value is the upper bound $\overline{\text{VaR}}_\alpha$, $\alpha = 90\%$.

As expected, we observe from Table 11.2 that, as compared to the case of moments that are known precisely, the upper bounds are larger, although the differences appear fairly limited. For $\alpha = 0.9$, in the Pareto–Clayton model, we find that the value of η^* given in (11.25) is constrained by the moment of order 4, so that the inequalities for higher-order moments are all satisfied. Therefore, the bounds do not change when we add moments from 5 to 10. The same phenomenon occurs with the moment of order 5 for the upper VaR bound at levels $\alpha = 0.95$ or $\alpha = 0.99$.

Hereafter, we provide upper and lower bounds on VaR, TVAR, and their generalization range value at risk (RVaR) under constraints on the first two moments for the case of an unbounded domain of the loss distribution. These have been extensively studied in previous literature; indicatively see Ghaoui et al. (2003) and Barrieu and

Scandolo (2015) for the VaR, Natarajan et al. (2010) for the TVaR, and Li et al. (2018) for the RVaR.

11.2   VaR Bounds: Unbounded Domain

When the domain of the loss distribution is unbounded, explicit formulas for the upper and lower bounds on VaR can often be obtained. By letting $a \to -\infty$ and $b \to \infty$ in Theorem 11.7, we obtain the following result:

Theorem 11.9   (VaR bounds, domain $\mathbb{R}$, known mean and variance) *The bounds* $\underline{\mathrm{VaR}}_\alpha$ *and* $\overline{\mathrm{VaR}}_\alpha$ *on* $\mathcal{F}((-\infty, +\infty), \mu_1, \mu_2)$ *are given as*

$$\underline{\mathrm{VaR}}_\alpha = \mu_1 - \sqrt{\frac{1-\alpha}{\alpha}(\mu_2 - \mu_1^2)} \quad and \quad \overline{\mathrm{VaR}}_\alpha = \mu_1 + \sqrt{\frac{\alpha}{1-\alpha}(\mu_2 - \mu_1^2)},$$

$$0 < \alpha < 1.$$

These bounds are also known as the Cantelli bounds. Furthermore, if in Theorem 11.7 we take $a = 0$ and let $b \to \infty$, we obtain the following bounds:

Theorem 11.10   (VaR bounds, domain $\mathbb{R}^+$, known mean and variance) *The bounds* $\underline{\mathrm{VaR}}_\alpha$ *and* $\overline{\mathrm{VaR}}_\alpha$ *on* $\mathcal{F}([0, +\infty), \mu_1, \mu_2)$ *are given as:*

- *For* $0 < \alpha < \frac{\mu_2 - \mu_1^2}{\mu_2}$,

$$\underline{\mathrm{VaR}}_\alpha = 0 \quad and \quad \overline{\mathrm{VaR}}_\alpha = \frac{\mu_1}{1-\alpha}.$$

- *For* $\frac{\mu_2 - \mu_1^2}{(a-\mu_1)^2 + \mu_2 - \mu_1^2} < \alpha < \frac{(b-\mu_1)^2}{(b-\mu_1)^2 + \mu_2 - \mu_1^2}$,

$$\underline{\mathrm{VaR}}_\alpha = \mu_1 - \sqrt{\frac{1-\alpha}{\alpha}(\mu_2 - \mu_1^2)} \quad and \quad \overline{\mathrm{VaR}}_\alpha = \mu_1 + \sqrt{\frac{\alpha}{1-\alpha}(\mu_2 - \mu_1^2)}.$$

In Theorem 11.8 we provide VaR bounds under statistical uncertainty, i.e., under moment inequalities. An inspection of the proof reveals that this theorem also holds when $b = \infty$. We obtain the following theorem:

Theorem 11.11   (VaR bounds under statistical uncertainty, domain is $\mathbb{R}^+$) *The bounds* $\underline{\mathrm{VaR}}_\alpha^\leqq$ *and* $\overline{\mathrm{VaR}}_\alpha^\leqq$ *in* $\mathcal{F}_\leq([0, \infty), \mu_1, d_2, \ldots, d_{s-1})$ *are explicitly given as*

$$\underline{\mathrm{VaR}}_\alpha^\leqq = \mathrm{VaR}_\alpha(Z_{\eta^*}) = \mu_1 - \eta^* \frac{1-\alpha}{\alpha}$$

$$and \quad \overline{\mathrm{VaR}}_\alpha^\leqq = \mathrm{VaR}_\alpha^+(Z_{\eta^*}) = \mu_1 + \eta^*, \tag{11.26}$$

where η^* *is the maximum value in the interval* $[0, \mu_1 \frac{\alpha}{1-\alpha}]$ *such that the inequalities*

$$(\mu_1 + \eta^*)^k (1-\alpha) + \left(\mu_1 - \eta^* \frac{1-\alpha}{\alpha}\right)^k \alpha \leq d_k \tag{11.27}$$

hold for $k = 1, \ldots, s - 1$.

11.3 TVaR and RVaR Bounds

VaR is the reference risk measure for risk quantification. Its main competitor is TVaR, also known as expected shortfall (ES), which indicates how much the expected loss is, given the loss exceeds VaR. We also study the so-called range value at risk (RVaR), proposed by Cont et al. (2010), as it can be seen as an interpolation between VaR and TVaR incorporating these two measures as special cases. The RVaR of a portfolio loss S is defined as

$$\text{RVaR}_{\alpha,\beta}(S) = \frac{1}{\beta - \alpha} \int_{\alpha}^{\beta} \text{VaR}_u(S) \, du, \qquad 0 < \alpha < \beta < 1. \tag{11.28}$$

Hence, $\text{RVaR}_{\alpha,\beta}(S)$ measures the average loss between confidendence level α and β. In this regard, Cont et al. (2010) argue that, while TVaR is coherent in the sense described in Artzner et al. (1999), it is not robust, in that small variations in the loss distribution, resulting either from estimation or misspecification, could result in significant variations in (the estimator for) TVaR. By contrast, VaR is robust but not coherent. Cont et al. (2010) point out that robustness and coherence are two conflicting objectives and that the use of RVaR makes it possible to balance these competing objectives appropriately.

Hürlimann (2002) provides an analytical expression for $\overline{\text{TVaR}}_\alpha$ on $\mathcal{F}([a, b], \mu_1, \mu_2)$ as given in the following theorem.

Theorem 11.12 (TVaR upper bounds, bounded domain, mean and variance are known) *The bound* $\overline{\text{TVaR}}_\alpha$ *on* $\mathcal{F}([a, b], \mu_1, \mu_2)$ *is given as:*

- *For* $\alpha < \frac{\mu_2 - \mu_1^2}{(a - \mu_1)^2 + \mu_2 - \mu_1^2}$,

$$\overline{\text{TVaR}}_\alpha = \frac{\mu_1 - a\alpha}{1 - \alpha}.$$

- *For* $\frac{\mu_2 - \mu_1^2}{(a - \mu_1)^2 + \mu_2 - \mu_1^2} < \alpha < \frac{(b - \mu_1)^2}{(b - \mu_1)^2 + \mu_2 - \mu_1^2}$,

$$\overline{\text{TVaR}}_\alpha = \mu_1 + \sqrt{\frac{\alpha}{1 - \alpha}(\mu_2 - \mu_1^2)}.$$

- *For* $\alpha > \frac{(b - \mu_1)^2}{(b - \mu_1)^2 + \mu_2 - \mu_1^2}$,

$$\overline{\text{TVaR}}_\alpha = b.$$

By taking in Theorem 11.12, $a = 0$ and letting $b \to \infty$, we obtain the following result.

Theorem 11.13 (TVaR bounds, domain is $\mathbb{R}^+$, mean and variance are known) *The bound* $\overline{\text{TVaR}}_\alpha$ *on* $\mathcal{F}((0, +\infty, \mu_1, \mu_2)$ *is given as:*

- *For $\alpha < \frac{\mu_2 - \mu_1^2}{\mu_2}$,*

$$\overline{\text{TVaR}}_\alpha = \frac{\mu_1}{1 - \alpha}.$$

- *For $\frac{\mu_2 - \mu_1^2}{(a - \mu_1)^2 + \mu_2 - \mu_1^2} < \alpha < 1$,*

$$\overline{\text{TVaR}}_\alpha = \mu_1 + \sqrt{\frac{\alpha}{1 - \alpha}(\mu_2 - \mu_1^2)}.$$

In the next theorem, we provide TVaR bounds in the presence of statistical uncertainty on the moment estimates.

Theorem 11.14 (TVaR bounds under statistical uncertainty)
The bounds $\underline{\text{TVaR}}_\alpha$ and $\overline{\text{TVaR}}_\alpha$ on $\mathcal{F}_\leq([a, b], \mu_1, d_2, \ldots, d_{s-1}), 0 < \alpha < 1$, are explicitly given as

$$\underline{\text{TVaR}}_\alpha = \mu_1 \ and \ \overline{\text{TVaR}}_\alpha = \text{VaR}_\alpha^+[Z_\psi] = \mu_1 + \psi, \tag{11.29}$$

where ψ is the maximum value in the interval $[0, \min\{b - \mu_1, (\mu_1 - a)\frac{\alpha}{1-\alpha}\}]$ such that the inequalities

$$(\mu_1 + \psi)^k (1 - \alpha) + \left(\mu_1 - \psi\frac{1 - \alpha}{\alpha}\right)^k \alpha \leq d_k \tag{11.30}$$

hold for $k = 1, \ldots, s - 1$.

Note that under statistical uncertainty, the TVaR upper bounds coincide with upper VaR bounds. Thus, a numerical illustration can be found in Table 11.2. The property that in the case of inequality constraints on moments, the upper bounds on VaR and TVaR coincide offers some new perspective on the discussion of the appropriateness of either risk measure.

Theorem 11.15 (TVaR bounds, domain $\mathbb{R}$, mean and variance are known) *The bounds $\underline{\text{TVaR}}_\alpha$ and $\overline{\text{TVaR}}_\alpha$ on $\mathcal{F}((-\infty, +\infty), \mu_1, \mu_2), 0 < \alpha < 1$, are given as*

$$\underline{\text{TVaR}}_\alpha = \mu_1 - \sqrt{\frac{1 - \alpha}{\alpha}(\mu_2 - \mu_1^2)} \quad and \quad \overline{\text{TVaR}}_\alpha = \mu_1 + \sqrt{\frac{\alpha}{1 - \alpha}(\mu_2 - \mu_1^2)}.$$

Theorem 11.16 (RVaR upper and lower bounds, domain is $\mathbb{R}$, mean and variance are known) *The bounds $\underline{\text{RVaR}}_{\alpha,\beta}$ and $\overline{\text{RVaR}}_{\alpha,\beta}$ on $\mathcal{F}((-\infty, +\infty), \mu_1, \mu_2), 0 < \alpha < \beta < 1$, are given as follows.*

$$\underline{\text{RVaR}}_{\alpha,\beta} = \mu - \sqrt{\frac{1 - \beta}{\beta}(\mu_2 - \mu_1^2)} \ and \ \overline{\text{RVaR}}_{\alpha,\beta} = \mu_1 + \sqrt{\frac{\alpha}{1 - \alpha}(\mu_2 - \mu_1^2)}.$$

11.4 Proofs

11.4.1 Proof of Proposition 11.3

Odd number of moments are fixed ($s = 2m$):
We define two particular elements in $\mathcal{F}([a, b], \mu_1, \mu_2, \ldots, \mu_{2m-1})$. The first one has the minimal number of support points, i.e., m mass points in (a, b). It puts positive masses $p_1, p_2, \ldots, p_m$ on the mass points $x_1, x_2, \ldots, x_m$, with $a < x_1 < x_2 < \cdots < x_m < b$. The mass points $x_1, x_2, \ldots, x_m$ are the m distinct roots of equation (11.2) with $i = 0$ and $j = m$. From Proposition 11.2 it follows that this distribution function is the smallest in the sense of the $2m$-convex order among all distribution functions in $\mathcal{F}([a, b], \mu_1, \mu_2, \ldots, \mu_{2m-1})$. We denote it by $F_{X_{\min}^{(2m)}}$.

The second element in $\mathcal{F}(\mu_1, \mu_2, \ldots, \mu_{2m-1})$ of interest has $m + 1$ mass points: a, b and $m - 1$ points strictly in (a, b). It puts positive masses $q_1, q_2, \ldots, q_{m+1}$ on the mass points $a, y_2, \ldots, y_m, b$, with $a < y_2 < \cdots < y_m < b$. The mass points $y_2, \ldots, y_m$ are the $m - 1$ distinct roots of equation (11.2) with $i = 2$, $\alpha_1 = a$, $\alpha_2 = b$, and $j = m - 1$. Proposition 11.2 implies that this distribution function is the largest in the sense of the $2m$-convex order among all the distribution functions in $\mathcal{F}([a, b], \mu_1, \mu_2, \ldots, \mu_{2m-1})$. It will be denoted below by $F_{X_{\max}^{(2m)}}$.

Even number of moments are fixed ($s = 2m + 1$):
We consider two particular discrete distribution functions in $\mathcal{F}([a, b], \mu_1, \mu_2, \ldots, \mu_{2m})$ having $(m + 1)$ mass points with one of the mass points given as either a or b.

The first one puts positive masses $p_1, p_2, \ldots, p_{m+1}$ on the mass points $a, x_2, \ldots, x_{m+1}$, with $a < x_2 < \cdots < x_{m+1} < b$. Here, the $x_2, \ldots, x_{m+1}$ are the s distinct roots of equation (11.2) with $i = 1$, $\alpha_1 = a$, and $j = m$. From Proposition 11.2 we know that the corresponding distribution has at least $2m - 1$ crossing points with any other distribution function in $\mathcal{F}([a, b], \mu_1, \mu_2, \ldots, \mu_{2m})$, and by construction it must then have exactly $2m$ such crossings. This distribution function is the smallest for the $(2m + 1)$-convex order among all distribution functions in $\mathcal{F}([a, b], \mu_1, \mu_2, \ldots, \mu_{2m})$. We denote it by $X_{\min}^{(2m+1)}$.

The second one puts positive masses $q_1, q_2, \ldots, q_{m+1}$ on the mass points $y_1, y_2, \ldots, y_m, b$, with $a < y_1 < y_2 < \cdots < y_m < b$, where $y_1, y_2, \ldots, y_m$ are the m distinct roots of (11.2) with $i = 1$, $\alpha_1 = b$, and $j = m$. From Proposition 11.2 it readily follows that this distribution function is the largest in the sense of the $(2m + 1)$-convex order among all distribution functions in $\mathcal{F}([a, b], \mu_1, \mu_2, \ldots, \mu_{2m})$. We denote it by $F_{X_{\max}^{(2m+1)}}$.

11.4.2 Derivation of Moment Bounds in Subsection 11.1.2

We consider a discrete distribution F_Z in $\mathcal{F}([a, b], \mu_1, \mu_2, \ldots, \mu_{s-1})$ with mass points that are either given as $z_1, \ldots, z_{m+2}$ with $a = z_1 < \cdots < z_{m+2} = b$, or as $z_1, \ldots, z_{m+1}$ with $a < z_1 < \cdots < z_{m+1} < b$. If the mass points are given as $z_1, \ldots, z_{m+2}$ with $z_1 = a$ and $z_{m+2} = b$, it readily shows that one has a total of exactly $2m + 1 = s$ crossings with any other element F_X in $\mathcal{F}([a, b], \mu_1, \mu_2, \ldots, \mu_{s-1})$, $F_X \neq F_{X_{\min}^{(s)}}$, $F_X \neq F_{X_{\max}^{(s)}}$:

- one crossing over each flat part (z_j, z_{j+1}), $j = 1, \ldots, m + 1$, i.e., $m + 1$ crossings;
- one crossing at each mass point $z_2, \ldots, z_{m+1}$, i.e., m crossings.

If the mass points are $z_1, \ldots, z_{m+1}$ with $a < z_1 < \cdots < z_{m+1} < b$, we also have a total of $2m + 1 = s$ crossings:
- one crossing over each flat part (z_j, z_{j+1}), $j = 1, \ldots, m$, i.e., m crossings;
- one crossing at each mass point $z_1, \ldots, z_{m+1}$, i.e., $m + 1$ crossings.

Depending on the location of t with respect to the mass points of the s-convex extrema $F_{X_{\min}^{(s)}}$ and $F_{X_{\max}^{(s)}}$, there is only one possible F_Z, including either a and b as mass points or neither of them. To explain this further let us remark that the mass points of $F_{X_{\min}^{(s)}}$ are $a = x_1 < x_2 < \cdots < x_{m+1} < b$, and the support points of $F_{X_{\max}^{(s)}}$ are $a < y_1 < y_2 < \cdots < y_m < y_{m+1} = b$. The x_j and y_j satisfy

$$a = x_1 < y_1 < x_2 < y_2 < \cdots < y_m < x_{m+1} < y_{m+1} = b.$$

At each mass point z_j of F_Z, F_Z must cross both $F_{X_{\min}^{(s)}}$ and $F_{X_{\max}^{(s)}}$ so that two cases are possible:
- If $t \in (x_j, y_j)$ for some $j = 1, \ldots, m + 1$, then each mass point of F_Z must belong to an interval (x_ℓ, y_ℓ), $\ell = 1, \ldots, m + 1$, and no probability mass is placed on the extremities a and b. The mass points of F_Z are thus given by
$$a < z_1 < z_2 < \cdots < z_m < z_{m+1} < b$$
with each $z_k \in (x_k, y_k)$ and $z_j = t$.
- If $t \in (y_j, x_{j+1})$ for some $j = 1, \ldots, m$, then each mass point of F_Z must belong to an interval $(y_\ell, x_{\ell+1})$, $\ell = 1, \ldots, m$, and a and b must both belong to the domain of Z. The mass points of F_Z are thus given by
$$a = z_1 < z_2 < \cdots < z_{m+1} < z_{m+2} = b$$
with each $z_{k+1} \in (y_k, x_{k+1})$ for $k = 1, \ldots, m$ and $z_{j+1} = t$. Thus, we see that the location of t with respect to the partition of $[a, b]$ created by the union of the supports of $F_{X_{\min}^{(s)}}$ and $F_{X_{\max}^{(s)}}$ satisfying (11.11) determines whether both a and b belong to the support of Z or neither of them do.

11.4.3   Proof of Theorem 11.7

We consider the moment space $\mathcal{F}([a, b], \mu_1, \mu_2)$ in which (see also Example 11.6)

$$\text{range}(F_{X_{\min}^{(3)}}) \cup \text{range}(F_{X_{\max}^{(3)}})$$
$$= \left\{ 0, \frac{\mu_2 - \mu_1^2}{(a - \mu_1)^2 + \mu_2 - \mu_1^2}, \frac{(b - \mu_1)^2}{(b - \mu_1)^2 + \mu_2 - \mu_1^2}, 1 \right\}.$$

- For $\alpha < \frac{\mu_2 - \mu_1^2}{(a - \mu_1)^2 + \mu_2 - \mu_1^2}$ we select Z^* with support $\{a, t, b\}$, and we impose $P(Z^* = a) = \alpha$ so that $t = \frac{\alpha a(a - b) + b\mu_1 - \mu_2}{b - \mu_1 + \alpha(a - b)}$. The bounds are then given by

$$\underline{\text{VaR}}_\alpha = a \quad \text{and} \quad \overline{\text{VaR}}_\alpha = \frac{\alpha a(a - b) + b\mu_1 - \mu_2}{b - \mu_1 + \alpha(a - b)}.$$

- For $\frac{\mu_2-\mu_1^2}{(a-\mu_1)^2+\mu_2-\mu_1^2} < \alpha < \frac{(b-\mu_1)^2}{(b-\mu_1)^2+\mu_2-\mu_1^2}$, we select Z^* with support $\{t, \mu_1 + \frac{\mu_2-\mu_1^2}{\mu_1-t}\}$ in (a, b) such that $P(Z^* = t) = \alpha$ so that $t = \mu_1 - \sqrt{\frac{1-\alpha}{\alpha}(\mu_2 - \mu_1^2)}$. The other support point of Z^* is then equal to $\mu_1 + \sqrt{\frac{\alpha}{1-\alpha}(\mu_2 - \mu_1^2)}$. The bounds on $\text{VaR}_\alpha(X)$ are then given by

$$\underline{\text{VaR}}_\alpha = \mu_1 - \sqrt{\frac{1-\alpha}{\alpha}(\mu_2 - \mu_1^2)}$$

$$\text{and} \quad \overline{\text{VaR}}_\alpha = \mu_1 + \sqrt{\frac{\alpha}{1-\alpha}(\mu_2 - \mu_1^2)}.$$

- For $\alpha > \frac{(b-\mu_1)^2}{(b-\mu_1)^2+\mu_2-\mu_1^2}$ we consider Z^* with support $\{a, t, b\}$ with $P(Z^* \le t) = \alpha$; thus $P(Z^* = b) = 1 - \alpha$ so that $t = \frac{(1-\alpha)b(b-a)+a\mu_1-\mu_2}{a-\mu_1+(1-\alpha)(b-a)}$. Then,

$$\underline{\text{VaR}}_\alpha = \frac{(1 - \alpha)b(b - a) + a\mu_1 - \mu_2}{a - \mu_1 + (1 - \alpha)(b - a)} \quad \text{and} \quad \overline{\text{VaR}}_\alpha = b.$$

11.4.4 Proof of Theorem 11.8

We prove the result for the VaR upper bound. The result for the lower bound can be proven in a similar way. We denote the set of variables Z_α defined as in (11.23) by $\mathcal{Z}([a, b], \mu_1)$. For convenience, we use the shorthand notation $\mathcal{F}_\le([a, b])$ for $\mathcal{F}_\le([a, b], \mu_1, d_2, \dots, d_{s-1})$. For F_{Z_η} to belong to $\mathcal{F}_\le([a, b])$, it must hold that $0 \le \eta \le \psi := \min\left(b - \mu_1, (\mu_1 - a)\frac{\alpha}{1-\alpha}\right)$, and note that $F_{Z_0} \in \mathcal{Z}([a, b], \mu_1)$. Since for all $k = 1, 2, \dots, s - 1$, $\eta \to E(Z_\eta^k)$ is continuously increasing on $[0, \psi]$, it follows that there is a maximum value for η in $[0, \psi]$ such that $F_{Z_\eta} \in \mathcal{F}_\le([a, b])$. We denote this maximum value by η^*. Clearly, among all distribution functions in $\mathcal{F}_\le([a, b]) \cap \mathcal{Z}([a, b], \mu_1)$, $F_{Z_{\eta^*}}$ is by construction the one with maximum VaR at the confidence level α.

Assume next that there exists a distribution $F_X \in \mathcal{F}_\le([a, b])$ such that $\text{VaR}_\alpha^+(X) > \text{VaR}_\alpha^+(Z_{\eta^*}) = \mu_1 + \eta^*$. Without loss of generality we can represent X as $X = F_X^{-1}(U)$ in which U is a standard uniformly distributed variable. The variable X can also be expressed as

$$X = F_X^{-1}(U)I_{U \le \alpha} + F_X^{-1}(U)I_{U > \alpha}.$$

Consider the distribution function $F_Y \in \mathcal{Z}([a, b], \mu_1)$ defined as

$$Y = \text{LTVaR}_\alpha(X)I_{U \le \alpha} + \text{TVaR}_\alpha(X)I_{U > \alpha},$$

in which $\text{LTVaR}_\alpha(Y)$ is defined as $\text{LTVaR}_\alpha(X) = \frac{1}{\alpha}\int_0^\alpha \text{VaR}_u(X) \, du$, $0 < \alpha < 1$. Since $F_X \in \mathcal{F}_\le([a, b])$, Jensen's inequality implies that also $F_Y \in \mathcal{F}_\le([a, b])$. Hence, $F_Y \in \mathcal{Z}([a, b], \mu_1) \cap \mathcal{F}_\le([a, b])$ and $\text{VaR}_\alpha^+(Y) > \text{VaR}_\alpha^+(Z_{\eta^*}) = \mu_1 + \eta^*$. However, this observation contradicts the fact that among all distribution functions in $\mathcal{Z}([a, b], \mu_1) \cap \mathcal{F}_s([a, b])$, $F_{Z_{\eta^*}}$ is by construction the one with maximum VaR.

11.4.5   Proof of Theorem 11.14

Since TVaR_α is increasing in α and coincides with the mean when $\alpha = 0$, we obtain that the TVaR lower bound is obtained for the constant random variable equal to μ_1; thus it is just equal to μ_1. For the upper bound, the proof is similar to the proof of Theorem 11.11. We consider again $F_{Z_{\eta^*}}$ as the distribution function in $\mathcal{F}_\le([a, b]) \cap \mathcal{Z}([a, b], \mu_1)$ that has the highest VaR (at the confidence level α). Assume next that there exists a distribution $F_X \in \mathcal{F}_\le([a, b])$ such that $\mathrm{TVaR}_\alpha(X) > \mathrm{TVaR}_\alpha^+(Z_{\eta^*}) = \mathrm{VaR}_\alpha^+(Z_{\eta^*}) = \mu_1 + \eta^*$. Without loss of generality we can represent X as $X = F_X^{-1}(U)$ in which U is a standard uniformly distributed variable. Consider the distribution function $F_Y \in \mathcal{Z}([a, b], \mu_1)$ defined as

$$Y = \mathrm{LTVaR}_\alpha(X) I_{U \le \alpha} + \mathrm{TVaR}_\alpha(X) I_{U > \alpha}.$$

Since $F_X \in \mathcal{F}_\le([a, b])$, Jensen's inequality implies that also $F_Y \in \mathcal{F}_\le([a, b])$. Hence, $F_Y \in \mathcal{Z}([a, b], \mu_1) \cap \mathcal{F}_\le([a, b])$, and thus it follows that $\mathrm{VaR}_\alpha^+(Y) = \mathrm{TVaR}_\alpha(Y) > \mathrm{VaR}_\alpha^+(Z_{\eta^*}) = \mu_1 + \eta^*$. However, this observation contradicts the fact that among all distribution functions in $\mathcal{Z}([a, b], \mu_1) \cap \mathcal{F}_s([a, b])$, $F_{Z_{\eta^*}}$ is by construction the one with maximum VaR.

11.4.6   Proof of Theorem 11.15

The result for the upper bound follows by letting $a \to -\infty$ and $b \to \infty$ in Theorem 11.12. As for the lower bound, note that for every variable X, $\mathrm{VaR}_\alpha(X) = -\mathrm{VaR}_\alpha^+(-X)$. This observation allows us to conclude that $\underline{\mathrm{TVaR}}_\alpha$ coincides with the negative of the RVaR upper bound at confidence interval $(0, 1 - \alpha)$ on $\mathcal{F}([-\infty, +\infty], -\mu_1, \mu_2)$. From Theorem 11.16, we obtain an expression for the bound $\overline{\mathrm{RVaR}}_{\varepsilon, 1-\alpha}$, $0 < \varepsilon < 1 - \alpha$, on $\mathcal{F}([-\infty, +\infty], -\mu_1, \mu_2)$. The result follows by letting ε approach zero.

11.4.7   Proof of Theorem 11.16

The upper bound follows from the fact that $\overline{\mathrm{VaR}}_\alpha \le \overline{\mathrm{RVaR}}_{\alpha, \beta} \le \overline{\mathrm{TVaR}}_\alpha$, $0 < \alpha < \beta$. Hence, as $\overline{\mathrm{VaR}}_\alpha = \overline{\mathrm{TVaR}}_\alpha$ (Theorem 11.9 and Theorem 11.15), the result follows. In a similar way as in the proof of Theorem 11.15, we find that $\underline{\mathrm{RVaR}}_{\alpha, \beta}$ coincides with the negative of the upper RVaR bound at confidence interval $(1 - \beta, 1 - \alpha)$ on $\mathcal{F}([-\infty, +\infty], -\mu_1, \mu_2)$, which implies the result.

12 Bounds for Distortion Risk Measures under Moment Information

In this chapter, we study bounds on a distortion risk measure of a portfolio sum under the sole knowledge of some of its moments. A subclass of distortion risk measures is the class of concave distortion risk measures. Its particular importance is due to the fact that a concave distortion risk measure is coherent, which is often seen as a desirable property. Moreover, by the representation result of Kusuoka (2001), concave distortion risk measures are exactly those law-invariant coherent risk measures that are comonotone additive. No set of axioms is all encompassing, however. In this regard, Dhaene et al. (2008) point out that VaR, which is not subadditive and thus not coherent, may arise as an optimal choice in that it is the minimizer of a cost function that appears suitable when determining the required capital of a financial institution. The use of a coherent risk measure definition encourages diversification. However, in the case of heavy-tailed portfolios with infinite means, Mainik and Rüschendorf (2010) show that diversification may make the portfolio worse.

The moment bounds we derive can also be used to numerically approximate a sharp upper bound for the risk of a portfolio, measured with a distortion risk measure, when both the marginal distribution functions of the portfolio components and some of the moments of the portfolio sum are known. Specifically, in Section 6.2 of Chapter 6, we discussed how a suitable modification of the (B)RA provides a simple-to-implement numerical tool to obtain such bounds numerically when the VaR is used as a risk measure. This approach can be adapted in a straightforward way to the case of general distortion risk measures.

12.1 Bounds for Distortion Risk Measures: Bounded Domain

We first deal with distribution functions having a bounded domain. We derive upper bounds for any strictly *concave* distortion risk measure when k, not necessarily consecutive, moments of the distribution function are known.

12.1.1 Problem Formulation

As we are dealing with distribution functions with a bounded domain, we can consider them after suitable rescaling on the unit interval $[0, 1]$. Hence, let $\mathcal{F}_0$ denote a set of distribution functions on $[0, 1]$ for which $k \in \mathbb{N}_0$ moments are given,

$$\mathcal{F}_0 = \left\{ F \in \mathcal{F}([0,1]) \middle| \int_0^1 x^i \, dF(x) = c_i, i \in \mathcal{I} \right\} := \mathcal{F}((c_i)_{i \in \mathcal{I}}), \qquad (12.1)$$

where $\mathcal{I} \subset \mathbb{N}_0$ and $card(\mathcal{I}) = k$. Note that in general $\mathcal{F}_0$ may correspond to a set of distribution functions with any k moments fixed, not necessarily the first k ones and not necessarily starting with the mean. We assume that $\mathcal{F}_0$ contains at least two different elements (and hence infinitely many, since $\mathcal{F}_0$ is convex).

A distortion risk measure of a random variable X having cumulative distribution F_X is defined as the Choquet integral (Choquet, 1954)

$$H_g(X) = \int_0^\infty g(1 - F_X(x)) \, dx - \int_{-\infty}^0 [1 - g(1 - F_X(x))] \, dx, \qquad (12.2)$$

where g is a distortion function, i.e., an increasing function on $[0,1]$ with $g(0) = 0$ and $g(1) = 1$ and which is well defined when at least one of the integrals is finite. Note that $H_g(X)$ depends solely on the distribution function F_X (law-invariance), and in what follows we also write $H_g(F_X)$ instead of $H_g(X)$. It is well known (Wirch and Hardy, 1999) that when g is concave, H_g is a coherent risk measure. Examples of concave distortion risk measures are the power distortion risk measure in which $g(x) = x^\alpha, \alpha \in (0,1)$, the dual-power distortion risk measure in which $g(x) = 1 - (1 - x)^\beta, \beta \in (1, \infty)$, and the Wang distortion risk measure in which $g(x) = \Phi(\Phi^{-1}(x) + \Phi^{-1}(p)), p \in (0.5, 1)$.

We aim to determine the upper bound for a distortion risk measure on the distributional uncertainty set $\mathcal{F}_0$, i.e., we study the problem

$$\overline{H}_g(\mathcal{F}_0) = \sup_{F \in \mathcal{F}_0} H_g(F). \qquad (12.3)$$

In addition to the worst case value, we also aim to determine the distribution function that attains this bound. When only one moment is specified, say the ith one with value c_i, it is easy to show that the solution is obtained by a discrete distribution function $F^\star$ that is concentrated on 0 and 1 and has ith moment equal to c_i. To see this, observe that since it crosses all other distribution functions exactly once from above and has the largest possible mean, namely c_i, $F^\star$ will dominate all other admissible distribution functions in the sense of stop-loss order, and moreover it is well known that concave distortion risk measures are consistent with stop-loss order (Bäuerle and Müller, 2006). Since little distributional information is used in the optimization, this case is not very useful in practice in that it leads to wide bounds. Therefore, in the remainder of this section we consider only the case in which $k \geq 2$. Unless otherwise stated, it is also assumed that g is strictly concave and twice differentiable.

The following results are based on Cornilly et al. (2018). We first provide necessary conditions that maximizing distribution functions must satisfy. As a consequence, the optimization problem we consider reduces to a problem in which we only need to perform an optimization with respect to some parameters. The conditions that we obtain are in general not sufficient to single out the maximizing distribution. However, when the mean and any other higher moment are known, we show that the feasible set of maximizing distribution functions becomes a singleton, and we explicitly obtain the maximizing distribution.

12.1.2 Necessary Conditions for the General Case

The following theorem provides necessary conditions that a worst case distribution $F^\star$ for problem (12.3) must satisfy.

Theorem 12.1 (Necessary conditions, bounded domain) *For $k \geq 2$, there is a unique worst case distribution $F^\star$ for problem (12.3). This worst case distribution $F^\star$ is continuous on $(0, 1)$, and on intervals where it is not constant, $F^\star$ coincides with some F_η of the form*

$$F_\eta(x) = \begin{cases} 0 & \text{if } \sum_{i \in I} \eta_i x^{i-1} < g'(1), \\ 1 - (g')^{-1} \left(\sum_{i \in I} \eta_i x^{i-1} \right) & \text{else,} \\ 1 & \text{if } \sum_{i \in I} \eta_i x^{i-1} \geq g'(0), \end{cases} \qquad (12.4)$$

where $\eta := (\eta_1, \ldots, \eta_k)$ is a vector of admissible parameters satisfying

$$\int_0^1 x^i dF^\star(x) = c_i, \qquad i \in I, \qquad (12.5)$$

and

$$\int_a^b \left[g' \left(1 - F^\star(x) \right) - \sum_{i \in I} \eta_i x^{i-1} \right] dx = 0 \qquad (12.6)$$

on intervals $(a, b] = (F^{\star-}(c), F^{\star+}(c)]$ for all $c \in (0, 1)$.

Due to the conditions on the distortion function g, it holds that $g'(1) \leq g'(0)$, and hence the function F_η in the theorem is well defined. Theorem 12.1 does not completely characterize the maximizing distribution function F^*. In fact, it reduces problem (12.3) to a parametric optimization problem over a set of admissible parameter vectors $(\eta_1, \eta_2, \ldots, \eta_k)$. Since this set is not readily known, this optimization problem is difficult to deal with in general.

It is shown in Cornilly et al. (2018), however, that when the mean and some other higher-order moment are known, we are able to determine $F^\star$. In this regard, we note that Rustagi (1957, 1976) claims that in the case that he considers (first two moments given), the unique solution is given by F_η, as defined in Theorem 12.1, provided F_η is a distribution function; see Rustagi (1976, pg. 104) and subsequent theorems, as well as the corollary in Rustagi (1957, pg. 318). However, there is no proof for this statement, and the argument is not obvious since there could be several admissible distribution functions of the form F_η.

12.1.3   Bounds when Mean and Variance (or another Higher Order Moment) Are Known

Let $\mathcal{F}_0 = \mathcal{F}(c_1, c_i)$ be the set of distribution functions with fixed first moment c_1 and fixed ith moment c_i. Note that the case $i = 2$ covers the case of a fixed mean and variance. We formulate the following theorem.

Theorem 12.2 (Sharp bounds, bounded domain)   *If $\mathcal{F} = \mathcal{F}(c_1, c_i)$, then the worst case distribution $F^\star$ for (12.3) is given as:*

$$F^\star(x) = \begin{cases} 0 & \text{if } x < \max\left(\left(\frac{g'(1)-\eta_1}{\eta_i}\right)^{\frac{1}{i-1}}, 0\right), \\ 1 - (g')^{-1}\left(\eta_1 + \eta_i x^{i-1}\right) & \text{else,} \\ 1 & \text{if } x \geq \min\left(\left(\frac{g'(0)-\eta_1}{\eta_i}\right)^{\frac{1}{i-1}}, 1\right). \end{cases} \quad (12.7)$$

Here, η_1, η_i are uniquely determined parameters satisfying $\eta_i > 0$, $\eta_1 \in (g'(1) - \eta_i, g'(0))$,

$$\int_0^1 x \, dF^\star(x) = c_1 \quad \text{and} \quad \int_0^1 x^i \, dF^\star(x) = c_i. \quad (12.8)$$

In particular, Theorem 12.2 implies that the maximizing distribution is either continuous or a mixture of a continuous distribution with point masses in zero, one, or both. Note that the domain of (η_1, η_i) follows directly from the condition that $F^\star$ is a non-degenerate distribution on $[0, 1]$; for more details, we refer to the proof in Section 12.3.

Algorithm 12.1 Solve for (η_1, η_i)

1: mu1 ← function(eta1, etai)
2: mui ← function(eta1, etai)
3: diffMoments ← function(eta1, etai) {
4:        **return** (mu1(eta1, etai) − c1, mui(eta1, etai) − ci)
5: }
6: (eta1, etai) = solve.system(diffMoments)

Pseudocode for finding the solution (η_1, η_i) for given moments (c_1, c_i) is given in Algorithm 12.1. First, two functions are defined that compute the moments (μ_1, μ_i) of F_η, given (η_1, η_i). Then, a non-linear system is solved to find the values of (η_1, η_i) for which the moments are equal to (c_1, c_i). Due to Theorem 12.2, the solution is then found.

So far, we have only considered distribution functions with a bounded domain. After rescaling, the domain was taken as the interval $[0, 1]$. The influence of the end points of this interval, i.e., zero and one, becomes apparent in the definition of the domain of $F^\star$ and has the consequence that it is not readily possible to express the coefficients η_1 and η_i as explicit functions of c_1 and c_i. A similar expression for the maximizing distribution $F^\star$ is obtained when considering as support $[-b, b]$, $b > 0$. Taking the limit of b to infinity, the case of an unbounded domain can be approached. In the case of given mean and variance, an elegant formula for the worst case distorted expectation is obtained. We formulate the following corollary.

Theorem 12.3 (Sharp bounds, unbounded domain) *Let $\mathcal{F}_0 = \mathcal{F}(\mu, s^2)$ denote the set of distributions with unbounded domain and having mean μ and variance s^2. Then, the worst case distribution for (12.3) is unique and given as follows:*
a) When $g'(0) < \infty$,

$$F^\star(x) = \begin{cases} 0 & \text{if } x < \mu + \frac{g'(1)-1}{\eta_2}, \\ 1 - (g')^{-1}(\eta_1 + \eta_2 x) & \text{else}, \\ 1 & \text{if } x > \mu + \frac{g'(0)-1}{\eta_2}. \end{cases} \quad (12.9)$$

b) When $g'(0) = \infty$,

$$F^\star(x) = \begin{cases} 0 & \text{if } x < \mu + \frac{g'(1)-1}{\eta_2}, \\ 1 - (g')^{-1}(\eta_1 + \eta_2 x) & \text{else}. \end{cases} \quad (12.10)$$

The constants η_1, η_2 are in both cases given in explicit form as

$$\eta_1 = 1 - \eta_2\mu \quad \text{and} \quad \eta_2 = \frac{1}{s}\sqrt{\int_0^1 (g')^2 (1-p)\, \mathrm{d}p} - 1. \quad (12.11)$$

Furthermore, the worst case value is given as

$$\overline{H}_g(\mathcal{F}_0) = H_g(F^\star) = \mu + w_g s, \quad (12.12)$$

with weight w_g given as

$$w_g = \sqrt{\int_0^1 (g')^2 (1-p)\, \mathrm{d}p} - 1. \quad (12.13)$$

Formula (12.12) shows that the worst case distorted expectation is a weighted sum of the mean μ and standard deviation s in which the weight w_g allocated to the standard deviation is driven entirely by the distortion function g. In fact, the formula can be seen as a generalization of the Cantelli bounds for VaR and TVaR at confidence level p, in which case the weight w_g is given as

$$w_g = \sqrt{\frac{p}{1-p}}. \quad (12.14)$$

Ghaoui et al. (2003) consider the problem of finding a robust portfolio that optimizes the worst case VaR when the distribution of returns is only partially known in that only a mean and covariance matrix are available. In their analysis, the Cantelli bound is crucial because it essentially allows a reformulation of their optimization problem as a mean-variance optimization problem à la Markowitz. Using formula (12.12), this approach to robust portfolio selection can thus be extended to include any concave distorted expectation as objective. Formula (12.12) first appeared in Li (2018). Theorem 12.9 generalizes formula (12.12) to the case of general distortion functions.

12.1.4　Bounds for the General Case

Solving problem (12.3) for a general choice of $\mathcal{F}_0 = \mathcal{F}((c_i)_{i \in I})$ appears difficult. However, from Theorem 12.2 one obtains a simple upper bound on the distorted expectation when $\mathcal{F}_0$ contains all distribution functions with given mean and $k-1$ given higher order moments; we denote this class by $\mathcal{F}_0 = \mathcal{F}(c_1, (c_i)_{i \in I \setminus \{1\}})$. This bound is a direct consequence of the observation that the value of the distorted expectation for each of the two-moment solutions, i.e., when the mean and some other higher order moment are fixed, as in Theorem 12.2, provides an upper bound for this case.

Corollary 12.4　*Let $\mathcal{F}_0 = \mathcal{F}(c_1, (c_i)_{i \in I \setminus \{1\}})$, and denote by $F^{\star}_{c_1, c_i}$ the maximizing distribution for (12.3) when optimizing over the set $\mathcal{F}_0 = \mathcal{F}(c_1, c_i)$, $i \in I \setminus \{1\}$. Then*

$$\overline{H}_g(\mathcal{F}_0) \le \min_{i \in I \setminus \{1\}} H_g(F^{\star}_{c_1, c_i}). \tag{12.15}$$

The bound obtained in Corollary 12.4 cannot be expected to be sharp. To show this, consider the distortion function $g(x) = 1 - (1-x)^2$, and consider the optimization problem (12.3) over $\mathcal{F}(c_1, c_2)$ and $\mathcal{F}(c_1, c_3)$ with $c_1 = 0.50$, $c_2 = 0.33$, and $c_3 = 0.24$. The maximizing distribution functions can be obtained numerically using Algorithm 12.1 and are given as

$$F^{*}_{c_1, c_2}(x) = \begin{cases} 0 & \text{if } x < 0.0101, \\ -0.0103 + 1.0206x & \text{if } 0.0101 \le x < 0.9899, \\ 1 & \text{if } x \ge 0.9899 \end{cases} \tag{12.16}$$

and

$$F^{*}_{c_1, c_3}(x) = \begin{cases} 0 & \text{if } x < 0, \\ 0.1615 + 1.0482x^2 & \text{if } 0 \le x < 0.8944, \\ 1 & \text{if } x \ge 0.8944, \end{cases} \tag{12.17}$$

respectively; see Figure 12.1.

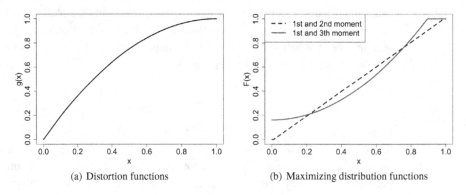

(a) Distortion functions　　　　(b) Maximizing distribution functions

Figure 12.1　Non-sharpness of the bound obtained in Corollary 12.4.

The values of the distorted expectations $H_g(F^{*}_{c_1, c_2})$ and $H_g(F^{*}_{c_1, c_3})$ are equal to 0.6633 and 0.6646, respectively. However, since the first three moments of these distribution functions are respectively $(0.5000, 0.3300, 0.2450)$ and $(0.5000, 0.3354, 0.2400)$, both distribution functions are not even admissible with respect to the the optimization problem (12.3) over the set $F_0 = \mathcal{F}(c_1, c_2, c_3)$. Hence, F^* can coincide with neither $F^*_{c_1, c_2}$ nor $F^*_{c_1, c_3}$, and $H_g(F^*)$ is strictly lower than $H_g(F^*_{c_1, c_2})$ and $H_g(F^*_{c_1, c_3})$.

12.1.5 Bounds for some Particular Distortion Risk Measures

We first use Algorithm 12.1 to numerically determine the sharp upper bound and corresponding maximizing distribution function for several concave distortion risk measures and moment spaces. Next, we provide an analytical solution when the distortion function is given by $g(x) = 1 - (1 - x)^2$ and the first two moments are fixed.

Example 12.5 (Power, dual-power, and Wang distortions)
We consider the power distortion function with $g(x) = x^\alpha$, where $\alpha = 0.5, 0.2, 0.1$, the dual-power distortion function with $g(x) = 1 - (1 - x)^\beta$, where $\beta = 2, 5, 10$, and the Wang distortion function with $g(x) = \Phi(\Phi^{-1}(x) + \Phi^{-1}(p))$, where $p = 0.8, 0.9, 0.95$. For the moment constraints we take those that arise from the uniform distribution function on $[0, 1]$, that is, we take $c_1 = 1/2$, $c_2 = 1/3$, $c_3 = 1/4$, and $c_4 = 1/5$. For $\mathcal{F}_0 = \mathcal{F}(c_1, c_2)$, $\mathcal{F}_0 = \mathcal{F}(c_1, c_3)$, and $\mathcal{F}_0 = \mathcal{F}(c_1, c_4)$, we use Algorithm 1 to determine the maximizing distribution functions numerically and display the distorted expectations in Table 12.1.

| | domain $[0, 1]$ | | |
Distortion function	$\mathcal{F}_{(c_1, c_2)}$	$\mathcal{F}_{(c_1, c_3)}$	$\mathcal{F}_{(c_1, c_4)}$
power ($\alpha = 0.5$)	0.6754	0.6711	0.6693
power ($\alpha = 0.2$)	0.8450	0.8407	0.8382
power ($\alpha = 0.1$)	0.9175	0.9148	0.9130
dual-power ($\beta = 2$)	0.6667	0.6714	0.6782
dual-power ($\beta = 5$)	0.8660	0.8472	0.8366
dual-power ($\beta = 10$)	0.9686	0.9540	0.9404
Wang ($p = 0.80$)	0.7330	0.7276	0.7273
Wang ($p = 0.90$)	0.8360	0.8270	0.8230
Wang ($p = 0.95$)	0.9012	0.8923	0.8866

Table 12.1 Maximum value $H_g(F^\star)$ for several choices of distortion functions and moment spaces ($c_1 = 0.50$, $c_2 = 0.33$, $c_3 = 0.24$).

To obtain more insight into the maximizing distribution functions and their link with the distortion functions, we further focus on the power distortion function with parameter $\alpha = 0.2$, the dual-power distortion function with parameter $\beta = 5$, and the Wang distortion function with $p = 0.8$. These distortion functions are displayed in Figure 12.2(a). Figures 12.2(b), 12.2(c), and 12.2(d) then show the corresponding maximizing distribution functions on $[0, 1]$ for some of the cases considered in Table 12.1. Interestingly, unlike in the case of VaR and TVaR, where maximizing distribution functions are known to be discrete, the maximizing distribution functions are either continuous or appear as a mixture of a continuous and a discrete distribution function.

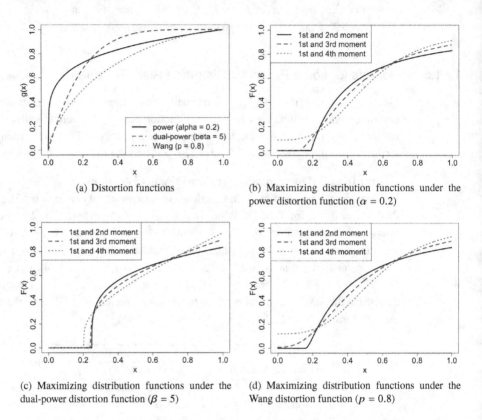

(a) Distortion functions

(b) Maximizing distribution functions under the power distortion function ($\alpha = 0.2$)

(c) Maximizing distribution functions under the dual-power distortion function ($\beta = 5$)

(d) Maximizing distribution functions under the Wang distortion function ($p = 0.8$)

Figure 12.2 Maximizing distribution functions under different distortion risk measures.

Example 12.6 (Analytical solution for the dual-power distortion risk measure)
Moment spaces for distribution functions with a compact support are compact in the topology of convergence in distribution; see, e.g., Karlin and Shapley (1953). However, the set E defined as

$$E = \left\{ \eta \mid F_\eta(x) \text{ is a distribution function} \right\} \tag{12.18}$$

is not necessarily compact. Therefore, it does not seem possible to always obtain an analytical description of the maximizing distribution function. In the case of the dual-power distortion function,

$$g(x) = 1 - (1 - x)^2, \tag{12.19}$$

it is, however, possible to explicitly describe the solution $\eta = (\eta_1, \eta_2)'$ as a function of given first two moments (c_1, c_2).

The set of pairs (c_1, c_2), such that at least one distribution function with those moments exists, is given by (Karlin and Shapley, 1953)

$$N = \{(x, y) \mid x \in [0, 1]; x^2 \le y \le x\}. \tag{12.20}$$

According to Theorem 12.2, the maximizing distribution function is of the following form:

$$F^\star(x) = F_\eta(x) = \begin{cases} 0 & \text{if } x < \max\left(0, \frac{-\eta_1}{\eta_2}\right), \\ \frac{\eta_1 + \eta_2 x}{2} & \text{else,} \\ 1 & \text{if } x \ge \min\left(1, \frac{2-\eta_1}{\eta_2}\right), \end{cases} \tag{12.21}$$

with $\eta_2 > 0$ and (η_1, η_2) determined by the moments (c_1, c_2). This distribution function is thus uniform on a certain interval (a, b), $0 \le a < b \le 1$, with potential mass points in zero, one, or both. Hence, if the given moments c_1 and c_2 allow for it, a uniform distribution is optimal. If this is not the case, we obtain a truncated uniform distribution, i.e., a mixture of a uniform distribution reaching the boundary with a point mass in one or both of the boundary points.

These four cases partition the moment set N into four sets $N_i, i = 1, \ldots, 4$, each corresponding to one of these cases. The set E is then also partitioned into four subsets $E_i, i = 1, \ldots, 4$, and from Theorem 12.2 we obtain analytic expressions relating the moments (c_1, c_2) to the optimal parameter vector (η_1, η_2). All these sets are shown in Figure 12.3(a) and Figure 12.3(b). We refer to Section 12.3 for their detailed description as well as for the corresponding exact analytic expressions. Note that the intersection of the four subsets $N_i, i = 1, \ldots, 4$, is given by $c_1 = 1/2, c_2 = 1/3$ and corresponds as expected to a maximizing distribution that is uniform on $(0, 1)$.

Finally, note that if we allow for an unbounded domain, we obtain from Theorem 12.3 that for $c_1 = 1/2$ and $c_2 = 1/3$ the maximizing distribution $F^\star$ is uniform on $(0,1)$. Moreover, since $w_g = \sqrt{1/3}$, we also obtain that $H_g(F^\star) = 2/3$, which is consistent with the value reported in Table 12.1 using Algorithm 1 for the case of a bounded domain.

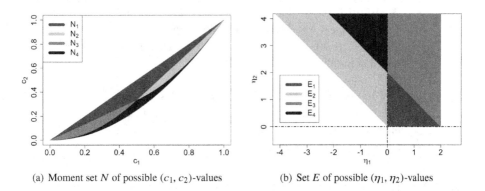

(a) Moment set N of possible (c_1, c_2)-values (b) Set E of possible (η_1, η_2)-values

Figure 12.3 Moment set and corresponding set for (η_1, η_2).

12.2   Bounds for Distortion Risk Measures: Unbounded Domain

The seminal Cantelli moment bound provides an upper bound on VaR when optimizing over a set of distribution functions with given mean and variance and without domain restrictions (see Eqs (12.12) and (12.14)).

Theorem 12.3 extends this result to the case of concave distortion risk measures. In this section, this result is further extended to the case of general distortion risk measures. We study both upper and lower bounds.

12.2.1   Setup

It is well known that when a *distortion function* $g: [0, 1] \to [0, 1]$ is absolutely continuous, the corresponding distortion risk measure H_g has a representation of the form

$$H_g(X) = \int_0^1 \gamma(u) F_X^{-1}(u) \, du, \tag{12.22}$$

where $\gamma(u) = \partial_- g(x)|_{x=1-u}$, $0 < u < 1$, such that $\int_0^1 \gamma(u) \, du = 1$ and where ∂_- denotes the derivative from the left; see, e.g., Föllmer and Schied (2004). We recall that we may sometimes write $H_g(F_X)$ or $H_g(F_X^{-1})$ instead of $H_g(X)$.

We study the *best case* and *worst case values* for a distortion risk measure H_g on a distributional uncertainty set $\mathcal{F}_0(\mu, \sigma)$, $\mu \in \mathbb{R}$ and $\sigma > 0$, i.e., we consider the problems

$$\inf_{F \in \mathcal{F}(\mu,\sigma)} H_g(F) \qquad (12.23a) \qquad\qquad \sup_{F \in \mathcal{F}(\mu,\sigma)} H_g(F). \qquad (12.23b)$$

Throughout this section, we make use of various technical assumptions that we summarize here:

Assumption 12.7   We assume that representation (12.22) holds and also that $\int_0^1 |\gamma(u)|^2 \, du < +\infty$. Furthermore, we assume that $\gamma(\cdot) \neq 1$; otherwise $H_g(G) = \mu$ for all $F \in \mathcal{F}(\mu, \sigma)$. Finally, we assume that $\mathcal{F}(\mu, \sigma)$ contains at least two elements and hence infinitely many.

The assumption that $\gamma(\cdot) \neq 1$ is needed since otherwise problems (12.23a) and (12.23b) are clearly obsolete. The assumption that $\mathcal{F}(\mu, \sigma)$ contains at least two elements is needed since otherwise both problems are either ill-posed or trivial.

12.2.2   Upper Bounds for Concave Distortion Risk Measures

For concave distortion risk measures, we obtain analytic solutions to problems (12.23b), which we present in the following theorem.

Theorem 12.8 (Upper bounds for concave distortion risk measures) *Let g be a concave distortion function. Then, it holds that*

$$\overline{H}_g(\mathcal{F}_0) = \mu + \sigma \operatorname{Std}(\gamma(U)) = \mu + \sigma \sqrt{\int_0^1 (\gamma(p) - 1)^2 \, dp}. \qquad (12.24)$$

The bound is sharp and is attained by a unique maximizing distribution function with quantile function h defined as

$$h(u) = \mu + \sigma \left(\frac{\gamma(u) - a}{b} \right), \qquad 0 < u < 1, \qquad (12.25)$$

in which $a = E\gamma(U)$ and $b = \operatorname{Std}(\gamma(U))$, where $U \sim U(0, 1)$.

Formula (12.24) is identical to formula (12.12) that appeared in Theorem 12.3. (Note that $\int_0^1 (\gamma(p) - 1)^2 \, dp = \int_0^1 \gamma^2(p) \, dp - 1$.) The proof for Theorem 12.8 is straightforward in that it follows from the representation (12.22) of a distortion risk measure as an inner product combined with the Cauchy–Schwarz inequality.

12.2.3 Upper Bounds for General Distortion Risk Measures

To analyze the worst case for general distortion risk measures, we make use of the L^2-projection of γ onto the set of square integrable non-decreasing functions, denoted by $\gamma^\uparrow$. That is, $\gamma^\uparrow$ is the unique solution to

$$\min_{k \in \mathcal{K}} \|\gamma - k\|^2, \qquad (12.26)$$

where $\| \cdot \|$ denotes the L^2 norm, and $\mathcal{K} = \{k \colon (0, 1) \to \mathbb{R} \mid k \text{ is non-decreasing,} \int k^2(u) du < \infty\}$. We also define the quantile function

$$h^\uparrow(u) := \mu + \sigma \left(\frac{\gamma^\uparrow(u) - a}{b} \right), \qquad 0 < u < 1, \qquad (12.27)$$

where $a = E(\gamma^\uparrow(U))$ and $b = \operatorname{Std}(\gamma^\uparrow(U))$. Note that $\gamma^\uparrow$ is constant whenever γ is decreasing, i.e., when g is convex.

Theorem 12.9 (Upper bounds for general distortion risk measures) *Let g be a distortion function. Then:*

1 If $\gamma^\uparrow$ is not constant, then

$$\overline{H}_g(\mathcal{F}_0) = \mu + \sigma \operatorname{Std}(\gamma^\uparrow(U)) = \mu + \sigma \sqrt{\int_0^1 (\gamma^\uparrow(p) - 1)^2 \, dp}. \qquad (12.28)$$

The bound is sharp (best possible) and attained by a maximizing distribution function with quantile function $h^\uparrow$, as defined in (12.27).

2 If $\gamma^\uparrow$ is constant, then

$$\overline{H}_g(\mathcal{F}_0) = \mu, \qquad (12.29)$$

and this bound cannot be attained.

The proof of this result builds on the result for the case of concave distortion risk measures (Theorem 12.8) combined with known properties of isotonic projections.

Remark 12.10   When g is concave, γ is an increasing function and thus equal to $\gamma^\uparrow$, i.e., Theorem 12.9 reduces to Theorem 12.8 in the case of concave distortion risk measures.   ◊

12.2.4   Lower Bounds for General Distortion Risk Measures

In a similar way as in Section 14.2.2, we define by $\gamma^\downarrow$ the projection of γ onto the set of square integrable non-increasing functions. We also define the quantile function

$$h^\downarrow(u) := \mu + \sigma \left(\frac{a - \gamma^\downarrow(u)}{b} \right), \quad 0 < u < 1, \tag{12.30}$$

where $a = \mathbb{E}(\gamma^\downarrow(U))$ and $b = \mathrm{Std}(\gamma^\downarrow(U))$.

The lower bounds for general distortion risk measures are presented in Theorem 12.11.

Theorem 12.11   (Lower bounds for general distortion risk measures) *Let g be a distortion function; then*

$$\underline{H}_g(\mathcal{F}_0) = \mu - \sigma \, \mathrm{Std}(\gamma(U)) \, \mathrm{Cor}\left(\gamma(U), \gamma^\downarrow(U)\right). \tag{12.31}$$

If $\gamma^\downarrow$ is not constant then the bound is attained by a distribution function with quantile function $h^\downarrow$. Otherwise, the infimum is equal to μ but is not attained.

Note in particular that when g is concave, $\gamma^\downarrow$ is constant and

$$\inf_{F \in \mathcal{F}(\mu,\sigma)} H_g(G) = \mu.$$

Remark 12.12   The results in this chapter can be applied to derive upper and lower bounds for RVaR on $\mathcal{F}(\mu, \sigma)$ as well as for its limiting cases VaR and TVaR. For given probability levels $0 < \alpha < \beta \le 1$, $\mathrm{RVaR}_{\alpha,\beta}$ is indeed a distortion risk measure determined by the distortion

$$g(x) = \min\left\{\max\left\{\frac{x + \beta - 1}{\beta - \alpha}, 0\right\}, 1\right\}, \quad \gamma(u) = \frac{1}{\beta - \alpha} 1_{(\alpha,\beta]}(u).$$

Doing so would provide an alternative for the approach pursued in Chapter 11 for deriving these bounds.   ◊

12.3   Proofs

In the first part of this section, we prove some useful lemmas based on adaptations and extensions of the results presented in Rustagi (1957, 1976). The proofs for Theorem 12.1 and Theorem 12.2 are provided thereafter.

Some Useful Lemmas

Let φ be a strictly convex and twice differentiable bounded function on $[0, 1]$. The results in Sections 12.1 and 12.2 concern a solution $F^\star$ of the optimization problem

$$F^\star = \arg\min_{F \in \mathcal{F}} \int_0^1 \varphi(F(x))\, dx, \tag{12.32}$$

where $\mathcal{F}_0$ is the moment class of distribution functions given in (12.1).

Lemma 12.13 (Existence and unicity)   *A solution $F^\star$ to (12.32) exists and is unique.*

Proof   The existence of a solution follows from Lemma 3.1 and Lemma 3.2 in Rustagi (1957). Specifically, let $I = \{i_1, \ldots, i_k\}$, and observe that $\int_0^1 x^{i_j}\, dF(x) = c_{i_j}$ if and only if $\int_0^1 x^{i_j - 1} F(x)\, dx = d_{i_j}$ for appropriate d_{i_j}. Consider the transformation

$$T : \mathcal{F} \to \mathbb{R}^{k+1} : \tag{12.33}$$

$$T(F) = \left(\int_0^1 \varphi(F(x))\, dx, \int_0^1 x^{i_1 - 1} F(x)\, dx, \ldots, \int_0^1 x^{i_k - 1} F(x)\, dx \right).$$

This transformation is continuous and convex in F, and since $\mathcal{F}$ is convex and compact in the topology of convergence in distribution, it maps $\mathcal{F}$ into a convex and compact set. The imposed moment restrictions on F thus yield a non-empty subset that is also bounded and closed, and hence a solution to (12.32) exists.

To prove uniqueness, assume that $F_0 \neq F_1$ are two admissible solutions. Let

$$M = \int_0^1 \varphi(F_0(x))\, dx = \int_0^1 \varphi(F_1(x))\, dx. \tag{12.34}$$

Since φ is strictly convex, it holds for any $\lambda \in (0, 1)$ that

$$\int_0^1 \varphi(\lambda F_0(x) + (1 - \lambda) F_1(x))\, dx \tag{12.35}$$

$$< \lambda \int_0^1 \varphi(F_0(x))\, dx + (1 - \lambda) \int_0^1 \varphi(F_1(x))\, dx = M,$$

which is a contradiction; see also Rustagi (1976, pg. 100). □

Define φ_y as the derivative of the function φ.

Lemma 12.14 (Linearisation)   *If $F^\star$ solves (12.32), then there exists $(\eta_i)_{i \in I}$ such that $F^\star$ minimizes over all $F \in \mathcal{F}$ the function*

$$\int_0^1 \left[\varphi_y(F^\star(x)) + \sum_{i \in I} \eta_i x^{i-1} \right] F(x)\, dx. \tag{12.36}$$

Proof   Let $F^\star$ be the solution to (12.32). Lemma 5.3.1 in Rustagi (1976) proves that $F^\star$ minimizes (12.32) if and only if

$$\int_0^1 \varphi_y(F^\star(x)) F(x)\, dx \geq \int_0^1 \varphi_y(F^\star(x)) F^\star(x)\, dx, \qquad F \in \mathcal{F}. \tag{12.37}$$

Denote $I = \{i_1, \ldots, i_k\}$. Consider $\mathcal{J} = T(\mathcal{F}_0)$ the set of points $(\zeta_0, \zeta_1, \ldots, \zeta_k)$ given as

$$\left(\int_0^1 \varphi_y(F^\star(x)) F(x)\, dx, \int_0^1 x^{i_1-1} F(x)\, dx, \ldots, \int_0^1 x^{i_k-1} F(x)\, dx \right),$$
$$F \in \mathcal{F}, \tag{12.38}$$

where the second coordinates onward are determined by the moments of the distribution function F, obtained by integration by parts. The set $\mathcal{J}$ is closed, bounded, and convex in $k+1$ dimensions. We show this based on Theorem 7.2 in Karlin and Shapley (1953) and in Rustagi (1957, pg. 313). Consider the transformation

$$T: \mathcal{F} \to \mathbb{R}^{k+1}: \tag{12.39}$$
$$T(F) = \left(\int_0^1 \varphi_y(F^\star(x)) F(x)\, dx, \int_0^1 x^{i_1-1} F(x)\, dx, \ldots, \int_0^1 x^{i_k-1} F(x)\, dx \right).$$

The transformation T is continuous and linear in F. Since $\mathcal{F}$ is convex and compact in the topology of convergence in distribution, its image is also convex and compact.

Because the set $\mathcal{J} = T(\mathcal{F}_0)$ is closed, bounded, and convex, there exists a boundary point $(z_0, \ldots, z_k)$ of $\mathcal{J}$ such that $F^\star$ corresponds to this boundary point. Therefore, there exists a supporting hyperplane of $\mathcal{J}$ at $(z_0, \ldots, z_k)$, i.e., there exist $\eta_0, \eta_{i_1}, \ldots, \eta_{i_k}, \eta$ such that

$$\eta_0 z_0 + \eta_{i_1} z_1 + \cdots + \eta_{i_k} z_k + \eta = 0 \tag{12.40}$$

and

$$\eta_0 \zeta_0 + \eta_{i_1} \zeta_1 + \cdots + \eta_{i_k} \zeta_k + \eta \geq 0, \qquad \text{for all } \zeta = (\zeta_0, \ldots, \zeta_k) \in \mathcal{J}. \tag{12.41}$$

Hence, it holds that

$$\eta_0 (\zeta_0 - z_0) + \cdots + \eta_{i_k} (\zeta_k - z_k) \geq 0, \qquad \text{for all } \zeta \in \mathcal{J}. \tag{12.42}$$

We now show that $\eta_0 > 0$ by eliminating the other possibilities. Let

$$\mathcal{J}^\star = \{ (\zeta_0^\star, \zeta_1, \ldots, \zeta_k) \mid \zeta_0^\star \geq \zeta_0; (\zeta_0, \zeta_1, \ldots, \zeta_k) \in \mathcal{J} \}. \tag{12.43}$$

The set $\mathcal{J}^\star$ is convex, and $\mathcal{J} \subseteq \mathcal{J}^\star$. The value z_0 is the minimum of ζ_0 subject to $\zeta_1 = z_1, \ldots, \zeta_k = z_k$. Hence $(z_0, \ldots, z_k)$ is also a minimum point of $\mathcal{J}^\star$ and thus a boundary point of $\mathcal{J}^\star$. Therefore, there exist $(\eta_0, \ldots, \eta_{i_k}) \neq (0, \ldots, 0)$ such that

$$\eta_0 (\zeta_0^\star - z_0) + \cdots + \eta_{i_k} (\zeta_k - z_k) \geq 0 \qquad \text{for all } (\zeta_0^\star, \zeta_1, \ldots, \zeta_k) \in \mathcal{J}^\star. \tag{12.44}$$

Hence, since $\mathcal{J} \subseteq \mathcal{J}^\star$,

$$\eta_0 (\zeta_0 - z_0) + \cdots + \eta_{i_k} (\zeta_k - z_k) \geq 0 \qquad \text{for all } (\zeta_0, \zeta_1, \ldots, \zeta_k) \in \mathcal{J}. \tag{12.45}$$

Suppose $\eta_0 < 0$. Since $(z_0, \ldots, z_k)$ is a minimum point of $\mathcal{J}^\star$, there exists some $\alpha > 0$ such that $(z_0 + \alpha, z_1, \ldots, z_k) \in \mathcal{J}^\star$. By Equation (12.44), $\eta_0 \alpha \geq 0$ would hold, which would imply $\eta_0 = 0$. Next, suppose that $\eta_0 = 0$. This corresponds to the boundary points of the set $\mathcal{J}$ where the supporting hyperplanes are parallel to the

ζ_0-axis, and hence $(z_1, \ldots, z_k)$ corresponds to the boundary of the projection of $\mathcal{J}$ on the $(\zeta_1, \ldots, \zeta_k)$ hyperplane. But the conditions on $\mathcal{F}_0$ are such that the given point $(z_1, \ldots, z_k)$ will be interior to the projection set, and hence $\eta_0 \neq 0$.

Since $\eta_0 > 0$ and we can normalize it to be equal to one, we obtain for any $F \in \mathcal{F}_0$,

$$
\int_0^1 \left[\varphi_y(F^\star(x)) + \sum_{i \in I} \eta_i x^{i-1} \right] F(x) \, dx
$$
$$
\geq \int_0^1 \left[\varphi_y(F^\star(x)) + \sum_{i \in I} \eta_i x^{i-1} \right] F^\star(x) \, dx. \tag{12.46}
$$

Equivalently, $F^\star$ minimizes (12.36). □

Remark 12.15   Another approach to the constrained optimization problem (12.32) consists in using Lagrange multipliers. In this regard, a result from Everett III (1963) can be used to reformulate the problem as an unconstrained problem and to obtain sufficient conditions for a solution. However, unlike the approach we propose here, the result in Everett III (1963) requires the Lagrange multipliers to be strictly positive, resulting in a severe restriction in that not all moment sequences are admissible. Moreover, one would still need to solve the unconstrained optimization problem, which would require the linearization step we dealt with in Lemma 12.14 as well as the steps we describe hereafter. ◇

Remark 12.16   In the literature, one finds early contributions to the related problem of obtaining a distribution F that maximizes or minimizes an expected utility $E_F u$ given as

$$
E_F u = \int u(x) \, dF(x), \tag{12.47}
$$

subject to the side constraints $c_1 = \int x \, dF(x)$ and $c_2 = \int x^2 \, dF(x)$, and in which u is some continuous function. For some specific choice for u, Gumbel (1954) considers a variational approach for maximization of $E_F u$ and derives sufficient conditions for obtaining a solution. Our basic minimization problem (12.32) can, after partial integration, also be recast as a maximum expected utility problem of the form (12.47); see also Chapter 6 of Rustagi (1957). By equating the first variation to zero, sufficient conditions for a maximizing distribution can be obtained. It is, however, not clear whether these conditions are also necessary. For general u, Chernoff and Reiter (1954) study the problem of minimizing $E_F u$ and derive necessary conditions that lead to a discrete minimizing distribution. In the case of a given mean and variance, one obtains a solution with at most three mass points. Their result is consistent with the results in Chapter 8 of Rustagi (1957), which considers the maximization counterpart for problem (12.32), i.e., the maximization of

$$
\int_0^1 \varphi(F(x)) \, dx, \tag{12.48}
$$

which after partial integration leads to an expected utility maximization problem of the form (12.47). ◇

Next we provide necessary conditions that a solution to (12.36) should satisfy. In the next lemma, we specifically consider

$$\varphi: [0, 1] \to [0, 1]: \varphi(y) = 1 - g(1 - y) \tag{12.49}$$

for a distortion function g satisfying the imposed conditions of strict concavity and twice differentiability.

Lemma 12.17 (Necessary conditions) *Consider the optimization problem*

$$F^\star = \arg\min_{F \in \mathcal{F}_0} \int_0^1 \left[\varphi_y(F^\star(x)) - \sum_{i \in I} \eta_i x^{i-1} \right] F(x)\, dx. \tag{12.50}$$

Then $F^\star$ is continuous on $(0, 1)$, and on intervals where $F^\star$ is not constant, it coincides with some F_η of the form

$$F_\eta(x) = \begin{cases} 0 & \text{if } z < g'(1), \\ 1 - (g')^{-1}(z) & \text{else,} \\ 1 & \text{if } z \geq g'(0), \end{cases} \tag{12.51}$$

with $z = \sum_{i \in I} \eta_i x^{i-1}$ and $(\eta_i)_{i \in I}$ such that

$$\int_0^1 x^i \, dF^\star(x) = c_i, \quad i \in I \tag{12.52}$$

and

$$\int_a^b \left[g'\left(1 - F^\star(x)\right) - \sum_{i \in I} \eta_i x^{i-1} \right] dx = 0 \tag{12.53}$$

on intervals $[a, b] = [F^{\star-}(c), F^{\star+}(c)]$ for all $c \in (0, 1)$.

Proof From Lemma 12.14 we obtain that properties of the solution $F^\star$ can be described in terms of $G_\eta(x)$ solving

$$\varphi_y(G_\eta(x)) - \sum_{i \in I} \eta_i x^{i-1} = 0. \tag{12.54}$$

Because φ_y is differentiable, this equation readily inverts on some set I to an expression for $G_\eta(x)$ in terms of the original distortion function g:

$$G_\eta : I \to \mathbb{R}, G_\eta(x) = 1 - (g')^{-1}\left(\sum_{i \in I} \eta_i x^{i-1}\right), \tag{12.55}$$

with $I := I_\eta$ defined as

$$I = \left\{ x \in [0, 1] \,\Big|\, \sum_{i \in I} \eta_i x^{i-1} \in [g'(1), g'(0)) \right\}. \tag{12.56}$$

We extend the function G_η to the interval $[0, 1]$ in a straightforward fashion:

$$F_\eta(x) = \begin{cases} 0 & \text{if } z < g'(1), \\ 1 - (g')^{-1}(z) & \text{if } x \in I, \\ 1 & \text{if } z \geq g'(0), \end{cases} \qquad (12.57)$$

with $z = \sum_{i \in I} \eta_i x^{i-1}$.

Next, some restrictions on how to transform F_η into a distribution function are derived. Define the function

$$A: (0, 1) \to \mathbb{R}, \, A(x) = \varphi_y(F^\star(x)) - \sum_{i \in I} \eta_i x^{i-1} \qquad (12.58)$$

and the set

$$\mathcal{D} = \{x \in (0, 1) \mid A(x) \neq 0\}. \qquad (12.59)$$

Theorem 5.1 of Rustagi (1957, pg. 316) states that if $F^\star$ minimizes (12.36), then the set $\mathcal{D}$ has $F^\star$-measure zero, i.e., $F^\star$ is constant on $\mathcal{D}$. Hence, $F^\star$ coincides with F_η on $(0, 1)$ when $F^\star$ is not constant.

A corollary of Rustagi (1957, pg. 316) gives restrictions on $F^\star$ where it is constant. On intervals $(a, b] = (F^-(c), F^+(c)]$ for all $c \in (0, 1)$, it holds that

$$\int_a^b A(x)\, dx = 0. \qquad (12.60)$$

Finally, the continuity of $F^\star$ on $(0, 1)$ follows from its right-continuity and the fact that it has no jumps, as proven in Theorem 5.2 in Rustagi (1957). This concludes the proof. □

Proof of Theorem 12.1

First, we rewrite the maximization problem (12.3) as a minimization problem and note that all distribution functions we consider are defined on $[0, 1]$. Hence, we study the problem

$$F^\star = \arg \min_{F \in \mathcal{F}_0} \int_0^1 (1 - g(1 - F(x)))\, dx. \qquad (12.61)$$

Define the function φ as

$$\varphi: [0, 1] \to [0, 1]: \varphi(y) = 1 - g(1 - y). \qquad (12.62)$$

This function is strictly convex, bounded, and twice differentiable under the assumptions made on the distortion function g. Thus, by Lemma 12.13, there exists a unique $F^\star$ minimizing (12.61) and thus maximizing (12.3).

By Lemma 12.14 it holds that $F^\star$ minimizes the function given in equation (12.36) over all $F \in \mathcal{F}_0$. Since this optimization problem is similar to the one in Lemma 12.17, the necessary conditions in Theorem 12.1 are shown.

Proof of Theorem 12.2

Under the moment conditions it follows that F_{η} as stated in Theorem 12.1 is monotonic as it is the composition of an increasing function and a linear one. Furthermore, F_{η} then has to be increasing, since otherwise $F^{\star}$ would be degenerate and thus not admissible (note that $\mathcal{F}_0$ contains at least two elements, and the degenerate distribution function is unique with respect to its moment sequence). Hence, since $F^{\star}$ is continuous on $(0, 1)$, it holds that the optimal solution $F^{\star}$ should be of following form:

$$
F_{\eta}(x) = \begin{cases} 0 & \text{if } x < \max\left(\left(\frac{g'(1)-\eta_1}{\eta_i}\right)^{\frac{1}{i-1}}, 0\right), \\ 1 - (g')^{-1}\left(\eta_1 + \eta_i x^{i-1}\right) & \text{else,} \\ 1 & \text{if } x \geq \min\left(\left(\frac{g'(0)-\eta_1}{\eta_i}\right)^{\frac{1}{i-1}}, 1\right), \end{cases} \tag{12.63}
$$

for some $\eta := (\eta_1, \eta_i)$ such that $\eta_i > 0$ and $\eta_1 \in (g'(1) - \eta_i, g'(0))$ and

$$
\mu_1(\eta) = \int_0^1 x\, dF_{\eta}(x) = c_1 \quad \text{and} \quad \mu_i(\eta) = \int_0^1 x^i\, dF_{\eta}(x) = c_i. \tag{12.64}
$$

The condition that F_{η} is a non-decreasing function implies that $\eta_i > 0$, whereas $\eta_1 \in (g'(1) - \eta_i, g'(0))$ follows from the observation that when $\eta_1 = g'(1) - \eta_i$ or $\eta_1 = g'(0)$, the resulting distribution function is degenerate in one or zero and thus not admissible, since by assumption $\mathcal{F}_0$ contains at least two elements. When $\eta_1 < g'(1) - \eta_i$ or $\eta_1 > g'(0)$, the resulting F_{η} is not a cumulative distribution function on $[0, 1]$ and hence not admissible.

We aim to show that there exists only one $\eta^{\star} = (\eta_1^{\star}, \eta_i^{\star})$ such that $F_{\eta^{\star}}$ satisfies the moment conditions (12.64).

First, we show that for any $\eta_i > 0$, there exists some η_1 such that $\mu_1(\eta_1, \eta_i) = c_1$. For any $\eta_i > 0$, the function $\mu_1(\eta_1, \eta_i)$ is strictly decreasing in η_1 on $(g'(1) - \eta_i, g'(0))$. Since $\lim_{\eta_1 \to g'(1)-\eta_i} \mu_1(\eta_1, \eta_i) = 1$ and $\lim_{\eta_1 \to g'(0)} \mu_1(\eta_1, \eta_i) = 0$, the full domain of possible mean values for any distribution function F on $[0, 1]$ can be reached by varying η_1. Hence, there exists a unique η_1 such that $\mu_1(\eta_1, \eta_i) = c_1$; write $\eta_1 = \eta_1(\eta_i)$.

Next, we show that there exists a unique $\eta_i^{\star} > 0$ such that $\mu_i(\eta_1(\eta_i), \eta_i) = c_i$. To this end, consider the function

$$
\widetilde{\mu}_i \colon (0, \infty) \to [0, 1] \colon \widetilde{\mu}_i(\eta_i) = \int_0^1 x^{i-1} F_{(\eta_1(\eta_i), \eta_i)}(x)\, dx. \tag{12.65}
$$

The derivative of $\widetilde{\mu}_i$ with respect to η_i is given as

$$
\frac{d\widetilde{\mu}_i(\eta_i)}{d\eta_i} = \int_0^1 x^{2i-2} H'(\eta_1 + \eta_i x^{i-1})\, dx
$$
$$
+ \left(\int_0^1 x^{i-1} H'(\eta_1 + \eta_i x^{i-1})\, dx\right) \frac{d\eta_1(\eta_i)}{d\eta_i}, \tag{12.66}
$$

where H is defined as

$$H(y) = \begin{cases} 0 & \text{if } y < \max\left(g'(1), 0\right), \\ 1 - (g')^{-1}(y) & \text{else,} \\ 1 & \text{if } y \geq \min\left(g'(0), 1\right), \end{cases} \qquad (12.67)$$

and with H' the derivative of H on $(0, 1)$ a.e., it holds that the derivative $d\eta_1(\eta_i)/d\eta_i$ equals

$$\frac{d\eta_1(\eta_i)}{d\eta_i} = -\left(\int_0^1 x^{i-1} H'(\eta_1 + \eta_i x^{i-1})\, dx\right) \Big/ \left(\int_0^1 H'(\eta_1 + \eta_i x^{i-1})\, dx\right). \quad (12.68)$$

The latter can be seen by setting the total derivative $d\tilde\mu_1$ equal to zero, since we keep the mean fixed when varying η_1 and η_i accordingly.

Note that $H'(\eta_1 + \eta_i x^{i-1})/ \int_0^1 H'(\eta_1 + \eta_i x^{i-1})\, dx$ is a density on $(0, 1)$. Because H' is not degenerate, the variance of X^{i-1}, where X is a random variable with this density, is strictly positive, and hence

$$\frac{d\tilde\mu_i(\eta_i)}{d\eta_i} > 0. \qquad (12.69)$$

Since integration by parts yields the relation $i\tilde\mu_i(\eta_i) = 1 - \mu_i(\eta_1(\eta_i), \eta_i)$, we thus also have that

$$\frac{d\mu_i(\eta_1(\eta_i), \eta_i)}{d\eta_i} < 0. \qquad (12.70)$$

Thus, if $\eta^\star$ exists, it is unique due to strict monotonicity.

Consider now the limits

$$\lim_{\eta_i \to 0} \int_0^1 x^i \, dF_{(\eta_1(\eta_i), \eta_i)}(x) = c_1$$

$$\text{and} \qquad \lim_{\eta_i \to \infty} \int_0^1 x^i \, dF_{(\eta_1(\eta_i), \eta_i)}(x) = c_1^i, \qquad (12.71)$$

which show that the full domain of possible values for the ith moment can be reached by varying η_i, implying that $\eta_i^\star$ exists. Hence, $\boldsymbol{\eta}^\star = (\eta_1^\star, \eta_i^\star)$ exists and is unique, and thus $F^\star = F_{\boldsymbol{\eta}^\star}$.

Proof of Corollary 12.3

Consider a random variable X on $[-b, b]$, $b > 0$, and let Y denote the transformed variable $Y = (X + b)/2b$ on $[0, 1]$. Denote $EX = \mu_X$ and $\text{Var}(X) = s_X^2$; then $EY = \mu_Y = (\mu_X + b)/2b$ and $\text{Var}(Y) = s_Y^2 = s_X^2/4b^2$. Due to Theorem 12.2, the distribution $F_Y^\star$ on $[0, 1]$ with this mean and variance, maximizing the distorted expectation $H_g(Y)$, is of the form (12.7):

$$F_Y^\star(y) = \begin{cases} 0 & \text{if } y < \max\left(\frac{g'(1)-\tilde\eta_1}{\tilde\eta_2}, 0\right), \\ 1 - (g')^{-1}(\tilde\eta_1 + \tilde\eta_2 y) & \text{else,} \\ 1 & \text{if } y \geq \min\left(\frac{g'(0)-\tilde\eta_1}{\tilde\eta_2}, 1\right), \end{cases} \qquad (12.72)$$

for $(\widetilde{\eta}_1, \widetilde{\eta}_2)$ such that $\widetilde{\eta}_2 > 0$ and $\widetilde{\eta}_1 \in (g'(1) - \widetilde{\eta}_2, g'(0))$ and the moment conditions are satisfied.

The quantile function of this distribution equals

$$
F_Y^{-1}(p) = \begin{cases} 0 & \text{if } p \le F_Y^\star(0), \\ \frac{1}{\widetilde{\eta}_2}\left(g'(1-p) - \widetilde{\eta}_1\right) & \text{else,} \\ 1 & \text{if } p > \lim_{y\uparrow 1} F_Y^\star(y). \end{cases} \tag{12.73}
$$

Using

$$
EY = \int_0^1 F_Y^{-1}(p)\, dp \quad \text{and} \quad EY^2 = \int_0^1 \left(F_Y^{-1}(p)\right)^2 dp, \tag{12.74}
$$

one obtains the following expressions for $(\widetilde{\eta}_1, \widetilde{\eta}_2)$ satisfying the mean and variance conditions on Y:

$$
\widetilde{\eta}_1 = \frac{1}{\beta - \alpha}\left(g(1-\alpha) - g(1-\beta) - \widetilde{\eta}_2(\mu_Y - (1-\beta))\right),
$$

$$
\widetilde{\eta}_2 = \sqrt{\frac{\int_\alpha^\beta \left(g'(1-p) - \frac{g(1-\alpha)-g(1-\beta)}{\beta-\alpha}\right)^2 dp}{s_Y^2 + \mu_Y^2 - (1-\beta) - (\beta-\alpha)^{-1}(\mu_Y - (1-\beta))^2}}, \tag{12.75}
$$

where $\alpha = F_Y^\star(0)$ and $\beta = \lim_{y\uparrow 1} F_Y^\star(y)$. We remark that these expressions are possibly implicit, depending on whether or not $\alpha = 0$ and $\beta = 1$. However, this is no concern since Theorem 12.2 guarantees the uniqueness of $(\widetilde{\eta}_1, \widetilde{\eta}_2)$.

Since $F_Y^\star$ maximizes the distorted expectation over all functions on $[0, 1]$ with mean $(\mu_X + b)/2b$ and variance $s_X^2/4b^2$, it follows that the distribution function $F_X^\star$, defined as

$$
F_X^\star(x) = \begin{cases} 0 & \text{if } x < \max\left(\frac{g'(1)-\eta_1}{\eta_2}, -b\right), \\ 1 - (g')^{-1}(\eta_1 + \eta_2 x) & \text{else,} \\ 1 & \text{if } x \ge \min\left(\frac{g'(0)-\eta_1}{\eta_2}, b\right) \end{cases} \tag{12.76}
$$

maximizes the distorted expectation over all distributions on $[-b, b]$ with mean μ_X and variance s_X^2, where

$$
\eta_1 = \frac{1}{\beta - \alpha}\left(g(1-\alpha) - g(1-\beta) - 2b\eta_2\left(\mu_Y - (1-\beta) - \frac{\beta-\alpha}{2}\right)\right),
$$

$$
\eta_2 = \frac{1}{2b}\sqrt{\frac{\int_\alpha^\beta \left(g'(1-p) - \frac{g(1-\alpha)-g(1-\beta)}{\beta-\alpha}\right)^2 dp}{s_Y^2 + \mu_Y^2 - (1-\beta) - (\beta-\alpha)^{-1}(\mu_Y - (1-\beta))^2}}. \tag{12.77}
$$

When $b \to \infty$, it holds that $\mu_Y \to 0.5$ and $s_Y^2 \to 0$ for μ_X and s_X^2 fixed. It follows that $b \to \infty$ implies that $\alpha \to 0$ and $\beta \to 1$, which implies $g(1-\alpha) \to 1$ and $g(1-\beta) \to 0$. The constants η_1 and η_2 then simplify to

$$
\eta_1 = 1 - \eta_2 \mu_X \quad \text{and} \quad \eta_2 = \frac{\sqrt{\int_0^1 (g')^2 (1-p)\, dp} - 1}{s_X}. \tag{12.78}
$$

When $g'(0) < \infty$, it is possible to take b large enough such that

$$F^\star(x) = \begin{cases} 0 & \text{if } x < \mu_X + \frac{g'(1)-1}{\eta_2}, \\ 1 - (g')^{-1}(\eta_1 + \eta_2 x) & \text{else,} \\ 1 & \text{if } x \geq \mu_X + \frac{g'(0)-1}{\eta_2} \end{cases} \qquad (12.79)$$

is independent of b and hence maximizes the distorted expectation over all distributions with mean μ_X and variance s_X^2 on an unbounded support.

In the case $g'(0) = \infty$, observe that

$$\lim_{x \to \infty} (1 - (g')^{-1}(\eta_1 + \eta_2 x)) = 1 - \lim_{x \to \infty} (g')^{-1}(x) = 1. \qquad (12.80)$$

This indicates that the function

$$F_X^\star(x) = \begin{cases} 0 & \text{if } x < \mu_X + \frac{g'(1)-1}{\eta_2}, \\ 1 - (g')^{-1}(\eta_1 + \eta_2 x) & \text{else} \end{cases} \qquad (12.81)$$

is a distribution function and hence maximizes the distorted expectation over all distributions with mean μ_X and variance s_X^2 on an unbounded support.

Furthermore, it is well known that $H_g(F^\star)$ can also be expressed as

$$H_g(F^\star) = \int_0^1 g'(1-p)(F^\star)^{-1}(p)\,dp, \qquad (12.82)$$

and by substituting $(F^\star)^{-1}(p)$ we thus get

$$H_g(F^\star) = \frac{\int_0^1 (g')^2 (1-p)\,dp - 1}{\eta_2} - \frac{\eta_1}{\eta_2}. \qquad (12.83)$$

Substituting the expressions for η_1 and η_2 delivers the result.

Proof of Corollary 12.4

Let $F^\star$ be the optimal solution to (12.3) when $\mathcal{F}_0 = \mathcal{F}(c_1, (c_i)_{i \in I \setminus \{1\}})$, and denote by $F_{c_1, c_i}^\star$ the solution to (12.3) when optimizing over the set $\mathcal{F}(c_1, c_i)$, $i \in I \setminus \{1\}$. Since $\mathcal{F}_0 \subset \mathcal{F}(c_1, c_i)$ for $i \in I \setminus \{1\}$, it follows that

$$H_g(F^\star) \leq \min_{i \in I \setminus \{1\}} H_g(F_{c_1, c_i}^\star). \qquad (12.84)$$

Analytical Expressions for Section 12.6

The solution is a mixture of a uniform distribution reaching the boundary with a point mass in one or both of the boundary points. These cases partition the moment set N into four subsets $N_i, i = 1, \ldots, 4$, given by

$$N_1 = \left\{ (c_1, c_2) \in N \mid c_2 \geq \max\left(\frac{1}{3}(4c_1 - 1), \frac{2}{3}c_1\right) \right\},$$

$$N_2 = \left\{ (c_1, c_2) \in N \mid \frac{1}{3}(4c_1^2 - 2c_1 + 1) \leq c_2 < \frac{1}{3}(4c_1 - 1) \right\},$$

$$N_3 = \left\{ (c_1, c_2) \in N \mid \frac{4}{3}c_1^2 \leq c_2 < \frac{2}{3}c_1 \right\}, \tag{12.85}$$

$$N_4 = \left\{ (c_1, c_2) \in N \mid c_2 < \min\left(\frac{4}{3}c_1^2, \frac{1}{3}(4c_1^2 - 2c_1 + 1)\right) \right\}.$$

For a graphical representation, see Figure 12.3(a).

For each of the four subsets N_i we obtain as a consequence of Theorem 12.2 analytical expressions relating the moments (c_1, c_2) to the optimal parameters (η_1, η_2). They are given by

$$(\eta_1, \eta_2) = \begin{cases} (-8c_1 + 6c_2 + 2, \, 12(c_1 - c_2)) & \text{if } (c_1, c_2) \in N_1, \\[2mm] \left(\frac{8(1-c_1)^2(1+3c_2-4c_1)}{9(1+c_2-2c_1)^2}, \, \frac{16(1-c_1)^3}{9(1+c_2-2c_1)^2}\right) & \text{if } (c_1, c_2) \in N_2, \\[2mm] \left(2 - \frac{8c_1^2}{3c_2}, \, \frac{16c_1^3}{9c_2^2}\right) & \text{if } (c_1, c_2) \in N_3, \\[2mm] \left(1 - \frac{c_1}{\sqrt{3(c_2-c_1^2)}}, \, \frac{1}{\sqrt{3(c_2-c_1^2)}}\right) & \text{if } (c_1, c_2) \in N_4. \end{cases} \tag{12.86}$$

This identification induces one-to-one mappings of the moment sets N_i to parameter sets $E_i, i = 1, \ldots, 4$, which yields a decomposition of the set E of admissible parameter vectors. The sets $E_i, i = 1, \ldots, 4$, are given by

$$\begin{aligned} E_1 &= \{(\eta_1, \eta_2) \mid \eta_1 \in [0, 2); \;\; 0 < \eta_2 \leq 2 - \eta_1\}, \\ E_2 &= \{(\eta_1, \eta_2) \mid \eta_1 \in (-\infty, 0); \;\; -\eta_1 < \eta_2 \leq 2 - \eta_1\}, \\ E_3 &= \{(\eta_1, \eta_2) \mid \eta_1 \in [0, 2); \;\; 2 - \eta_1 < \eta_2\}, \\ E_4 &= \{(\eta_1, \eta_2) \mid \eta_1 \in (-\infty, 0); \;\; 2 - \eta_1 < \eta_2\}. \end{aligned} \tag{12.87}$$

This decomposition is shown in Figure 12.3(b).

Throughout, we denote by $\langle \cdot, \cdot \rangle$ the inner product on the space of square integrable functions on $(0, 1)$, that is, $\langle \ell, k \rangle = \int_0^1 \ell(u)k(u)\, du$, $\ell, k \colon (0, 1) \to \mathbb{R}$. Further, we define the set $\mathcal{F}^{-1}(\mu, \sigma) = \{G^{-1} \mid F \in \mathcal{F}(\mu, \sigma)\}$, the corresponding set of quantile functions of $\mathcal{F}(\mu, \sigma)$.

Proof of Theorem 12.8

Note that the quantile function h has mean μ and variance σ^2, i.e., $h \in \mathcal{F}^{-1}(\mu, \sigma)$. Let $k \in \mathcal{F}^{-1}(\mu, \sigma)$; then by the Cauchy–Schwarz inequality it holds that $\langle \gamma, k \rangle \leq \langle \gamma, h \rangle$, with equality if and only if h is linear in γ and thus given as

$$h(u) = \mu + \sigma\left(\frac{\gamma(u) - a_0}{b_0}\right), \quad 0 < u < 1,$$

in which $a_0 = E\gamma(U)$ and $b_0 = \mathrm{Std}(\gamma(U))$ are such that $h \in \mathcal{F}^{-1}(\mu, \sigma)$ and with distorted expectation given as

$$H_g(h) = \mu + \sigma \, \mathrm{Std}\,(\gamma(U)) = \mu + \sigma \sqrt{\int_0^1 (\gamma(p) - 1)^2 \, dp}.$$

Proof of Theorem 12.9: Upper Bounds

Note that the projection $\gamma^\uparrow$ of γ is non-constant, which implies that $h^\uparrow \in \mathcal{F}^{-1}(\mu, \sigma)$. Let $\beta \in \mathcal{F}^{-1}(\mu, \sigma)$; then $k(u) = b_0(\beta(u) - \mu)/\sigma + a_0$ is a non-decreasing function. Moreover, it holds that $||\gamma^\uparrow||^2 = ||k||^2$. Thus, the following inequalities are equivalent:

$$||\gamma - \gamma^\uparrow||^2 \leq ||\gamma - k||^2$$
$$\Longleftrightarrow \quad \langle \gamma, \gamma^\uparrow \rangle \geq \langle \gamma, k \rangle$$
$$\Longleftrightarrow \quad \langle \gamma, h^\uparrow \rangle \geq \langle \gamma, h \rangle.$$

Note that unless $\gamma^\uparrow = k$, the inequalities are strict, which implies unicity of the solution.

Proof of Theorem 12.11: Lower Bounds

Note that the isotonic projection, $\gamma^\downarrow$ of γ onto the set of non-increasing function is non-constant. Hence, $h^\downarrow \in \mathcal{F}^{-1}(\mu, \sigma)$. Let $\beta \in \mathcal{F}^{-1}(\mu, \sigma)$; then $k(u) = (-\beta(u) + \mu)/b\sigma + a$ is a non-increasing function. Moreover, it holds that $||\gamma^\downarrow||^2 = ||k||^2$. Thus, the following inequalities are equivalent:

$$||\gamma - \gamma^\downarrow||^2 \leq ||\gamma - k||^2$$
$$\Longleftrightarrow \quad \langle \gamma, \gamma^\downarrow \rangle \geq \langle \gamma, k \rangle$$
$$\Longleftrightarrow \quad \langle \gamma, h^\downarrow \rangle \geq \langle \gamma, h \rangle.$$

Note that unless $k = \gamma^\downarrow$, the inequalities are strict, which implies unicity of the solution.

13 Bounds for VaR, TVaR, and RVaR under Unimodality Constraints

In this chapter, we derive bounds for the VaR, RVaR, and TVaR of a portfolio loss S under the assumption that in addition to knowledge of its mean and variance, we also know that its loss distribution function F_S is unimodal. The assumption of unimodality is very reasonable from a practical point of view as it is typically backed by observed data. The use of unimodal distributions is also common in insurance and financial risk modeling (e.g., exponential, Pareto, Gamma, normal, log-normal, logistic, Beta, Weibull, and Student t-distributions are all unimodal).

Definition 13.1 (Khintchine (1938))   A distribution function F is said to be uni-modal if it is convex-concave, i.e., there exists $m \in \mathbb{R}$, called the mode, such that F is convex on $(-\infty, m)$ and concave on $(m, +\infty)$. We also denote $p_m := F(m)$ for the mode level of F.

A standard uniform distribution is unimodal. More generally, a distribution function F with density f is unimodal if the density is unimodal, i.e., increasing before the mode m and decreasing after the mode. Note that a distribution function F is convex-concave if and only if its quantile function F^{-1} is concave-convex. We study risk bounds for distribution functions that belong to the following set:

$$\mathcal{F}_{\leq}^u(\mu, s) = \{F \in \mathcal{F}_{\leq}(\mu, s); F \text{ is unimodal}\},$$

where $\mathcal{F}_{\leq}(\mu, s)$ is the set of all distribution functions with mean $\mu \in \mathbb{R}$ and a variance that is bounded by $s^2 \in \mathbb{R}^+$.

To derive the bounds, we introduce the set $\mathcal{F}_R^{pl}$ as a set of distribution functions having a quantile function that is piecewise linear with a first piece that is constant. Formally,

$$\mathcal{F}_R^{pl} = \left\{ F \in \mathcal{F}; F^{-1}(p) = \begin{cases} a & \text{for } p \in [0, b), \\ c(p - b) + a & \text{for } p \in [b, 1], \end{cases} \right. \tag{13.1}$$

$$\left. a \in \mathbb{R}, b \in [0, 1], c \in \mathbb{R}^+ \right\}.$$

In a similar way, we define $\mathcal{F}_L^{pl}$ as the set of distribution functions having a quantile function that is piecewise linear with a second piece that is constant, i.e.,

$$\mathcal{F}_L^{pl} = \left\{ F \in \mathcal{F} ; F^{-1}(p) = \begin{cases} c(p-b) + a & \text{for } p \in [0, b), \\ a & \text{for } p \in [b, 1], \end{cases} \right. \tag{13.2}$$

$$a \in \mathbb{R}, b \in [0, 1], c \in \mathbb{R}^+ \Big\}.$$

From Definition 13.1, we deduce that all elements of $\mathcal{F}_R^{pl}$ and $\mathcal{F}_L^{pl}$ are unimodal. Thus, it is clear that

$$\mathcal{F}_R^{pl} \cap \mathcal{F}_{\leq}(\mu, s) \subset \mathcal{F}_{\leq}^u(\mu, s). \tag{13.3}$$

In a similar way, we have that

$$\mathcal{F}_L^{pl} \cap \mathcal{F}_{\leq}(\mu, s) \subset \mathcal{F}_{\leq}^u(\mu, s) \tag{13.4}$$

and thus also that

$$\mathcal{F}^{pl} \cap \mathcal{F}_{\leq}(\mu, s) \subset \mathcal{F}_{\leq}^u(\mu, s), \tag{13.5}$$

in which $\mathcal{F}^{pl} := \mathcal{F}_L^{pl} \cup \mathcal{F}_R^{pl}$.

13.1   VaR Bounds

We consider the following problem for $\alpha \in (0, 1)$:

Problem 1 $\overline{\mathrm{VaR}}_\alpha^u := \sup \left\{ \mathrm{VaR}_\alpha(X); F_X \in \mathcal{F}_{\leq}^u(\mu, s) \right\},$ $\qquad$ (13.6)

$\underline{\mathrm{VaR}}_\alpha^u := \inf \left\{ \mathrm{VaR}_\alpha(X); F_X \in \mathcal{F}_{\leq}^u(\mu, s) \right\}.$ $\qquad$ (13.7)

To solve Problem 1, we proceed in two steps. In a first step, we reduce the optimization problem to a parametric one. This is the topic of the following proposition. Note that the proof of it only uses some basic results on convex order.

Proposition 13.2   (Reduction to a parametric optimization problem) *The following holds for the upper bound $\overline{\mathrm{VaR}}_\alpha^u$ on $\mathcal{F}_{\leq}^u(\mu, s)$.*

$$\overline{\mathrm{VaR}}_\alpha^u = \sup \left\{ \mathrm{VaR}_\alpha(X); F_X \in \mathcal{F}^{pl} \cap \mathcal{F}_{\leq}(\mu, s) \right\}. \tag{13.8}$$

Proof By (13.3) and (13.4) we have that

$$\sup \left\{ \mathrm{VaR}_\alpha(Y); F_Y \in \mathcal{F}^{pl} \cap \mathcal{F}_{\leq}(\mu, s) \right\} \leq \overline{\mathrm{VaR}}_\alpha^u, \tag{13.9}$$

and we only need to prove the reverse inequality. To do so, consider an arbitrary element $F_X \in \mathcal{F}_{\leq}^u(\mu, s)$ with mode m and mode level p_m. We will show that there exists $F_Y \in \mathcal{F}^{pl} \cap \mathcal{F}_{\leq}(\mu, s)$ such that $\mathrm{VaR}_\alpha(Y) = \mathrm{VaR}_\alpha(X)$, as in this case

$$\overline{\mathrm{VaR}}_\alpha^u \leq \sup \left\{ \mathrm{VaR}_\alpha(Y); F_Y \in \mathcal{F}^{pl} \cap \mathcal{F}_{\leq}(\mu, s) \right\}.$$

The rest of the proof is split into two cases, depending on whether the level α is higher or lower than the mode level p_m of F_X.

i) First case: $\alpha \in [p_m, 1)$.

Let $Y_c = F_{Y_c}^{-1}(U)$ with the quantile function $F_{Y_c}^{-1}$ given as

$$F_{Y_c}^{-1}(p) = \begin{cases} F_X^{-1}(p) & \text{for } p \in [0, \alpha), \\ c(p - \alpha) + F_X^{-1}(\alpha) & \text{for } p \in [\alpha, 1], \end{cases} \tag{13.10}$$

where $c \in \mathbb{R}^+$ is chosen such that $EY_c = \mu$. By construction, we also have that $\text{VaR}_\alpha(Y_c) = \text{VaR}_\alpha(X)$.

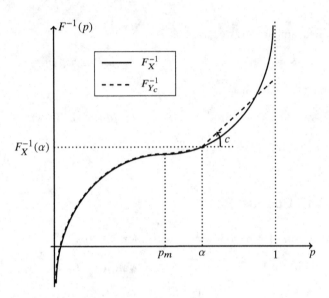

Figure 13.1 Quantile functions of X and Y_c.

In other words, on the interval $[0, \alpha)$, the quantile functions $F_{Y_c}^{-1}$ and F_X^{-1} are identical, whereas on the interval $[\alpha, 1]$ we have that $F_{Y_c}^{-1}$ is linear and F_X^{-1} is convex (see also Figure 13.1). As a consequence, F_X^{-1} up-crosses $F_{Y_c}^{-1}$ once, and since $EY_c = EX$, we obtain from the cut criterion of Karlin and Novikoff (1963) that Y_c is smaller than X in the sense of convex order, i.e., $Y_c \leq_{cx} X$. Hence, $\text{Var}(Y_c) \leq \text{Var}(X) \leq s^2$, and we thus obtain that $F_{Y_c} \in \mathcal{F}_{\leq}(\mu, s)$. Next, we define $Y_l = F_{Y_l}^{-1}(U)$ with a linear quantile function having slope c, i.e.,

$$F_{Y_l}^{-1}(p) = c(p - \alpha) + F_X^{-1}(\alpha), \quad \text{for } p \in [0, 1]. \tag{13.11}$$

We compare the mean of Y_l with the mean of Y_c and proceed further based on this comparison:

- If $EY_l \leq EY_c$, then we consider $Y_R = F_{Y_R}^{-1}(U)$ with quantile function $F_{Y_R}^{-1}$ given as

$$F_{Y_R}^{-1}(p) = \begin{cases} c(b - \alpha) + F_X^{-1}(\alpha) & \text{for } p \in [0, b), \\ c(p - \alpha) + F_X^{-1}(\alpha) & \text{for } p \in [b, \alpha), \\ F_{Y_c}^{-1}(p) & \text{for } p \in [\alpha, 1], \end{cases} \tag{13.12}$$

in which $b \in [0, \alpha]$ is determined such that $EY_R = EY_c$. The quantile function $F_{Y_R}^{-1}$ is thus first constant and then becomes linear in such a way that it corresponds on the interval $[\alpha, 1]$ with the quantile function $F_{Y_c}^{-1}$ (see also Figure 13.2).

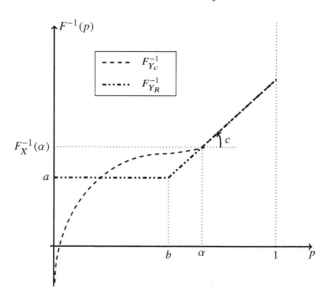

Figure 13.2 Quantile functions of Y_c and Y_R.

Clearly, $F_{Y_c}^{-1}$ up-crosses $F_{Y_R}^{-1}$ exactly once. Since $EY_R = EY_c$, we thus obtain that $Y_R \leq_{cx} Y_c$. As a result, $\mathrm{Var}(Y_R) \leq \mathrm{Var}(Y_c)$ and $F_{Y_R} \in \mathcal{F}_R^{pl} \cap \mathcal{F}_{\leq}(\mu, s)$, and by construction we obtain that $\mathrm{VaR}_\alpha(Y_R) = \mathrm{VaR}_\alpha(Y_c) = \mathrm{VaR}_\alpha(X)$.

- If $EY_l > EY_c$, we consider $Y_{c'} = F_{Y_{c'}}^{-1}(U)$ with a quantile function that is linear with a slope c' on $[0, \alpha)$ and identical to F_X^{-1} on $[\alpha, 1]$ such that $EY_{c'} = \mu$ (see also Figure 13.3 and note that for the sake of clarity of the graph, the shape of F_X^{-1} is slightly modified). In this case $c' > c$ must hold, which implies that F_X^{-1} up-crosses $F_{Y_{c'}}^{-1}$ exactly once. Hence, $Y_{c'} \leq_{cx} X$ and $F_{Y_{c'}} \in \mathcal{F}_{\leq}(\mu, s)$.

Since $c' > c$, we can consider a random variable $Y_L = F_{Y_L}^{-1}(U)$ having a quantile function $F_{Y_L}^{-1}$ with the property that it is identical to $F_{Y_{c'}}^{-1}$ on $[0, \alpha]$, linear on $[0, b)$ with slope c', flat on $[b, 1]$, and such that $EY_L = \mu$ (see also Figure 13.4). $F_{Y_{c'}}^{-1}$ necessarily up-crosses $F_{Y_L}^{-1}$ exactly once. Since $EY_L = EY_{c'}$, we deduce that $Y_L \leq_{cx} Y_{c'}$ and $Y_L \in \mathcal{F}_L^{pl} \cap \mathcal{F}_{\leq}(\mu, s)$. Moreover, by construction, we have that $\mathrm{VaR}_\alpha(X) = \mathrm{VaR}_\alpha(Y_c) = \mathrm{VaR}_\alpha(Y_L)$.

In summary, for every $F_X \in \mathcal{F}_{\leq}^u(\mu, s)$ we can determine a distribution function $F_Y \in \mathcal{F}^{pl} \cap \mathcal{F}_{\leq}(\mu, s)$ such that $\mathrm{VaR}_\alpha(X) = \mathrm{VaR}_\alpha(Y)$. Thus, we finally get

$$\mathrm{VaR}_\alpha(X) \leq \sup_{F_Y \in \mathcal{F}^{pl} \cap \mathcal{F}_{\leq}(\mu, s)} \mathrm{VaR}_\alpha(Y). \qquad (13.13)$$

ii) Second case: $\alpha \in (0, p_m)$.

This case is analogous to the first case. Specifically, for any arbitrary element $F_X \in \mathcal{F}_{\leq}^u(\mu, s)$, we construct a distribution function $F_{Y_{c'}}$ having a quantile function

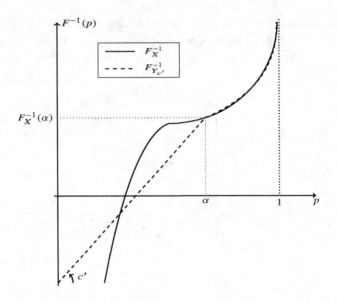

Figure 13.3 Quantile functions of X and $Y_{c'}$.

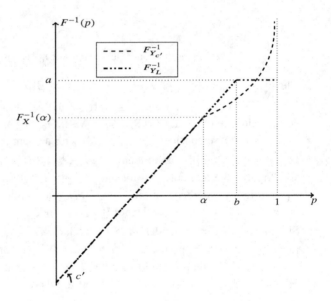

Figure 13.4 Quantile functions of $Y_{c'}$ and Y_L.

that is linear on $[0, \alpha)$ and identical to F_X^{-1} on $[\alpha, 1]$, and such that $EY_{c'} = \mu$. We then proceed in a similar way as in i) and reach the inequality (13.13). □

From Proposition 13.2, in order to optimize VaR bounds one thus only needs to consider quantile functions that are fully parametrized by three parameters. We deal with such a parametric optimization problem using standard techniques and formulate the results in the following theorem.

Theorem 13.3 (**VaR upper bounds for unimodal distributions with given mean and variance**) *The following holds for the upper bound* $\overline{\text{VaR}}_\alpha^u$ *on* $\mathcal{F}_\leq^u(\mu, s)$.

i) We have that

$$\overline{\text{VaR}}_\alpha^u = \begin{cases} \mu + s\sqrt{\frac{4}{9(1-\alpha)} - 1} & \text{for } \alpha \in [\frac{5}{6}, 1), \\ \mu + s\sqrt{\frac{3\alpha}{4-3\alpha}} & \text{for } \alpha \in (0, \frac{5}{6}). \end{cases} \tag{13.14}$$

ii) For $\alpha \in [\frac{5}{6}, 1)$, *a maximizing distribution function is given by* $F^* \in \mathcal{F}_R^{pl}$, *defined in (13.1) with* $a = \mu - s\sqrt{\frac{1-\alpha}{\alpha - 5/9}}$, $b = 3\alpha - 2$, *and* $c = \frac{2s}{9\sqrt{(1-\alpha)^3(\alpha - 5/9)}}$.

For $\alpha \in (0, \frac{5}{6})$, *a maximizing distribution function is given by* $F^* \in \mathcal{F}_L^{pl}$, *defined as in (13.2), with* $a = \mu + s\sqrt{\frac{3\alpha}{4-3\alpha}}$, $b = \alpha$, *and* $c = \frac{2\sqrt{3}s}{\sqrt{\alpha^3(4-3\alpha)}}$. *Furthermore, in both cases it holds that* $\text{Var}(X) = s^2$.

Proof Proposition 13.2 shows that the maximization of VaR over $\mathcal{F}_\leq^u(\mu, s)$ can be reduced to a parametric optimization over $\mathcal{F}^{pl} \cap \mathcal{F}_\leq(\mu, s)$. In what follows we split the optimization into two parts, namely over the sets $\mathcal{F}_R^{pl} \cap \mathcal{F}_\leq(\mu, s)$ and $\mathcal{F}_L^{pl} \cap \mathcal{F}_\leq(\mu, s)$, respectively. We first obtain for both cases bounds and maximizing distribution functions, which we denote by $F_{Y_R^*}$ and $F_{Y_L^*}$, respectively. The solution to our problem follows by comparing these distribution functions and the bounds they yield.

Case 1 Maximization of VaR over $\mathcal{F}_R^{pl} \cap \mathcal{F}_\leq(\mu, s)$.

We denote by F_{Y_R} a distribution function that belongs to $\mathcal{F}_R^{pl} \cap \mathcal{F}_\leq(\mu, s)$. Its quantile function $F_{Y_R}^{-1}$ is expressed as in (13.1):

$$F_{Y_R}^{-1}(p) = \begin{cases} a & \text{for } p \in [0, b), \\ c(p - b) + a & \text{for } p \in [b, 1], \end{cases} \quad a \in \mathbb{R}, b \in [0, 1], c \in \mathbb{R}^+.$$

Using the equations $EY_R = \mu$ and $\text{Var}(Y_R) = \sigma^2 \leq s^2$, we are able to express a and c as functions of b, μ, and σ. We obtain $a = \mu - \sigma\sqrt{\frac{1-b}{1/3+b}}$ and $c = \frac{2\sigma}{\sqrt{(1-b)^3(1/3+b)}}$, which yield the following expression for the quantile function $F_{Y_R}^{-1}(p)$:

$$F_{Y_R}^{-1}(p) = \begin{cases} \mu - \sigma\sqrt{\frac{1-b}{1/3+b}} & \text{for } p \in [0, b), \\ \mu + \sigma\frac{(2p-1-b^2)}{\sqrt{(1-b)^3(1/3+b)}} & \text{for } p \in [b, 1]. \end{cases} \tag{13.15}$$

Next, we maximize $F_{Y_R}^{-1}(\alpha)$ as a function of b for a given $\sigma \in [0, s]$ (note that by construction, $b \in [0, \alpha]$ must hold; see also Figure 13.2 for an illustration). Finally, we maximize over $\sigma \in [0, s]$ to find $F_{Y_R^*} \in \mathcal{F}_R^{pl} \cap \mathcal{F}_\leq(\mu, s)$ having the property

$$\text{VaR}_\alpha(Y_R^*) = \sup_{F_{Y_R} \in \mathcal{F}_R^{pl} \cap \mathcal{F}_\leq(\mu, s)} \text{VaR}_\alpha(Y_R).$$

By our procedure, the problem of determining $F_{Y_R^*}$ in explicit form is split into two cases depending on the value of α. First, assume that $\alpha \in (\frac{2}{3}, 1)$. Given $\sigma \in [0, s]$, $F_{Y_R}^{-1}(\alpha)$ is a concave function of b on $[0, \alpha]$. It is straightforward to show that it attains its maximum at $b = 3\alpha - 2$. Furthermore, when $b = 3\alpha - 2$, $F_{Y_R}^{-1}(\alpha)$ becomes an increasing function of σ that attains its maximum for $\sigma = s$. Hence, the maximizing distribution function $F_{Y_R^*}$ is obtained for $b = 3\alpha - 2$ and $\sigma = s$, and

$$\text{VaR}_\alpha(Y_R^*) = \mu + s\sqrt{\frac{4}{9(1-\alpha)} - 1}. \tag{13.16}$$

Second, assume that $\alpha \in (0, \frac{2}{3})$. For $\sigma \in [0, s]$, we find that $F_{Y_R}^{-1}(\alpha)$ is a decreasing function of b on $[0, \alpha]$ that attains its maximum for $b = 0$. In this case, $F_{Y_R}^{-1}(\alpha)$ is an increasing function of σ for $\alpha \in (\frac{1}{2}, \frac{2}{3}]$ but a decreasing function for $\alpha \in (0, \frac{1}{2}]$. Hence:

- For $\alpha \in (\frac{1}{2}, \frac{2}{3}]$, a maximizing distribution function $F_{Y_R^*}$ is obtained for $b = 0$ and $\sigma = s$, and

$$\text{VaR}_\alpha(Y_R^*) = \mu + s\sqrt{3}(2\alpha - 1). \tag{13.17}$$

- For $\alpha \in (0, \frac{1}{2}]$, a maximizing distribution function $F_{Y_R^*}$ is obtained for $b = 0$ and $\sigma = 0$, and

$$\text{VaR}_\alpha(Y_R^*) = \mu. \tag{13.18}$$

Case 2 Maximization of VaR over $\mathcal{F}_L^{pl} \cap \mathcal{F}_\le(\mu, s)$.

Consider an arbitrary choice for $F_{Y_L} \in \mathcal{F}_L^{pl} \cap \mathcal{F}_\le(\mu, s)$.

In a similar way as in the first case, we use the conditions $EY_L = \mu$ and $\text{Var}(Y_L) = \sigma^2 \le s^2$ to obtain the following expression for the quantile function $F_{Y_L}^{-1}(p)$:

$$F_{Y_L}^{-1}(p) = \begin{cases} \mu + \sigma\sqrt{3} \, \dfrac{(2p-2b+b^2)}{\sqrt{b^3(4-3b)}} & \text{for } p \in [0, b), \\ \mu + \sigma\sqrt{\dfrac{3b}{4-3b}} & \text{for } p \in [b, 1]. \end{cases} \tag{13.19}$$

We maximize $F_{Y_L}^{-1}(\alpha)$ first in terms of $b \in [\alpha, 1]$ and next in terms of $\sigma \in [0, s]$. We obtain that for all $\alpha \in (0, 1)$, a maximizing distribution function $F_{Y_L^*}$ is obtained for $b = \alpha$ and $\sigma = s$, and

$$\text{VaR}_\alpha(Y_L^*) = \mu + s\sqrt{\frac{3\alpha}{4 - 3\alpha}}. \tag{13.20}$$

Finally, to determine $\overline{\text{VaR}}_\alpha^u$ we just need to compare $\text{VaR}_\alpha(Y_R^*)$ and $\text{VaR}_\alpha(Y_L^*)$. As a result, we obtain that

$$\overline{\text{VaR}}_\alpha^u = \begin{cases} \mu + s\sqrt{\dfrac{4}{9(1-\alpha)} - 1} & \text{for } \alpha \in [\frac{5}{6}, 1), \\ \mu + s\sqrt{\dfrac{3\alpha}{4-3\alpha}} & \text{for } \alpha \in (0, \frac{5}{6}). \end{cases} \qquad \square$$

We point out that the above theorem conforms well with the results formulated in Theorem 1 in Li et al. (2018) for $\alpha \ge \frac{5}{6}$ (by letting β tend to α in their formula of

$\mathrm{RVaR}_{\alpha,\beta}$). Proposition 13.3 provides upper bounds for VaR. The lower bounds for VaR follow as a corollary.

Corollary 13.4 (**VaR lower bounds for unimodal distributions with given mean and variance**) *The lower bound* $\underline{\mathrm{VaR}}_{\alpha}^{u}$ *on* $\mathcal{F}_{\leq}^{u}(\mu, s)$ *is given as*

$$\underline{\mathrm{VaR}}_{\alpha}^{u} = \begin{cases} \mu - s\sqrt{\frac{1-\alpha}{1/3+\alpha}} & \text{for } \alpha \in (1/6, 1), \\ \mu - s\sqrt{\frac{4}{9\alpha} - 1} & \text{for } \alpha \in (0, 1/6]. \end{cases} \tag{13.21}$$

Proof As for any distribution function F_X and $\alpha \in (0, 1)$, $\mathrm{VaR}_{\alpha}(X) = -\mathrm{VaR}_{1-\alpha}^{+}(-X)$, we have that

$$\inf_{F_X \in \mathcal{F}_{\leq}^{u}(\mu, s)} \mathrm{VaR}_{\alpha}(X) = - \sup_{F_X \in \mathcal{F}_{\leq}^{u}(\mu, s)} -\mathrm{VaR}_{\alpha}^{+}(F_X)$$

$$= - \sup_{F_X \in \mathcal{F}_{\leq}^{u}(\mu, s)} \mathrm{VaR}_{1-\alpha}^{+}(-X)$$

$$= - \sup_{F_X \in \mathcal{F}_{\leq}^{u}(\mu, s)} \mathrm{VaR}_{1-\alpha}^{+}(-X). \tag{13.22}$$

Moreover, it is clear that if F_X is unimodal, then F_{-X} is also unimodal. Hence,

$$\inf_{F_X \in \mathcal{F}_{\leq}^{u}(\mu, s)} \mathrm{VaR}_{\alpha}(X) = - \sup_{F_{-X} \in \mathcal{F}_{\leq}^{u}(-\mu, s)} \mathrm{VaR}_{1-\alpha}(-X). \tag{13.23}$$

Using Theorem 13.3, we conclude that,

$$\underline{\mathrm{VaR}}_{\alpha}^{u} = \begin{cases} \mu - s\sqrt{\frac{1-\alpha}{\frac{1}{3}+\alpha}} & \text{for } \alpha \in (\frac{1}{6}, 1), \\ \mu - s\sqrt{\frac{4}{9\alpha} - 1} & \text{for } \alpha \in (0, \frac{1}{6}]. \end{cases} \tag{13.24}$$

$\square$

One common robust estimator of the mean is the symmetric quantile average, which can be defined as the average of the values obtained by the quantile function at the two complementary probability levels α and $1 - \alpha$ (Benson, 1949). In the following corollary, we make use of the bounds derived so far to find the maximum distance between this robust estimator and the mean of a unimodal distribution with known first two moments.

Corollary 13.5 (**Maximum distance between the quantile average and mean**) *Let* $\mu \in \mathbb{R}$ *and* $s \in \mathbb{R}^{+}$. *For any distribution function* $F \in \mathcal{F}_{\leq}^{u}(\mu, s)$, *we have that*

$$\frac{\left| \frac{q_{\alpha}+q_{1-\alpha}}{2} - \mu \right|}{s} \leq f(\alpha), \tag{13.25}$$

where q_{α} *denotes the quantile of* F *at a probability level* α, *and*

$$f(\alpha) = \begin{cases} \dfrac{\sqrt{\frac{4}{9(1-\alpha)}-1}+\sqrt{\frac{1-\alpha}{1/3+\alpha}}}{2} & \text{for } \alpha \in [\frac{5}{6}, 1), \\ \dfrac{\sqrt{\frac{3\alpha}{4-3\alpha}}+\sqrt{\frac{1-\alpha}{1/3+\alpha}}}{2} & \text{for } \alpha \in (\frac{1}{6}, \frac{5}{6}), \\ \dfrac{\sqrt{\frac{3\alpha}{4-3\alpha}}+\sqrt{\frac{4}{9\alpha}-1}}{2} & \text{for } \alpha \in (0, \frac{1}{6}]. \end{cases} \tag{13.26}$$

The proof of this result follows in a direct manner from Theorem 13.3 and Corollary 13.4 and is thus omitted.

Remark 13.6 The distance $f(\alpha)$ derived in Corollary 13.5 is minimized at $\alpha = 0.5$, which motivates the common choice of the median as a robust estimator for the mean. Moreover, when $\alpha = 0.5$ in (13.26), we find that $f(\alpha) = \sqrt{\frac{3}{5}}$, which is the maximum distance between the median and the mean of a unimodal distribution (see, e.g., Basu and DasGupta, 1997). ◊

13.2   RVaR and TVaR Bounds

In this section, we derive bounds on range value-at-risk (RVaR) and obtain as a corollary also bounds on TVaR. We consider the following problem for $0 < \alpha < \beta < 1$:

Problem 2
$$\overline{\text{RVaR}}_{\alpha,\beta}^{u} := \sup \left\{ \text{RVaR}_{\alpha,\beta}(X); F_X \in \mathcal{F}_{\leq}^{u}(\mu, s) \right\}, \qquad (13.27)$$

$$\underline{\text{RVaR}}_{\alpha,\beta} := \inf \left\{ \text{RVaR}_{\alpha,\beta}(X); F_X \in \mathcal{F}_{\leq}^{u}(\mu, s) \right\}. \qquad (13.28)$$

In a similar way as in the case of VaR, in a first step we focus on the upper bound and reduce the optimization problem to a parametric one.

Proposition 13.7 (**Reduction to a parametric optimization problem**) *The following holds for the upper bound $\overline{\text{RVaR}}_{\alpha,\beta}^{u}$ on $\mathcal{F}_{\leq}^{u}(\mu, s)$.*

$$\overline{\text{RVaR}}_{\alpha,\beta}^{u} = \sup \left\{ \text{RVaR}_{\alpha,\beta}(X); F_X \in \mathcal{F}^{pl} \cap \mathcal{F}_{\leq}(\mu, s) \right\}.$$

Proof   Relations (13.3) and (13.4) directly imply that

$$\sup \left\{ \text{RVaR}_{\alpha,\beta}(Y); F_Y \in \mathcal{F}^{pl} \cap \mathcal{F}_{\leq}(\mu, s) \right\} \leq \overline{\text{RVaR}}_{\alpha,\beta}^{u},$$

and we only need to prove the reverse inequality. Consider an arbitrary element $F_X \in \mathcal{F}_{\leq}^{u}(\mu, s)$ with mode m and model level p_m. We consider two cases:

i) First case: $\alpha \in [p_m, 1)$. We define the random variable Y_c in a similar way as in the proof of case i) in Proposition 13.2 (see Figure 13.1). We obtain that $F_{Y_c} \in \mathcal{F}_{\leq}(\mu, s)$ and $Y_c \leq_{cx} X$.

If the quantile functions $F_{Y_c}^{-1}$ and F_X^{-1} are identical then $\text{RVaR}_{\alpha,\beta}(X) = \text{RVaR}_{\alpha,\beta}(Y_c)$. If they are not identical, we denote by i the abscissa of the intersection point. Since $Y_c \leq_{cx} X$, we can deduce the following:

  • When $\beta \leq i$, then for all $u \in [\alpha, \beta]$, $\text{VaR}_u(X) \leq \text{VaR}_u(Y_c)$. Hence, $\text{RVaR}_{\alpha,\beta}(X) \leq \text{RVaR}_{\alpha,\beta}(Y_c)$.

- When $\beta > i$, then for all $u \in [\beta, 1]$, $\mathrm{VaR}_u(X) \geq \mathrm{VaR}_u(Y_c)$, and

$$\mathrm{RVaR}_{\alpha,\beta}(X) = \frac{1}{\beta - \alpha}\left[EX - \int_0^\alpha \mathrm{VaR}_u(X)\,du - \int_\beta^1 \mathrm{VaR}_u(X)\,du\right]$$

$$= \frac{1}{\beta - \alpha}\left[EY_c - \int_0^\alpha \mathrm{VaR}_u(Y_c)\,du - \int_\beta^1 \mathrm{VaR}_u(X)\,du\right]$$

$$\leq \mathrm{RVaR}_{\alpha,\beta}(Y_c).$$

We conclude that $\mathrm{RVaR}_{\alpha,\beta}(X) \leq \mathrm{RVaR}_{\alpha,\beta}(Y_c)$. Consider the variable Y_l, in a similar way as in the proof of case i) in Proposition 13.2. We compare the mean of Y_l with the mean of Y_c.

- If $EY_l \leq EY_c$, then we consider a random variable Y_R defined in a similar way as in the proof of Proposition 13.2 (see Figure 13.2). Since for $p \in [\alpha, 1]$, $F_{Y_R}^{-1}(p) = F_{Y_c}^{-1}(p)$ and $F_{Y_R} \in \mathcal{F}_R^{pl} \cap \mathcal{F}_\leq(\mu, s)$, we obtain that $\mathrm{RVaR}_{\alpha,\beta}(Y_R) = \mathrm{RVaR}_{\alpha,\beta}(Y_c)$. Hence,

$$\mathrm{RVaR}_{\alpha,\beta}(X) \leq \sup_{F_Y \in \mathcal{F}_R^{pl} \cap \mathcal{F}_\leq(\mu,s)} \mathrm{RVaR}_{\alpha,\beta}(Y).$$

- If $EY_l > EY_c$, we define $Y_{c'}$ as in the proof of Proposition 13.2 (see also Figure 13.3). Thus, $F_{Y_{c'}} \in \mathcal{F}_\leq(\mu, s)$. Since $F_{Y_{c'}}^{-1}(p) = F_X^{-1}(p)$ for $p \in [\alpha, 1]$, we obtain $\mathrm{RVaR}_{\alpha,\beta}(Y_{c'}) = \mathrm{RVaR}_{\alpha,\beta}(X)$.

We then consider Y_L as defined in the proof of Proposition 13.2. Hence, $F_{Y_L} \in \mathcal{F}_L^{pl} \cap \mathcal{F}_\leq(\mu, s) \subset \mathcal{F}_\leq^u(\mu, s)$, and one can readily show that $\mathrm{RVaR}_{\alpha,\beta}(Y_c) \leq \mathrm{RVaR}_{\alpha,\beta}(Y_L)$. Thus,

$$\mathrm{RVaR}_{\alpha,\beta}(X) \leq \sup_{F_Y \in \mathcal{F}_L^{pl} \cap \mathcal{F}_\leq(\mu,s)} \mathrm{RVaR}_{\alpha,\beta}(Y).$$

We can thus conclude that

$$\mathrm{RVaR}_{\alpha,\beta}(X) \leq \sup\left\{\mathrm{RVaR}_{\alpha,\beta}(Y); F_Y \in \mathcal{F}^{pl} \cap \mathcal{F}_\leq(\mu, s)\right\},$$

and as this holds for any choice of F_X, we obtain that

$$\overline{\mathrm{RVaR}}_{\alpha,\beta}^u \leq \sup\left\{\mathrm{RVaR}_{\alpha,\beta}(Y); F_Y \in \mathcal{F}^{pl} \cap \mathcal{F}_\leq(\mu, s)\right\}.$$

ii) Second case: $\alpha \in (0, p_m)$. This case is symmetric to case i) and is thus omitted. □

Theorem 13.8 (**RVaR upper bounds for unimodal distributions with given mean and variance**) *Let* $\mu \in \mathbb{R}$, $s \in \mathbb{R}^+$, *and* $0 < \alpha < \beta < 1$. *The following holds for the upper bound* $\overline{\mathrm{RVaR}}_{\alpha,\beta}^u$ *on* $\mathcal{F}_\leq^u(\mu, s)$.

i) *When* $2\alpha + \beta \neq 1$, *we have*

$$\overline{\mathrm{RVaR}}_{\alpha,\beta}^{u} = \begin{cases} \mu + s\sqrt{\frac{8}{9(2-\alpha-\beta)} - 1} & \text{for } \alpha \in \left[\frac{5}{6}, 1\right), \\ \max\left\{\mu + s\sqrt{\frac{8}{9(2-\alpha-\beta)} - 1}, Q\right\} & \text{for } \alpha \in \left(\frac{1}{2}, \frac{5}{6}\right) \\ & \text{and } \beta \in (\beta^*, 1), \\ Q & \text{otherwise,} \end{cases}$$

(13.29)

where for a given $\alpha \in (\frac{1}{2}, \frac{5}{6})$, β^* *is the unique root of the equation*[1]

$$27\alpha^3 + 54\alpha^2\beta^2 - 27\alpha^2\beta - 54\alpha^2 + 36\alpha\beta^3 - 135\alpha\beta^2$$
$$+ 108\alpha\beta - 42\beta^4 + 95\beta^3 - 54\beta^2 = 0,$$

and

$$Q = \mu + \frac{s\sqrt{3}}{\beta - \alpha} \frac{b^{*2}(\beta - \alpha - 1) + 2b^*\alpha - \alpha^2}{\sqrt{b^{*3}(4 - 3b^*)}},$$

in which

$$b^* = \alpha \frac{3\alpha + 2 - \sqrt{(3\alpha - 2)^2 + 12(1 - \beta)}}{2(2\alpha + \beta - 1)}.$$

ii) *In the first equality, the upper bound is attained for* $F^* \in \mathcal{F}_R^{pl}$ *defined as in* (13.1), *with* $a = \mu - 3s\sqrt{\frac{2-(\alpha+\beta)}{9(\alpha+\beta)-10}}$, $b = \frac{3}{2}(\alpha + \beta) - 2$, *and* $c = \frac{8s}{3\sqrt{(2-(\alpha+\beta))^3(9(\alpha+\beta)-10)}}$. *The upper bound in the third equality is attained for* $F^* \in \mathcal{F}_L^{pl}$ *defined as in* (13.2), *with* $a = \mu + s\sqrt{\frac{3b^*}{4-3b^*}}$, $b = b^*$, *and* $c = \frac{2\sqrt{3}s}{\sqrt{b^{*3}(4-3b^*)}}$. *In all equalities,* $\mathrm{Var}(X^*) = s^2$.

iii) *When* $2\alpha + \beta = 1$, *we have*

$$\overline{\mathrm{RVaR}}_{\alpha,\beta}^{u} = \mu + \frac{s}{3}\sqrt{\alpha(3\alpha + 8)}.$$

(13.30)

This bound is attained for $F^* \in \mathcal{F}_L^{pl}$ *defined as in* (13.2), *with* $a = \mu + s\sqrt{\frac{9\alpha}{3\alpha+8}}$, $b = \frac{3\alpha}{3\alpha+2}$, *and* $c = \frac{2(3\alpha+2)^2 s}{3\alpha\sqrt{\alpha(3\alpha+8)}}$. *Additionally,* $\mathrm{Var}[X^*] = s^2$.

Corollary 13.9 **(RVaR lower bounds for unimodal distributions with given mean and variance)** *Let* $\mu \in \mathbb{R}$, $s \in \mathbb{R}^+$ *and* $0 < \alpha < \beta < 1$. *The following holds for the lower bound* $\underline{\mathrm{RVaR}}_{\alpha,\beta}^{u}$ *on* $\mathcal{F}_{\leq}^{u}(\mu, s)$.

i) *When* $\alpha + 2\beta \neq 2$, *we have*

$$\underline{\mathrm{RVaR}}_{\alpha,\beta}^{u} = \begin{cases} \mu - s\sqrt{\frac{8}{9(\alpha+\beta)} - 1} & \text{for } \beta \in \left(0, \frac{1}{6}\right), \\ \min\left\{\mu - s\sqrt{\frac{8}{9(\alpha+\beta)} - 1}; W\right\} & \text{for } \beta \in \left(\frac{1}{6}, \frac{1}{2}\right) \\ & \text{and } \alpha \in (0, \alpha^*), \\ W & \text{otherwise,} \end{cases}$$

(13.31)

[1] In fact, for a given $\alpha \in (\frac{1}{2}, \frac{5}{6})$, $\beta \in (\beta^*, 1)$ if and only if
$27\alpha^3 + 54\alpha^2\beta^2 - 27\alpha^2\beta - 54\alpha^2 + 36\alpha\beta^3 - 135\alpha\beta^2 + 108\alpha\beta - 42\beta^4 + 95\beta^3 - 54\beta^2 > 0$.

where for a given $\beta \in (\frac{1}{6}, \frac{1}{2})$, α^* is the only admissible root of the equation

$$-42\alpha^4 + 36\alpha^3\beta + 37\alpha^3 + 54\alpha^2\beta^2 - 81\alpha^2\beta + 6\alpha^2 - 81\alpha\beta^2$$
$$+ 108\alpha\beta - 36\alpha - 27\beta^3 + 54\beta^2 - 36\beta + 8 = 0,$$

and

$$w = \mu - \frac{s\sqrt{3}}{\beta - \alpha} \frac{b^{*2}(\beta - \alpha - 1) + 2b^*(1 - \beta) - (1 - \beta)^2}{\sqrt{b^{*3}(4 - 3b^*)}},$$

in which

$$b^* = (1 - \beta)\frac{5 - 3\beta - \sqrt{(1 - 3\beta)^2 + 12\alpha}}{2(2 - 2\beta - \alpha)}.$$

ii) When $\alpha + 2\beta = 2$, we have

$$\underline{\mathrm{RVaR}}^u_{\alpha,\beta} = \mu - \frac{s}{3}\sqrt{(1 - \beta)(11 - 3\beta)}. \tag{13.32}$$

As TVaR_α can be obtained as a limiting case of $\mathrm{RVaR}_{\alpha,\beta}$, we immediately get the following result.

Corollary 13.10 **(TVaR bounds for unimodal distributions with given mean and variance)** *Let $\mu \in \mathbb{R}$, $s \in \mathbb{R}^+$, and $\alpha \in (0, 1)$. The following holds for the bounds $\overline{\mathrm{TVaR}}^u_\alpha$ and $\underline{\mathrm{TVaR}}^u_\alpha$ on $\mathcal{F}^u_\le(\mu, s)$.*

i) It holds that

$$\overline{\mathrm{TVaR}}^u_\alpha = \begin{cases} \mu + s\sqrt{\frac{8}{9(1-\alpha)} - 1} & \text{for } \alpha \in \left[\frac{1}{2}, 1\right), \\ \mu + \frac{s}{3}\frac{\sqrt{\alpha(8-9\alpha)}}{1-\alpha} & \text{for } \alpha \in \left(0, \frac{1}{2}\right). \end{cases}$$

ii) Furthermore,

$$\underline{\mathrm{TVaR}}^u_\alpha = \mu. \tag{13.33}$$

Proof   From Lemma 2 in Li et al. (2018),

$$\sup_{F_X \in \mathcal{F}^u_\le(\mu,s)} \mathrm{TVaR}_\alpha(X) = \sup_{F_X \in \mathcal{F}^u_\le(\mu,s)} \lim_{\beta \to 1} \mathrm{RVaR}_{\alpha,\beta}(X)$$

$$= \lim_{\beta \to 1} \sup_{F_X \in \mathcal{F}^u_\le(\mu,s)} \mathrm{RVaR}_{\alpha,\beta}(X).$$

A similar result is obtained for the case of the infimum. The stated results now follow in a straightforward way.   □

---

[1] In fact, for a given $\beta \in (\frac{1}{6}; \frac{1}{2})$, $\alpha \in (0, \alpha^*)$ is equivalent to
$-42\alpha^4 + 36\alpha^3\beta + 37\alpha^3 + 54\alpha^2\beta^2 - 81\alpha^2\beta + 6\alpha^2 - 81\alpha\beta^2 + 108\alpha\beta - 36\alpha - 27\beta^3 + 54\beta^2 - 36\beta + 8 > 0.$

13.3   Application to Model Risk Assessment of a Credit Risk Portfolio

We assess the uncertainty inherent in a model that is used for assessing the risk of a portfolio of credit loans in a commercial bank.

We consider a credit risk portfolio containing n dependent risky loans. The risky loans are described by dependent default indicators $\mathbb{I}_1 \cdots \mathbb{I}_n$, with $P(\mathbb{I}_i = 1) = PD_i$, i.e., PD_i, $i = 1, \ldots, n$ is the probability of default of the ith loan. Furthermore, the maximum amount of loss (exposure-at-default = EAD) that can occur due to a default is denoted by EAD_i, and the percentage of the loss in the case of a default (loss-given-default = LGD) is denoted by LGD_i. In what follows we express the aggregate portfolio loss S as a percentage of the maximum possible loss (given as $\sum_{i=1}^{n} EAD_i \, LGD_i$), i.e.,

$$S = \frac{\sum_{i=1}^{n} \mathbb{I}_i \, EAD_i \, LGD_i}{\sum_{i=1}^{n} EAD_i \, LGD_i}.$$

There are a few reference models that are used in the industry for computing the (R)VaRs of S at various probability levels. These are the Moody's KMV model, the CreditRisk$^+$ model, and the Beta model; see also Gordy (2000) and Vandendorpe et al. (2008). However, it is known that these models are closely linked to each other (Dhaene et al., 2003). Therefore, we proceed by using the Beta model as our reference model, as it is simpler to implement. Hence, under the reference model we have that

$$S \sim \text{Beta}(a, b), \text{ in which } a > 0 \text{ and } b > 0.$$

Recall that the density function of $X \sim \text{Beta}(a, b)$ with $a > 0$ and $b > 0$ is given by

$$f_X(x) = \frac{\Gamma(a + b)}{\Gamma(a)\Gamma(b)} x^{a-1}(1 - x)^{b-1} \text{ for } 0 < x < 1,$$

where

$$\Gamma(x) = \int_0^{+\infty} t^{x-1} e^{-t} dt \text{ for } x > 0.$$

The first two central moments of X are $EX = \frac{a}{a+b}$ and $\text{Var}(X) = \frac{ab}{(a+b)^2(a+b+1)}$.

Next, we discuss the model risk that is inherent in this distribution when it is calibrated to real-world credit data.

We consider a corporate portfolio of a major European bank. The portfolio contains 4495 loans to corporate clients. The total exposure (EAD) is 18642.7 (million euros), and the top 10 % of the portfolio (in terms of EAD) accounts for 70.1 % of it. In Table 13.1, we provide some summary statistics. The bank also has models in place for getting estimates for the PDs, LGDs, and EADs, and these will be used for the further analysis.

From these basic components, the expected loss μ of the portfolio loss S follows in a straightforward way, and we obtain that $\mu = 0.00675$. The bank also has a correlation model that provides estimates for all default correlations among the different loans. Once the correlations are estimated, the portfolio standard deviation s can be computed, and we obtain that $s = 0.01824$. The parameters a and b of the Beta

Summary statistics of a corporate portfolio			
	Minimum	Maximum	Average
PD	0.0001	0.15	0.0119
EAD	0	750.2	116.7
LGD	0	0.90	0.41

Table 13.1 Summary statistics of a corporate portfolio containing 4495 loans of a major European bank.

distribution can now be directly computed by moment matching, and we conclude that $a = 0.1293$ and $b = 19.0225$.

α	Beta model	$(\underline{\text{VaR}}_\alpha; \overline{\text{VaR}}_\alpha)$	$(\underline{\text{VaR}}_\alpha^u; \overline{\text{VaR}}_\alpha^u)$
50 %	0.016	(−1.149; 2.499)	(−0.738; 2.088)
75 %	0.383	(−0.378; 3.834)	(−0.201; 2.743)
85 %	1.129	(−0.091; 5.017)	(0.026; 3.231)
90 %	1.987	(0.067; 6.147)	(0.156; 4.060)
92.5 %	2.709	(0.156; 7.081)	(0.230; 4.723)
95 %	3.851	(0.257; 8.626)	(0.315; 5.798)
99.5 %	11.746	(0.546; 26.406)	(0.563; 17.775)

Table 13.2 Bounds for **value-at-risk**. The first column depicts the VaR under the Beta model. The second column depicts the Cantelli bounds for VaR. The third column depicts the bounds provided in Theorem 13.3 and Corollary 13.4. All numbers are expressed as percentages.

We present the results of our calculations in Table 13.2 (the case of VaR), Table 13.3 (the case of TVaR), and Table 13.4 (the case of RVaR). In all these tables, the bounds presented in the second column refer to bounds that are derived under the sole assumption of having information only on the mean μ and the variance s^2. The bounds presented in the third column refer to the case in which we have information on the mean, the variance, and the unimodal shape of the aggregate portfolio loss.

In Table 13.2, by comparing the results in the second and third columns, we observe that the addition of the unimodality assumption has a great impact on the Cantelli bounds of the value-at-risk when evaluated at any probability level.

Similar conclusions as for Table 13.2 can be drawn for Tables 13.3 and 13.4. Specifically, the unimodality improves the Cantelli bounds for TVaR and RVaR, even if not to the same extent as in the case of the VaR.

13.4   Proofs

Proof of Theorem 13.8
The optimization is split into two parts, over the sets $\mathcal{F}_R^{pl} \cap \mathcal{F}_\le(\mu, s)$ and $\mathcal{F}_L^{pl} \cap \mathcal{F}_\le(\mu, s)$. We determine for both cases the upper bound and the maximizing distribution function. A comparison argument then yields the upper bounds presented in Theorem 13.8.

α	Beta model	$(\underline{\text{TVaR}}_\alpha; \overline{\text{TVaR}}_\alpha)$	$(\underline{\text{TVaR}}_\alpha^u; \overline{\text{TVaR}}_\alpha^u)$
75 %	2.573	(0.675; 3.834)	(0.675; 3.591)
85 %	3.827	(0.675; 5.017)	(0.675; 4.723)
90 %	4.984	(0.675; 6.147)	(0.675; 5.798)
95 %	7.191	(0.675; 8.626)	(0.675; 8.146)
99.5 %	15.423	(0.675; 26.406)	(0.675; 24.927)

Table 13.3 Bounds for **tail value-at-risk**. The first column depicts TVaR under the Beta model. The second column depicts the bounds provided in Theorem 11.15. The third column depicts the bounds provided in Corollary 13.10. All numbers are expressed as percentages.

α	β	Beta model	$(\underline{\text{RVaR}}_{\alpha,\beta}; \overline{\text{RVaR}}_{\alpha,\beta})$	$(\underline{\text{RVaR}}_{\alpha,\beta}^u; \overline{\text{RVaR}}_{\alpha,\beta}^u)$
75 %	90 %	0.965	(0.067; 3.834)	(0.121; 2.938)
90 %	95 %	2.778	(0.257; 6.147)	(0.287; 4.723)
95 %	99.5 %	7.191	(0.546; 8.626)	(0.553; 7.454)
99.5 %	99.9 %	13.952	(0.617; 26.406)	(0.621; 22.801)

Table 13.4 Bounds for **range value-at-risk**. The first column depicts the RVaR under the Beta model. The second column depicts the bounds provided in Theorem 11.16. The third column depicts the bounds provided in Theorem 13.8 and Corollary 13.9. All numbers are expressed as percentages.

Case 1 Maximization of $\text{RVaR}_{\alpha,\beta}(S)$ over $\mathcal{F}_R^{pl} \cap \mathcal{F}_\preceq(\mu, s)$.

We denote by F_{Y_R} an arbitrary element of $\mathcal{F}_R^{pl} \cap \mathcal{F}_\preceq(\mu, s)$ and by $F_{Y_R^*}$ the element of $\mathcal{F}_R^{pl} \cap \mathcal{F}_\preceq(\mu, s)$ such that

$$\text{RVaR}_{\alpha,\beta}(Y_R^*) = \sup_{F_{Y_R} \in \mathcal{F}_R^{pl} \cap \mathcal{F}_\preceq(\mu,s)} \text{RVaR}_{\alpha,\beta}(Y_R).$$

Applying the two equations $EY_R = \mu$ and $\text{Var}(Y_R) = \sigma^2$, $F_{Y_R}^{-1}$ can be expressed as in (13.15). We then express $\text{RVaR}_{\alpha,\beta}(Y_R)$ as

$$\text{RVaR}_{\alpha,\beta}(Y_R) = \mu + \sigma \frac{\alpha + \beta - 1 - b^2}{\sqrt{(1-b)^3(1/3 + b)}}, \quad 0 \le b \le \alpha < \beta < 1. \tag{13.34}$$

We maximize the above function in terms of b and σ over the intervals $[0, \alpha]$ and $[0, s]$, respectively. The optimization leads to $F_{Y_R^*}$ with the following characteristics.
- If $4/3 < \alpha + \beta < 2$, $F_{Y_R^*}$ is obtained for $b = \frac{3}{2}(\alpha + \beta) - 2$ and $\sigma = s$, with

$$\text{RVaR}_{\alpha,\beta}(Y_R^*) = \mu + s\sqrt{\frac{8}{9(2 - \alpha - \beta)} - 1}. \tag{13.35}$$

- If $1 < \alpha + \beta \le 4/3$, $F_{Y_R^*}$ is obtained for $b = 0$ and $\sigma = s$, with

$$\text{RVaR}_{\alpha,\beta}(Y_R^*) = \mu + s\sqrt{3}(\alpha + \beta - 1). \tag{13.36}$$

- If $0 < \alpha + \beta \le 1$, $F_{Y_R^*}$ is obtained for $b = 0$ and $\sigma = 0$, with

$$\text{RVaR}_{\alpha,\beta}(Y_R^*) = \mu. \tag{13.37}$$

Case 2 Maximization of $\text{RVaR}_{\alpha,\beta}(S)$ over $\mathcal{F}_L^{pl} \cap \mathcal{F}_\le(\mu, s)$.

Let F_{Y_L} be an arbitrary element of $\mathcal{F}_L^{pl} \cap \mathcal{F}_\le(\mu, s)$. In this case, $F_{Y_L}^{-1}$ can be expressed as in (13.19). In this case, we have two possibilities for the position of b, i.e., the probability level at which $F_{Y_L}^{-1}$ changes from linear increasing to flat, either between α and β or between β and 1. Since we need to maximize $\text{RVaR}_{\alpha,\beta}(Y_L)$ in terms of b over the full domain $[\alpha, 1]$, we study each possibility individually and then perform a comparison to find the absolute maximum.

Consider first the case when $0 < \alpha < \beta \le b \le 1$; then $\text{RVaR}_{\alpha,\beta}(Y_L)$ can be written as

$$\text{RVaR}_{\alpha,\beta}(Y_L) = \mu + \sigma\sqrt{3}\,\frac{\alpha + \beta - 2b + b^2}{\sqrt{b^3(4 - 3b)}}, \quad b \in [\beta, 1]. \tag{13.38}$$

We first maximize $\text{RVaR}_{\alpha,\beta}(Y_L)$ in terms of b for $b \in [\beta, 1]$. The maximum in terms of b is obtained at $b = 1$ in the case $\alpha < \beta(1 - \beta)$ and $3\alpha^2\beta^2 + 2\alpha^2\beta + \alpha^2 + 6\alpha\beta^3 - 2\alpha\beta^2 - 2\alpha\beta + 3\beta^4 - 4\beta^3 + \beta^2 > 0$, and at $b = \beta$ otherwise.

We then maximize in terms of σ over the interval $[0, s]$. The maximum in terms of σ is obtained at $\sigma = 0$ in the case $\alpha < \beta(1 - \beta)$ and at $\sigma = s$ otherwise. If we denote by $F_{Y_{L,1}^*}$ a distribution function of $\mathcal{F}_L^{pl} \cap \mathcal{F}_\le(\mu, s)$ such that

$$\text{RVaR}_{\alpha,\beta}(Y_{L,1}^*) = \sup_{Y_L \in \mathcal{F}_L^{pl} \cap \mathcal{F}_\le(\mu,s)} \text{RVaR}_{\alpha,\beta}(Y_L), 0 < \alpha < \beta \le b \le 1,$$

we can then express the result as follows:

$$\text{RVaR}_{\alpha,\beta}(Y_{L,1}^*) = \begin{cases} \mu, & \text{if } \alpha < \beta(1 - \beta), \\ \mu + s\sqrt{3}\,\frac{\alpha - \beta + \beta^2}{\sqrt{\beta^3(4 - 3\beta)}}, & \text{otherwise.} \end{cases} \tag{13.39}$$

Now, consider the case when $0 < \alpha \le b < \beta < 1$. In this case, $\text{RVaR}_{\alpha,\beta}$ can be expressed as

$$\text{RVaR}_{\alpha,\beta}(Y_L) = \mu + \frac{\sigma\sqrt{3}}{\beta - \alpha}\,\frac{b^2(\beta - \alpha - 1) + 2b\alpha - \alpha^2}{\sqrt{b^3(4 - 3b)}}, \quad b \in [\alpha, \beta). \tag{13.40}$$

If we denote by $F_{Y_{L,2}^*}$ a distribution function of $\mathcal{F}_L^{pl} \cap \mathcal{F}_\le(\mu, s)$ such that

$$\text{RVaR}_{\alpha,\beta}(Y_{L,2}^*) = \sup_{Y_L \in \mathcal{F}_L^{pl} \cap \mathcal{F}_\le(\mu,s)} \text{RVaR}_{\alpha,\beta}(Y_L), \quad 0 < \alpha \le b < \beta < 1, \text{ then:}$$

- If $2\alpha + \beta = 1$, then $F_{Y_{L,2}^*}$ is obtained at $b = \frac{3\alpha}{3\alpha+2}$ and $\sigma = s$ with

$$\text{RVaR}_{\alpha,\beta}(Y_{L,2}^*) = \mu + \frac{s}{3}\sqrt{\alpha(3\alpha + 8)}.$$

- If $2\alpha + \beta \ne 1$, then $F_{Y_{L,2}^*}$ is obtained at

$$b = \alpha\,\frac{3\alpha + 2 - \sqrt{(3\alpha - 2)^2 + 12(1 - \beta)}}{2(2\alpha + \beta - 1)} \text{ and } \sigma = s.$$

Finally, we compare $\text{RVaR}_{\alpha,\beta}(Y_R^*)$, $\text{RVaR}_{\alpha,\beta}(Y_{L,1}^*)$, and $\text{RVaR}_{\alpha,\beta}(Y_{L,2}^*)$ to find the upper bounds presented in Theorem 13.8, i.e.,

$$\overline{\text{RVaR}}_{\alpha,\beta}^u = \max\left\{\text{RVaR}_{\alpha,\beta}(Y_R^*); \text{RVaR}_{\alpha,\beta}(Y_{L,1}^*); \text{RVaR}_{\alpha,\beta}(Y_{L,2}^*)\right\}. \quad \square$$

Proof of Corollary 13.9

For any $(\alpha, \beta) \in (0, \beta) \times (0, 1)$, it holds that

$$\inf_{F_X \in \mathcal{F}_\le^u(\mu,s)} \text{RVaR}_{\alpha,\beta}(X) = - \sup_{F_X \in \mathcal{F}_\le^u(\mu,s)} -\text{RVaR}_{\alpha,\beta}(X)$$

$$= - \sup_{F_X \in \mathcal{F}_\le^u(\mu,s)} \text{RVaR}_{1-\beta,1-\alpha}(-X)$$

$$= - \sup_{F_{-X} \in \mathcal{F}_\le^u(-\mu,s)} \text{RVaR}_{1-\beta,1-\alpha}(-X). \qquad (13.41)$$

Hence, Corollary 13.9 follows from Theorem 13.8.

14 Moment Bounds in Neighborhood Models

In this chapter, we study sharp upper and lower bounds of distortion risk measures when, in addition to moment information on the underlying loss distribution, it is also known that the loss distribution lies in an ε-neighborhood of a reference distribution with respect to the Wasserstein distance. For sufficiently small distance constraints, i.e., when ε is small enough, this additional information on the reference model is expected to lead to a significant reduction of the worst resp. best case moment bound from Section 12.2. When $\varepsilon \to \infty$, we recover these moment bounds as limiting cases.

14.1   Problem Formulation

We study sharp bounds on distortion risk measures when the loss distribution is known to belong to the neighborhood of a reference model with respect to the Wasserstein distance of order 2, see also Bernard et al. (2023). To formalize this, we first recall the definition of the Wasserstein distance.

Definition 14.1 (Wasserstein distance of order 2)   The Wasserstein distance of order 2 between $G_1, G_2 \in \mathcal{M}^2$ is (Rachev and Rüschendorf, 1998a)

$$d_W(G_1, G_2) = \inf \left\{ \left[\mathbb{E}((X_1 - X_2)^2) \right]^{\frac{1}{2}} \middle| X_1 \sim G_1, \ X_2 \sim G_2 \right\},$$

where the infimum is taken over all bivariate distributions with marginals G_1 and G_2.

For distributions on the real line, the Wasserstein distance admits the following representation:

$$d_W(G_1, G_2) = \left(\int_0^1 \left(G_1^{-1}(u) - G_2^{-1}(u) \right)^2 du \right)^{\frac{1}{2}}.$$

Hence, the Wasserstein distance between G_1 and G_2 is uniquely determined by their corresponding quantile functions, and we may write $d_W(G_1^{-1}, G_2^{-1})$ instead of $d_W(G_1, G_2)$.

We denote the reference distribution by $F \in \mathcal{F}$ and its first two moments by $\int x \, dF(x) = \mu_F \in \mathbb{R}$ and $\int x^2 \, dF(x) = \mu_F^2 + \sigma_F^2$, $\sigma_F > 0$, respectively. Throughout, we fix the reference distribution F and consider distributional sets of the type

$$\mathcal{F}_\varepsilon(\mu, \sigma) = \left\{ G \in \mathcal{F}(\mu, \sigma) \middle| d_W(F, G) \le \sqrt{\varepsilon} \right\},$$

where $\mu \in \mathbb{R}$, $\sigma > 0$, and $0 \le \varepsilon \le +\infty$. The set $\mathcal{F}_\varepsilon(\mu, \sigma)$ thus contains all distribution functions whose first two moments are μ and $\mu^2 + \sigma^2$, respectively, and that lie within a $\sqrt{\varepsilon}$-Wasserstein ball around the reference distribution F. The ε-constraint in the definition of $\mathcal{F}_\varepsilon(\mu, \sigma)$ is redundant when $\varepsilon = +\infty$, and in this special case, we find that the set $\mathcal{F}_\infty(\mu, \sigma)$ coincides with $\mathcal{F}(\mu, \sigma)$.

We aim to determine sharp bounds of a distortion risk measure H_g over the distributional uncertainty set $\mathcal{F}_\varepsilon(\mu, \sigma) := \mathcal{F}_0$, i.e., we aim to determine

$$\underline{H}_g(\mathcal{F}_0) = \inf_{G \in \mathcal{F}_0} H_g(G) \quad (14.1a) \qquad \overline{H}_g(\mathcal{F}_0) = \sup_{G \in \mathcal{F}_0} H_g(G). \quad (14.1b)$$

In addition to the worst and best case values, we also study *best case* and *worst case distribution functions* if they exist; that is, the distribution functions attaining (14.1a) and (14.1b) respectively.

We recall from Section 12.2 that if g is absolutely continuous, a distortion risk measure H_g has the following representation:

$$H_g(G) = \int_0^1 \gamma(u) G^{-1}(u)\, du, \quad (14.2)$$

with *weight function* $\gamma(u) = \partial_- g(x)|_{x=1-u}$, $0 < u < 1$, that satisfies $\int_0^1 \gamma(u)\, du = 1$ and where ∂_- denotes the derivative from the left.

We also make the the following assumptions.

Assumption 14.2 Representation (14.2) holds and $\int_0^1 |\gamma(u)|^2\, du < +\infty$. We also assume that $\gamma(\cdot) \ne 1$ and that $\mathcal{F}_\varepsilon(\mu, \sigma)$ contains at least two elements, and hence infinitely many.

In this regard, the assumption that $\mathcal{F}_\varepsilon(\mu, \sigma)$ contains at least two elements is equivalent to assuming that $\varepsilon > (\mu_F - \mu)^2 + (\sigma_F - \sigma)^2$. To see this, note that for any $G \in \mathcal{F}_\varepsilon(\mu, \sigma)$,

$$d_W(F, G)^2 = \int_0^1 \left(G_1^{-1}(u) - G_2^{-1}(u) \right)^2 du$$

$$= \mu_F^2 + \sigma_F^2 + \mu^2 + \sigma^2 - 2\,\mathrm{Cov}\left(F^{-1}(U), G^{-1}(U)\right) - 2\mu\mu_F$$

$$= (\mu_F - \mu)^2 + (\sigma_F - \sigma)^2 + 2\sigma\sigma_F(1 - \mathrm{Cor}(F^{-1}(U), G^{-1}(U)))$$

$$\ge (\mu_F - \mu)^2 + (\sigma_F - \sigma)^2.$$

Hence, if $\varepsilon < (\mu_F - \mu)^2 + (\sigma_F - \sigma)^2$ then $\mathcal{F}_\varepsilon(\mu, \sigma) = \phi$, and if $\varepsilon = (\mu_F - \mu)^2 + (\sigma_F - \sigma)^2$ then $\mathcal{F}_\varepsilon(\mu, \sigma)$ is a singleton, containing only one distribution with quantile function

$$G^{-1}(u) = \mu + \sigma \left(\frac{F^{-1}(u) - \mu_F}{\sigma_F} \right), \quad 0 < u < 1. \quad (14.3)$$

Moreover, in this special case, G coincides with the reference distribution F if and only if $\mu = \mu_F$ and $\sigma = \sigma_F$. Thus, we further assume that $\varepsilon > (\mu_F - \mu)^2 + (\sigma_F - \sigma)^2$.

14.2   Bounds on Distortion Risk Measures

We first study sharp upper bounds for the class of *concave distortion risk measures*, which are distortion risk measures with concave distortion function g. The corresponding results for the case of general distortion risk measures are presented in Section 14.2.2 and those for sharp lower bounds in Section 14.2.3.

14.2.1   Upper Bounds on Concave Distortion Risk Measures

For the special case of concave distortion risk measures, we provide in the following theorem analytic solutions to the upper bound problem (14.1b).

Theorem 14.3 (Sharp upper bounds)   *Let g be a concave distortion function and denote $c_0 = \mathrm{Cor}(F^{-1}(U), \gamma(U))$.[1] Then, the following statements hold:*

1. *If $(\mu_F - \mu)^2 + (\sigma_F - \sigma)^2 < \varepsilon < (\mu_F - \mu)^2 + (\sigma_F - \sigma)^2 + 2\sigma\sigma_F(1 - c_0)$, then the solution to (14.1b) is unique and has a quantile function given by*

$$h_\lambda(u) = \mu + \sigma\left(\frac{\gamma(u) + \lambda F^{-1}(u) - a_\lambda}{b_\lambda}\right), \quad 0 < u < 1,$$

where $\lambda > 0$ denotes the unique positive solution to $d_W(F^{-1}, h_\lambda) = \sqrt{\varepsilon}$, which is explicitly given by

$$\lambda = \frac{K}{\sigma_F^2}\sqrt{\frac{C_{\gamma,F}^2 - V\sigma_F^2}{K^2 - \sigma^2\sigma_F^2}} - \frac{C_{\gamma,F}}{\sigma_F^2}, \tag{14.4}$$

where $V = \mathrm{var}(\gamma(U))$, $C_{\gamma,F} = \mathrm{Cov}\left(F^{-1}(U), \gamma(U)\right)$, $a_\lambda = \mathbb{E}(\gamma(U) + \lambda F^{-1}(U))$, $b_\lambda = \mathrm{std}(\gamma(U) + \lambda F^{-1}(U))$, and where

$$K = \frac{\mu_F^2 + \sigma_F^2 + \mu^2 + \sigma^2 - 2\mu\mu_F - \varepsilon}{2} \geq 0.$$

The corresponding worst case value, i.e., the solution to (14.1b), is

$$H_g(h_\lambda) = \mu + \frac{\sigma}{b_\lambda}\left(V + \lambda C_{\gamma,F}\right)$$
$$= \mu + \sigma\, \mathrm{std}\left(\gamma(U)\right) \mathrm{Cor}\left(\gamma(U), \gamma(U) + \lambda F^{-1}(U)\right),$$

and $H_g(h_\lambda)$ is continuous in ε.
2. *If $(\mu_F - \mu)^2 + (\sigma_F - \sigma)^2 + 2\sigma\sigma_F(1 - c_0) \leq \varepsilon$, then case 1 applies with $\lambda = 0$.*

The sharp upper bound provided in Theorem 14.3 is attained by the worst case quantile function h_λ. We observe that h_λ is a weighted average of the reference quantile function F^{-1} and the (non-decreasing) weight function γ from the distortion risk measure. As the worst case bound $H_g(h_\lambda)$ is non-decreasing in the tolerance distance ε, we obtain that $\mathrm{Cor}\left(\gamma(U), \gamma(U) + \lambda F^{-1}(U)\right)$ is non-decreasing in ε, which in turn implies that λ is non-increasing in ε. Hence, we obtain that the influence of the reference

---
[1] If $c_0 = 1$, then only case 2 applies.

distribution on the worst case quantile function h_λ diminishes with increasing tolerance distance ε. Furthermore, for ε sufficiently large, i.e., under case 2 of Theorem 14.3, λ is zero, and the worst case quantile function is independent of the reference distribution.

Applying Theorem 14.3 with $\varepsilon = +\infty$, we obtain sharp upper bounds on concave distortion risk measures H_g under the knowledge of the first two moments of the distribution, i.e., without considering an ε-Wasserstein distance constraint or a reference distribution.

Corollary 14.4 (Sharp upper bounds, $\varepsilon = +\infty$) *Let g be a concave distortion function. Then,*

$$\sup_{G \in \mathcal{F}(\mu,\sigma)} H_g(G) = \mu + \sigma \operatorname{Std}(\gamma(U)) = \mu + \sigma\sqrt{\int_0^1 (\gamma(p) - 1)^2 \, \mathrm{d}p}. \tag{14.5}$$

The bound is attained by a unique maximizing distribution function with quantile function h_0 defined in Theorem 14.3.

Note that this corollary also appeared as Theorem 12.9 in Section 12.2.

Example 14.5 We illustrate the sharp upper bounds of Theorem 14.3 for three concave distortion risk measures: the dual power distortion with parameter $\beta > 0$,

$$g(x) = 1 - (1 - x)^\beta, \quad \text{and} \quad \gamma(u) = \beta u^{\beta-1}; \tag{14.6}$$

the Wang transform (Wang, 1996) with parameter $0 < q_0 < 1$,

$$g(x) = \Phi\left(\Phi^{-1}(x) + \Phi^{-1}(q_0)\right), \quad \text{and} \quad \gamma(u) = \frac{\varphi\left(\Phi^{-1}(1 - u) + \Phi^{-1}(q_0)\right)}{\varphi\left(\Phi^{-1}(1 - u)\right)}, \tag{14.7}$$

where Φ and φ denote the standard normal distribution and its density, respectively; and the tail value-at-risk (TVaR_α) with

$$g(x) = \min\left\{\frac{x}{1 - \alpha}, 1\right\} \quad \text{and} \quad \gamma(u) = \frac{1}{1 - \alpha}\mathbb{1}_{(\alpha,1)}(u). \tag{14.8}$$

The reference distribution F is chosen to be standard normal, and we further set $\mu = \mu_F = 0$ and $\sigma = \sigma_F = 1$. In the left panels of Figure 14.1, we observe that the upper bounds are non-decreasing and continuous functions of ε. The vertical lines display $\varepsilon^* = (\mu_F - \mu)^2 + (\sigma_F - \sigma)^2 + 2\sigma\sigma_F(1 - c_0) = 2(1 - c_0)$. The parameter value ε^* indicates the transition from case 1 to case 2 in Theorem 14.3; indeed for $\varepsilon \geq \varepsilon^*$, case 2 of Theorem 14.3 applies, and the upper bound is independent of ε. For TVaR_α with $\alpha = 0.7$ (bottom left panel), the worst case bound is equal to $\mu + \sigma\sqrt{\frac{\alpha}{1-\alpha}} = \sqrt{\frac{\alpha}{1-\alpha}} = 1.53$, and we recover the well-known Cantelli upper bound for VaR and TVaR.

The right panels of Figure 14.1 display the worst case quantile functions for different values of ε. We observe that for $\varepsilon = 0$, the worst case quantile functions are equal to the reference quantile functions (solid line). When the Wasserstein distance ε increases, the influence of the reference distribution diminishes and, if $\varepsilon > 2(1 - c_0)$, the worst case quantile function (dash-dotted line) is independent of the reference distribution.

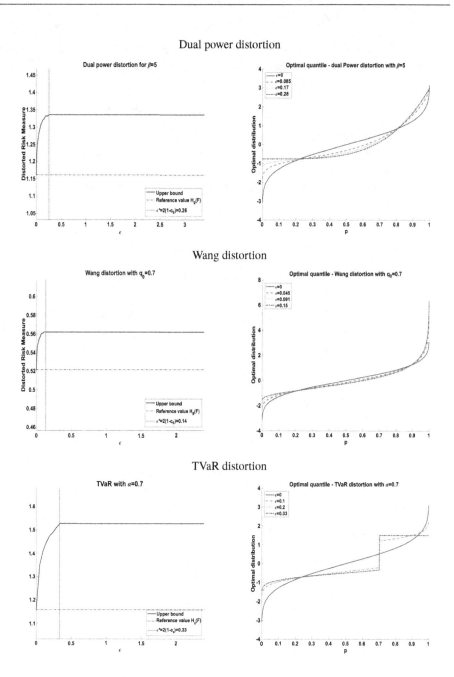

Figure 14.1 Sharp upper bounds and corresponding worst case quantile functions of three concave distortion risk measures: the dual power distortion (upper two figures), the Wang transform (middle figures), and the TVaR (bottom two figures). The left panels display the upper bounds as a function of ε; the right panels display the corresponding worst case quantile functions for different values of ε.

This can clearly be seen for TVaR, where the dash-dotted quantile function corresponds to a two-point distribution.

14.2.2   Upper Bounds on General Distortion Risk Measures

To characterize the sharp upper bounds and corresponding worst case quantile functions of general distortion risk measures, we make use of L^2-projections. Hence, for $\lambda \geq 0$, denote by $k_\lambda^\uparrow$ the L^2 projection of $\gamma + \lambda F^{-1}$ onto the space of square-integrable non-decreasing functions on $(0, 1)$, that is, the unique solution to

$$k_\lambda^\uparrow = \arg\inf_{k \in \mathcal{K}} ||\gamma + \lambda F^{-1} - k||^2, \tag{14.9}$$

where $|| \cdot ||$ denotes the L^2 norm and where

$$\mathcal{K} = \left\{ k : (0, 1) \to \mathbb{R} \,\middle|\, \int_0^1 k(u)^2 \, du < +\infty, \; k \text{ non-decreasing} \right\}. \tag{14.10}$$

Whenever $\lambda = 0$, we write $\gamma^\uparrow$ instead of $k_0^\uparrow$, as $k_0^\uparrow$ is indeed the L^2 projection of γ alone. In Section 14.6 we provide further details on these projections and their properties.

We further define

$$\lambda^\uparrow = \inf \left\{ \lambda \geq 0 \,\middle|\, k_\lambda^\uparrow \text{ is not constant} \right\} \tag{14.11}$$

and set $\inf \phi = +\infty$. Finally, we define the quantile function

$$h_\lambda^\uparrow(u) = \mu + \sigma \left(\frac{k_\lambda^\uparrow(u) - a_\lambda}{b_\lambda} \right), \quad 0 < u < 1, \tag{14.12}$$

with $a_\lambda = E\left(k_\lambda^\uparrow(U)\right)$ and $b_\lambda = \mathrm{Std}\left(k_\lambda^\uparrow(U)\right)$, which is well defined whenever $\lambda > \lambda^\uparrow$.

Before stating the sharp upper bound of general distortion risk measures, we need a further assumption.

Assumption 14.6   We assume that $\lambda^\uparrow = 0$.

Notice that Assumption 14.6 is a requirement on the weight function and the reference distribution. Assumption 14.6 is, for instance, satisfied when γ is bounded and the reference distribution has a quantile function that is unbounded to the right $(F^{-1}(u) \to +\infty$ for $u \to +\infty)$.

Theorem 14.7 (Sharp upper bounds)   *Let g be a distortion function, and denote $c_0 = \lim_{\lambda \searrow 0} \mathrm{Cor}\left(F^{-1}(U), k_\lambda^\uparrow(U)\right)$. Then,*

1. *If $(\mu_F - \mu)^2 + (\sigma_F - \sigma)^2 < \varepsilon < (\mu_F - \mu)^2 + (\sigma_F - \sigma)^2 + 2\sigma\sigma_F(1 - c_0)$, then there is a maximizing distribution to (14.1b) that is unique and has quantile function $h_\lambda^\uparrow$ defined in (14.12), where $\lambda > 0$ is the unique positive solution to $d_W(F^{-1}, h_\lambda^\uparrow) = \sqrt{\varepsilon}$. Furthermore, the sharp upper bound, i.e., the solution to (14.1b), is*

$$H_g(h_\lambda^\uparrow) = \mu + \sigma \, \mathrm{Std}(\gamma(U)) \, \mathrm{Cor}\left(\gamma(U), k_\lambda^\uparrow(U)\right),$$

and $H_g(h_\lambda^\uparrow)$ is continuous in ε.

2. *If $(\mu_F - \mu)^2 + (\sigma_F - \sigma)^2 + 2\sigma\sigma_F(1 - c_0) \le \varepsilon$ and $\gamma^\uparrow$ is not constant, then case 1) applies with $\lambda = 0$. Otherwise the solution to (14.1b) is equal to μ but cannot be attained.*

When g is a concave function, then $\gamma + \lambda F^{-1}$ is non-decreasing and thus equal to its isotonic projection. Hence, Theorem 14.7 reduces to Theorem 14.3 in the case of concave distortion risk measures. Applying Theorem 14.7 with $\varepsilon = +\infty$, we obtain sharp upper bounds on general distortion risk measures H_g under the knowledge of the first two moments of the distribution.

Corollary 14.8 (Sharp upper bounds and quantiles for $\varepsilon = +\infty$)   *Let g be a distortion function; then the following statements hold:*

1. *If $\gamma^\uparrow$ is not constant, then*

$$\sup_{G \in \mathcal{F}(\mu,\sigma)} H_g(G) = \mu + \sigma \, \mathrm{Std}\left(\gamma^\uparrow(U)\right) = \mu + \sigma\sqrt{\int_0^1 (\gamma^\uparrow(p) - 1)^2 \, dp} \,. \quad (14.13)$$

The bound is attained by a distribution with quantile function $h_0^\uparrow$ defined in (14.12).
2. *If $\gamma^\uparrow$ is constant, then*

$$\sup_{G \in \mathcal{F}(\mu,\sigma)} H_g(G) = \mu, \quad (14.14)$$

and the bound cannot be attained.

14.2.3 Lower Bounds on General Distortion Risk Measures

In a similar way as in Section 14.2.2, we use the tool of projections to derive the sharp lower bounds and corresponding best case quantile functions of general distortion risk measures.

For $\lambda \ge 0$, denote by $k_\lambda^\downarrow$ the isotonic projection of $\gamma - \lambda F^{-1}$ onto the set of square-integrable non-increasing functions, that is, the unique solution to

$$k_\lambda^\downarrow = \arg\inf_{-k \in \mathcal{K}} ||\gamma - \lambda F^{-1} - k||^2 \,. \quad (14.15)$$

Whenever $\lambda = 0$, we write $\gamma^\downarrow$ instead of $k_0^\downarrow$, as $k_0^\downarrow$ is the L^2 projection of γ onto the non-increasing functions. Further, we define

$$\lambda^\downarrow = \inf\left\{\lambda \ge 0 \mid k_\lambda^\downarrow \text{ is not constant}\right\} \quad (14.16)$$

and the quantile function

$$h_\lambda^\downarrow(u) = \mu + \sigma\left(\frac{a_\lambda - k_\lambda^\downarrow(u)}{b_\lambda}\right), \quad 0 < u < 1, \quad (14.17)$$

with $a_\lambda = E(k_\lambda^\downarrow(U))$ and $b_\lambda = \mathrm{Std}(k_\lambda^\downarrow(U))$, which is well defined when $\lambda > \lambda^\downarrow$.

Assumption 14.9 We assume that $\lambda^{\downarrow} = 0$.

The next theorem states the sharp lower bounds and corresponding best case quantile functions of general distortion risk measures.

Theorem 14.10 (Sharp lower bounds) *Let g be a distortion function, and denote* $c_0 = \lim_{\lambda \downarrow 0} \mathrm{Cor}(F^{-1}(U), -k_{\lambda}^{\downarrow}(U))$. *Then, the following statements hold:*

1. *If* $(\mu_F - \mu)^2 + (\sigma_F - \sigma)^2 \le \varepsilon < (\mu_F - \mu)^2 + (\sigma_F - \sigma)^2 + 2\sigma\sigma_F(1 - c_0)$, *then there is a minimizing distribution function to (14.1a) that is unique and given by the quantile function* $h_{\lambda}^{\downarrow}$ *as defined in (14.17), where* $\lambda > 0$ *is the unique positive solution to* $d_W(F^{-1}, h_{\lambda}^{\downarrow}) = \sqrt{\varepsilon}$. *The corresponding sharp lower bound, i.e., the solution to (14.1a), is given as*

$$H_g(h_{\lambda}^{\downarrow}) = \mu - \sigma \, \mathrm{Std}(\gamma(U)) \, \mathrm{Cor}\left(\gamma(U), k_{\lambda}^{\downarrow}(U)\right),$$

   and H_g *is continuous in* ε.
2. *If* $(\mu_F - \mu)^2 + (\sigma_F - \sigma)^2 + 2\sigma\sigma_F(1 - c_0) \le \varepsilon$ *and* $\gamma^{\downarrow}$ *is not constant, then case 1) applies with* $\lambda = 0$. *Otherwise the infimum is equal to* μ *but not attained.*

The next corollary collects the results for $\varepsilon = +\infty$, that is, when no reference distribution is considered.

Corollary 14.11 (Sharp lower bounds for $\varepsilon = +\infty$) *Let g be a distortion function.*

1. *If* $\gamma^{\downarrow}$ *is not constant, then*

$$\inf_{G \in \mathcal{F}(\mu,\sigma)} H_g(G) = \mu - \sigma \, \mathrm{Std}(\gamma(U)) \, \mathrm{Cor}\left(\gamma(U), \gamma^{\downarrow}(U)\right). \tag{14.18}$$

   The sharp bound is attained by a distribution with quantile function $h_0^{\downarrow}$ *defined in (14.17).*
2. *If* $\gamma^{\downarrow}$ *is constant, then*

$$\inf_{G \in \mathcal{F}(\mu,\sigma)} H_g(G) = \mu, \tag{14.19}$$

   and the bound cannot be attained.

Remark 14.12 Distortion risk measures have an alternative representation that allows us to write minimization problems in terms of maximization problems. Specifically,

$$\inf_{G_X \in \mathcal{F}(\mu,\sigma)} H_g(G_X) = - \sup_{G_{-X} \in \mathcal{M}(-\mu,\sigma)} H_{\bar{g}}(G_{-X}), \tag{14.20}$$

where $\bar{g}$ is the dual distortion of g, given by $\bar{g}(x) = 1 - g(1-x)$, and G_X, G_{-X} denote the distribution functions of the random variables X and $-X$, respectively. Equation (14.20) follows from the fact that $H_g(G_X) = -H_{\bar{g}}(G_{-X})$. This observation makes it possible to obtain Corollary 14.11 from Corollary 14.8 in a more direct manner. However, such reasoning cannot be extended when dealing with the Wasserstein distance constraints considered in this chapter.   ◊

Example 14.13 Figure 14.2 illustrates the sharp upper and lower bounds (Theorems 14.7 and 14.10) of three concave distortion risk measures: the dual power distortion, the Wang transform, and TVaR, with a standard normal reference distribution, $\mu_F = \mu = 0$, and $\sigma_F = \sigma = 1$ (see also Example 14.5). In the left panels of Figure 14.2, the solid line corresponds to the upper bound, and the dashed line corresponds to the lower bound. The dash-dotted horizontal line depicts the reference risk measure $H_g(F)$, and the vertical lines display the critical ε value for the transition between case 1 and case 2 in Theorem 14.7 (upper bound, solid line) and Theorem 14.10 (lower bound, dashed line).

As all three risk measures are concave, their lower bounds – for ε sufficiently large (case 2 in Theorem 14.10) – are all equal to $\mu = 0$ but not attained. That the sharp lower bounds are not attained can be seen in the right panels, where the corresponding best case quantile functions are displayed. The best case quantile functions become flatter for larger ε, and for ε sufficiently large, the quantile functions are no longer defined.

14.3 Bounds on Range VaR

The following two propositions provide upper and lower bounds on RVaR when, in addition to the moment constraints, the distributions lie within an ε-Wasserstein ball of the reference distribution F.

Proposition 14.14 (Sharp upper bounds for RVaR)   *Under the assumptions of Theorem 14.7 Case 1, a maximizing quantile function to* (14.1b) *with* $\mathrm{RVaR}_{\alpha,\beta}$ *as risk measure is given by* h_λ *defined in* (14.12)*, where* $k_\lambda^\uparrow$ *is*

$$
k_\lambda^\uparrow(u) = \begin{cases} \lambda F^{-1}(u) & 0 < u \le \alpha, \\ \frac{1}{\beta-\alpha} + \lambda F^{-1}(u) & \alpha < u \le w_0, \\ \lambda F^{-1}(w_1) & w_0 < u \le w_1, \\ \lambda F^{-1}(u) & w_1 < u < 1, \end{cases} \tag{14.21}
$$

and $\alpha \le w_0 \le \beta \le w_1$ *are solutions to*

$$
F^{-1}(w_0) = F^{-1}(w_1) - \frac{1}{\lambda(\beta - \alpha)} \quad and \tag{14.22a}
$$

$$
\lambda F^{-1}(w_1) = \frac{1}{w_1 - w_0} \frac{\beta - w_0}{\beta - \alpha} + \frac{\lambda}{w_1 - w_0} \int_{w_0}^{w_1} F^{-1}(u)\, du, \tag{14.22b}
$$

provided these exist. Otherwise $w_0 = \alpha$*, and* w_1 *is the solution to* (14.22b).

Proposition 14.15 (Sharp lower bounds for RVaR)   *Under the assumptions of Theorem 14.10 case 1, a maximizing quantile function to* (14.1a) *with* $\mathrm{RVaR}_{\alpha,\beta}$ *as the risk measure is unique and given by* $h_\lambda^\downarrow$ *defined in Theorem 14.10, where* $k_\lambda^\downarrow$ *is*

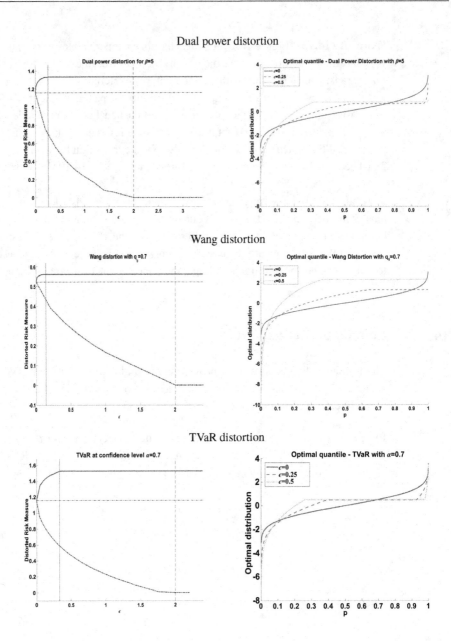

Figure 14.2 Sharp upper and lower bounds and quantile functions of three concave risk measures: the dual power distortion (upper two figures), the Wang transform (middle two figures), and TVaR (bottom two figures). The left panels display the upper and lower bounds as a function of ε; the right panels display the best case quantile functions for different values of ε. The vertical lines in the left panels correspond to the transition of case 1 to case 2, in the upper bound (Theorem 14.7, solid line) and lower bound (Theorem 14.10, dashed line).

$$
k_\lambda^\downarrow(u) = \begin{cases} -\lambda F^{-1}(u) & 0 < u \le z_0, \\ -\lambda F^{-1}(z_0) & z_0 < u \le z_1, \\ -\lambda F^{-1}(u) + \frac{1}{\beta - \alpha}, & z_1 < u \le \beta, \\ -\lambda F^{-1}(u) & \beta < u < 1, \end{cases} \tag{14.23}
$$

and $z_0 \le \alpha \le z_1 \le \beta$ are solutions to

$$
F^{-1}(z_1) = F^{-1}(z_0) + \frac{1}{\lambda(\beta - \alpha)} \quad \text{and} \tag{14.24a}
$$

$$
\lambda F^{-1}(z_0) = -\frac{z_1 - \alpha}{z_1 - z_0} \frac{1}{\beta - \alpha} + \frac{\lambda}{z_1 - z_0} \int_{z_0}^{z_1} F^{-1}(u)\, du, \tag{14.24b}
$$

provided these exist. Otherwise $z_1 = \beta$, and z_0 is the solution to (14.24a).

The subsequent corollary provides the best and worst case quantile function of VaR and TVaR, as limiting cases of the RVaR. The results are also summarized in Table 14.1.

Corollary 14.16 (Best and worst case quantiles of VaR and TVaR) *Under the assumptions of Theorems 14.7 and 14.10 case 1, respectively, it holds that for* VaR_α *and* VaR_α^+:

1. *A worst case quantile function is given by Proposition 14.14, with $w_0 = \alpha$ and w_1 the solution to (14.22b).*
2. *A best case quantile function is given by Proposition 14.15 with $z_1 = \beta$ and z_0 the solution to (14.24a) with $\beta = \alpha$.*

Furthermore, it holds that for TVaR_α:

3. *A worst case quantile function is given by Proposition 14.14 with $\beta = 1$ and $w_0 = w_1 = 1$.*
4. *A best case quantile function is given by Proposition 14.15 with $\beta = 1$.*

Risk measure	Best case quantile	Worst case quantile
$\text{RVaR}_{\alpha,\beta}$	Proposition 14.15	Proposition 14.14
VaR_α & VaR_α^+	$z_1 = \beta$, and z_0 solves (14.24a) with $\beta = \alpha$	$w_0 = \alpha$, and w_1 solves (14.22b)
TVaR_α	$\beta = 1$	$\beta = 1$, and $w_0 = w_1 = 1$

Table 14.1 Best and worst case quantile functions of VaR_α, VaR_α^+, $\text{RVaR}_{\alpha,\beta}$, and TVaR_α for case 1 of Theorems 14.7 and 14.10, respectively, that is, for ε sufficiently small.

Example 14.17 Figures 14.3 and 14.4 illustrate the L^2 projections for the worst and best case quantile function of $\text{RVaR}_{\alpha,\beta}$ with $\alpha = 0.6$ and different values of $\beta \in \{0.61, 0.85, 0.99\}$. Specifically, Figure 14.3 displays $\gamma(u) + \lambda F^{-1}(u)$, $u \in (0,1)$, its isotonic projection $k_\lambda^\uparrow$ onto the set of non-decreasing functions (derived in Proposition 14.14), and the numerically obtained isotonic projection. We observe that the isotonic projection $k_\lambda^\uparrow$ (black dashed line) matches the numerically obtained isotonic projection (solid gray line) well. Note that we chose ε sufficiently small such that the worst case quantile functions of $\mathrm{RVaR}_{\alpha,\beta}$ are not two point distributions, and case 1 of Theorem 14.7 applies. Figure 14.4 displays the corresponding graphs for the case of best case quantile functions.

The left plot of Figure 14.5 displays the lower and upper bounds on VaR_α, $\mathrm{RVaR}_{\alpha,\beta}$, and TVaR_α as a function of the Wasserstein distance ε. When ε is sufficiently large (case 2 in Theorems 14.7 and 14.10), the Wasserstein distance no longer affects the bounds; these bounds coincide with the well-known Cantelli bounds. The normalized lengths of the bounds on VaR_α, $\mathrm{RVaR}_{\alpha,\beta}$, and TVaR_α as a function of ε are displayed in the right plot of Figure 14.5. The normalized lengths of the bounds are the differences between (14.1b) and (14.1a) divided by the risk measure evaluated at the reference distribution, and these were introduced to assess model risk by Barrieu and Scandolo (2015). The relative length of the bounds on $\mathrm{RVaR}_{\alpha,\beta}$, for example, is

$$
\frac{\sup\limits_{G \in \mathcal{F}_\varepsilon(\mu,\sigma)} \mathrm{RVaR}_{\alpha,\beta}(G) - \inf\limits_{G \in \mathcal{F}_\varepsilon(\mu,\sigma)} \mathrm{RVaR}_{\alpha,\beta}(G)}{\mathrm{RVaR}_{\alpha,\beta}(F)}. \tag{14.25}
$$

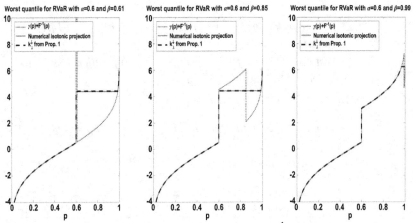

Figure 14.3 For the risk measure $\mathrm{RVaR}_{\alpha,\beta}$, we plot $\gamma(u) + \lambda F^{-1}(u)$, $u \in (0,1)$, its isotonic projection $k_\lambda^\uparrow$ onto the set of non-decreasing functions (derived in Proposition 14.14), and the numerically obtained isotonic projection. The reference distribution F is chosen to be standard normal with $\mu = \mu_F = 0$ and $\sigma = \sigma_F = 1$, and we set $\varepsilon = 0.2$.

We observe on Figure 14.4 that the normalized lengths of the bounds on all three risk measures increase with respect to ε. Moreover, the normalized lengths of the VaR bounds are significantly larger than in the case of TVaR, a fact that is well known for the case when $\varepsilon = +\infty$ (Embrechts et al., 2015).

14.4   Application to an Insurance Portfolio

This section illustrates the best and worst case bounds of VaR on a simulated insurance portfolio. We provide a comparison with VaR bounds in the literature, such as, for

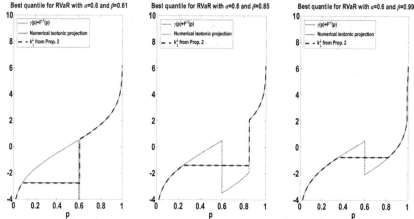

Figure 14.4 For the risk measure $\text{RVaR}_{\alpha,\beta}$, we plot $\gamma(u) - \lambda F^{-1}(u)$, $u \in (0,1)$, its isotonic projection k_{λ}^{i} onto the set of non-increasing functions (derived in Proposition 14.15), and the numerically obtained isotonic projection. The reference distribution F is chosen to be standard normal with $\mu = \mu_F = 0$ and $\sigma = \sigma_F = 1$, and we set $\varepsilon = 0.2$.

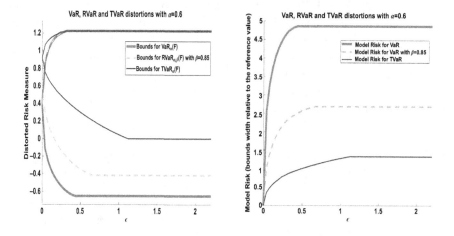

Figure 14.5 Left: Best and worst case bounds of VaR_{α}, $\text{RVaR}_{\alpha,\beta}$, and TVaR_{α} as a function of the tolerance level ε. VaR and TVaR are at level $\alpha = 0.6$ and RVaR is at levels $\alpha = 0.6$ and $\beta = 0.85$. Right: Length of the normalized bounds of VaR, RVaR, and TVaR (as a measure for model risk), given in (14.25).

example, VaR bounds subject to constraints on higher moments, and give rationales for choosing the Wasserstein distance by means of model uncertainty. We consider the Pareto–Clayton model, which offers a flexible way of modeling portfolios with dependent risks; see Oakes (1989), Albrecher et al. (2011), and Yeh (2007) for applications in insurance and Dacorogna et al. (2016) and Bernard et al. (2018a) for applications in finance. In the Pareto–Clayton model, the portfolio components $X_1, \ldots, X_d$ are, given $\Theta = \vartheta > 0$, independent and exponentially distributed with parameter ϑ that is drawn from a Gamma distribution, i.e., $\Theta \sim \mathrm{Gamma}(a, b)$. The aggregate portfolio risk $S = \sum_{i=1}^{n} X_i$ thus follows a $\mathrm{Gamma}(d, \vartheta)$ distribution, conditionally on $\Theta = \vartheta$. The first nine moments and the quantile function of the aggregate portfolio risk S are explicitly given by

$$ES^k = \frac{b^k \, B(a - k, d + k)}{B(d, a)}, \quad \text{for} \quad k = 1, \ldots, 9 \quad \text{and} \tag{14.26}$$

$$F_S^{-1}(u) = b \frac{G_B^{-1}(u)}{1 - G_B^{-1}(u)}, \tag{14.27}$$

where $B(\cdot, \cdot)$ denotes the beta function and G_B^{-1} the quantile function of the beta distribution with parameters d and a.

We consider the situation where modelers or experts may not fully agree on the reference Pareto–Clayton model, as available data or estimation may induce uncertainties, or the distribution of Θ might not seem appropriate. As alternative models for the distribution of the aggregate portfolio risk S, we consider two-parameter distributions with support in $(0, +\infty)$ that are commonly utilized for modeling insurance portfolios. Specifically, we consider a lognormal model (LN), a Gamma model (Γ), a Weibull model (W), an inverse Gaussian model (IG), an inverse Gamma model (IΓ), an inverse Weibull model (IW), and a log-logistic model (LL). The different models are described in Table 14.2, and additional information is provided in Appendix 14.7. For the numerical simulations of the Pareto–Clayton model, we use $a = 10$, $b = 1$, and $d = 100$, as in Bernard et al. (2018a). The parameters of the alternative models, reported in Table 14.2, are then obtained by matching the first two moments to those of the reference model.

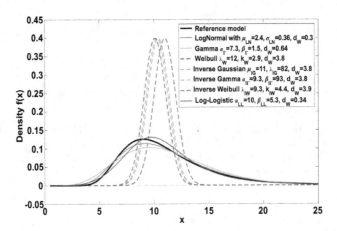

Figure 14.6 Densities of the reference model (solid black), the Pareto–Clayton model, and the alternative models. The parameters of the alternative models are calibrated by matching of the first two moments to those of the reference model and are reported in the legend.

We illustrate the reference Pareto–Clayton model (solid black) and the alternative models in Figure 14.6 via their densities. It is apparent that the alternative models

Model	Abb.	Parameters		d_W
lognormal model	LN	$\mu_{LN} = 2.4$	$\sigma_{LN} = 0.36$	0.298
Gamma	Γ	$\alpha_\Gamma = 7.3$	$\beta_\Gamma = 1.5$	0.637
Weibull	W	$\lambda_W = 12$	$k_W = 2.9$	3.787
inverse Gaussian	IG	$\mu_{IG} = 11$	$\lambda_{IG} = 82$	3.802
inverse Gamma	IΓ	$\alpha_{I\Gamma} = 9.3$	$\beta_{I\Gamma} = 93$	3.792
inverse Weibull	IW	$\lambda_{IW} = 9.3$	$k_{IW} = 4.4$	3.868
log-logistic	LL	$\alpha_{LL} = 10$	$\beta_{LL} = 5.3$	0.345

Table 14.2 Alternative models for the aggregate portfolio risk. The parameters are calculated such that the first two moments are equal to those of the reference Pareto–Clayton model, i.e., $\mathbb{E}S = 11.11$ and $\mathbb{E}S^2 = 140.28$. The last column reports the Wasserstein distance between the Pareto–Clayton model and the alternative model.

considered can be grouped into light-tailed models (W, IG, IΓ, IW) and heavy-tailed models (LN, Γ, LL).

Next, we discuss the choice of Wasserstein distance by means of model uncertainty. Specifically, we let ε be the maximum of all Wasserstein distances between the reference and a selection of alternative models. The constructed Wasserstein ball then includes all distribution functions whose Wasserstein distance is less than or equal to the maximum of all Wasserstein distances between the reference and considered alternative models; in particular, it contains all alternative models. We report the Wasserstein distances between the Pareto–Clayton model and all alternative models in Table 14.2 and observe that the Wasserstein distances between the reference distribution and the alternatives from the heavy-tailed models are significantly smaller compared to alternatives from the light-tailed models. This illustrates that model uncertainty is considered with respect to the reference distribution, which in our case is heavy-tailed. Thus, in a model uncertainty context, changing from the heavy-tailed reference model, i.e., Pareto–Clayton, to a light-tailed model, e.g., Weibull, results in a higher level of uncertainty compared to other models from the heavy-tailed family, e.g., lognormal.

Next, we calculate the best and worst case bounds of VaR_α, for $\alpha = 0.9, 0.95, 0.99$, for different sets and report them in Table 14.3. The values of VaR_α, for $\alpha = 0.9, 0.95, 0.99$, of the Pareto–Clayton model are respectively $\text{VaR}_{0.90} = 16.29$, $\text{VaR}_{0.95} = 18.75$, and $\text{VaR}_{0.99} = 24.79$. The sets we consider are $\mathcal{F}_\varepsilon(\mu, \sigma)$ with $\mu = \mu_F = 11.11$, $\sigma_F^2 = \sigma^2 = 16.82$, $\varepsilon = 0.637$, and $\varepsilon = 3.868$. The distance $\varepsilon = 0.637$ corresponds to model uncertainty with respect to heavy-tailed alternative models (LN, Γ, LL), whereas $\varepsilon = 3.868$ allows for alternative models that are light-tailed (W, IG, IΓ, IW). To compare these bounds with results in the literature, we calculate the lower and upper bounds on VaR when only the first k, $k = 1, \ldots, 5$ moments are known (Bernard et al., 2018a). Notice that the fourth row in Table 14.3 ($k = 2$) corresponds to $\mathcal{F}(\mu, \sigma)$, i.e., with Wasserstein distance $\varepsilon = +\infty$. Furthermore, we calculate the bounds on VaR for a set containing distribution functions with fixed mean, bounded standard deviation, and where moreover the distribution functions are

Set	$\alpha = 0.9$	$\alpha = 0.95$	$\alpha = 0.99$
$\mathcal{F}_\varepsilon(\mu,\sigma)$ with $\varepsilon = 0.637$	(12.8; 18.8)	(14.6; 22.8)	(19.0; 35.1)
$\mathcal{F}_\varepsilon(\mu,\sigma)$ with $\varepsilon = 3.868$	(10.7; 21.6)	(12.1; 26.1)	(15.0; 41.5)
# of moments $k = 1$	(0.0; 111.1)	(0.0; 222.2)	(1.1; 1000)
# of moments $k = 2$	(9.7; 23.4)	(10.2; 29.0)	(10.7; 51.9)
# of moments $k = 3$	(10.1; 22.5)	(10.8; 26.2)	(11.7; 38.7)
# of moments $k = 4$	(10.1; 22.5)	(11.5; 25.6)	(15.1; 32.9)
# of moments $k = 5$	(10.1; 22.4)	(11.5; 25.5)	(15.1; 32.9)
$E = \mu$, Std $\leq s$, unimodal	(9.9; 18.7)	(10.3; 22.6)	(10.7; 38.1)

Table 14.3 Best and worst case bounds of VaR$_\alpha$, for $\alpha = 0.9$, 0.95, 0.99, for different sets. The VaR bounds are reported as (;) in which the first value is the lower and the second is the upper bound. The quantiles of the Pareto–Clayton model are VaR$_{0.90}$ = 16.29, VaR$_{0.95}$ = 18.75, and VaR$_{0.99}$ = 24.79.

known to be unimodal (Bernard et al., 2020). We observe in Table 14.2, that adding the Wasserstein constraint significantly reduces the bounds and that information on higher moments only affect the bounds in a minor way. Moreover, when only heavy-tailed models are considered as alternatives ($\varepsilon = 0.637$), the lengths of the bounds are $\frac{18.8-12.8}{21.6-10.7} = 55\%$ of the bounds when both light- and heavy-tailed are valid alternative models ($\varepsilon = 0.3868$).

Worst and best case risk measures over sets that are constructed via a Wasserstein distance are conceptually distinct from bounds subject to moment constraints, as in the former case the distribution of the considered alternative models are close to a reference distribution. Therefore, the alternative models, over which we seek the best and worst case bounds, share features with the reference distribution, such as similar tail behavior.

Next, in Figure 14.7, we plot the VaR$_{0.95}$ bound with respect to the set $\mathcal{F}_\varepsilon(\mu,\sigma)$ as a function of the Wasserstein distance ε. Note that for $\varepsilon \to +\infty$, the lower and upper bounds on VaR$_{0.95}$ converge to 10.2 and 29.0, respectively, which correspond to the bounds on VaR$_{0.95}$ with constraints on the first two moments only; see also the fourth row of Table 14.3. It is apparent in Figure 14.7 that the distance and thus the bounds corresponding to heavy-tailed alternative models ($\varepsilon = 0.637$, dash-dotted vertical line) is significantly smaller than the one corresponding to both heavy-tailed and light-tailed alternative models ($\varepsilon = 3.868$, dotted vertical line). This stems from the fact that the reference model is itself heavy-tailed and that model uncertainty using the Wasserstein distance pertains to deviation from the reference model.

We also derive the worst case distribution function for RVaR and show that for small Wasserstein distances, the worst case distribution functions of VaR and TVaR are no longer two-point distributions, thus making the bounds attractive for risk management applications.

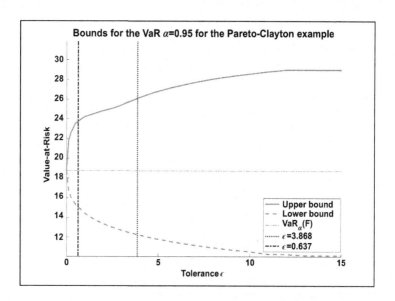

Figure 14.7 VaR_α bounds with $\alpha = 0.95$ for the Pareto–Clayton model. The vertical lines at $\varepsilon = 0.637$ and $\varepsilon = 3.868$ indicate the Wasserstein distance with respect to heavy-tailed and light-tailed model uncertainty; see also Table 14.2.

14.5   Proofs

We first collect a series of lemmas that we need in the proofs. We denote $\mathcal{F}_\varepsilon^{-1}(\mu, \sigma) = \{G^{-1} \mid G \in \mathcal{F}_\varepsilon(\mu, \sigma)\}$ and $\mathcal{F}^{-1}(\mu, \sigma) = \{G^{-1} \mid G \in \mathcal{F}(\mu, \sigma)\}$.

For the proof of Theorem 14.3, we need the following lemma.

Lemma 14.18   *Let $\lambda \geq 0$ and γ be non-decreasing; then*

$$\sup_{h \in \mathcal{F}^{-1}(\mu, \sigma)} \langle \gamma + \lambda F^{-1}, h \rangle$$

is uniquely attained by

$$h_\lambda(u) = \mu + \sigma \left(\frac{\gamma(u) + \lambda F^{-1}(u) - a_\lambda}{b_\lambda} \right), \quad 0 < u < 1, \qquad (14.28)$$

where $a_\lambda = E(\gamma(U) + \lambda F^{-1}(U))$, and $b_\lambda = \text{Std}(\gamma(U) + \lambda F^{-1}(U))$. Moreover, $\text{Cor}(F^{-1}(U), h_\lambda(U))$ is continuous in λ, for $\lambda \geq 0$.

Proof   The function h_λ, defined in (14.28), belongs to $\mathcal{F}^{-1}(\mu, \sigma)$. Let $h \in \mathcal{F}^{-1}(\mu, \sigma)$; then by the Cauchy–Schwarz inequality it holds that $\langle \gamma + \lambda F^{-1}, h \rangle \leq \langle \gamma + \lambda F^{-1}, h_\lambda \rangle$, with equality if and only if h is a linear combination of $\gamma + \lambda F^{-1}$. Continuity of $\text{Cor}(F^{-1}(U), h_\lambda(U))$ is immediate.   □

The following lemma is needed to prove Theorem 14.7.

Lemma 14.19 *Let $\lambda > 0$ and g be a distortion function with weight function γ. Then*

$$\sup_{h \in \mathcal{F}^{-1}(\mu, \sigma)} \langle \gamma + \lambda F^{-1}, h \rangle$$

is uniquely attained by

$$h_\lambda^\uparrow(u) = \mu + \sigma \left(\frac{k_\lambda^\uparrow(u) - a_\lambda}{b_\lambda} \right), \quad 0 < u < 1, \tag{14.29}$$

where $a_\lambda = E(k_\lambda^\uparrow(U))$, $b_\lambda = \text{Std}(k_\lambda^\uparrow(U))$, and $k_\lambda^\uparrow$ denotes the isotonic projection of $\gamma + \lambda F^{-1}$ onto the set of non-decreasing functions. Moreover, $\text{Cor}(F^{-1}(U), h_\lambda^\uparrow(U))$ is continuous in λ, for $\lambda > 0$.

Proof For $\lambda > 0$, note that the projection $k_\lambda^\uparrow$ is non-constant, which implies that $h_\lambda^\uparrow$, as defined in (14.29), fulfills $h_\lambda^\uparrow \in \mathcal{F}^{-1}(\mu, \sigma)$. Let $h \in \mathcal{F}^{-1}(\mu, \sigma)$; then $k(u) = b_\lambda(h(u) - \mu)/\sigma + a_\lambda$ is a non-decreasing function. Moreover, it holds that $\|k_\lambda^\uparrow\|^2 = \|k\|^2$. Thus, the following inequalities are equivalent:

$$\|\gamma + \lambda F^{-1} - k_\lambda^\uparrow\|^2 \le \|\gamma + \lambda F^{-1} - k\|^2$$
$$\Longleftrightarrow \quad \langle \gamma + \lambda F^{-1}, k_\lambda^\uparrow \rangle \ge \langle \gamma + \lambda F^{-1}, k \rangle$$
$$\Longleftrightarrow \quad \langle \gamma + \lambda F^{-1}, h_\lambda^\uparrow \rangle \ge \langle \gamma + \lambda F^{-1}, h \rangle.$$

Note that unless $k_\lambda^\uparrow = k$, the inequalities are strict, which implies uniqueness of the solution. Finally, continuity of the correlations

$$\text{Cor}(F^{-1}(U), h_\lambda^\uparrow(U)) = \text{Cor}(F^{-1}(U), k_\lambda^\uparrow(U))$$

follows from continuity of the projection k_λ, property 2 in Proposition 14.23.   □

Lemma 14.20   *The solution λ to $d_W(F^{-1}, h_\lambda)^2 = \mathbb{E}((F^{-1}(U) - h_\lambda(U))^2) = \varepsilon$, where $h_\lambda(\cdot)$ is defined in Lemma 14.18, is explicit and is given by*

$$\lambda = \frac{\sqrt{\Delta} - C_{\gamma,F}}{\sigma_F^2}, \quad \text{where} \quad \Delta = \frac{K^2(C_{\gamma,F}^2 - V\sigma_F^2)}{K^2 - \sigma^2 \sigma_F^2}, \tag{14.30}$$

with $V = \text{var}(\gamma(U))$, $C_{\gamma,F} = \text{Cov}\left(F^{-1}(U), \gamma(u)\right)$, and

$$K = \tfrac{1}{2}\left(\mu_F^2 + \sigma_F^2 + \mu^2 + \sigma^2 - 2\mu\mu_F - \varepsilon \right) \ge 0. \tag{14.31}$$

Proof Let $h_\lambda(u) \in \mathcal{M}^{-1}(\mu, \sigma)$ defined in (14.28). Solving for λ such that $d_W(F^{-1}, h_\lambda)^2 = \varepsilon$ amounts to solving

$$\varepsilon = \mu_F^2 + \sigma_F^2 + \mu^2 + \sigma^2 - 2 \, \text{Cov}\left(F^{-1}(U), h_\lambda(U)\right) - 2\mu\mu_F. \tag{14.32}$$

Using the expression of K in (14.31), equation (14.32) is simply

$$\text{Cov}\left(F^{-1}(U), h_\lambda(U)\right) = K, \tag{14.33}$$

which ensures that $K \geq 0$. Using the expression of h_λ,

$$\text{Cov}\left(F^{-1}(U), h_\lambda(U)\right) = \frac{\sigma}{b_\lambda} \text{Cov}\left(F^{-1}(U), \gamma(U) + \lambda F^{-1}(U)\right)$$

$$= \frac{\sigma}{b_\lambda}\left(\text{Cov}\left(F^{-1}(U), \gamma(U)\right) + \lambda \sigma_F^2\right).$$

Rearranging, we obtain

$$Kb_\lambda = \sigma \, \text{Cov}\left(F^{-1}(U), \gamma(u)\right) + \lambda \sigma \sigma_F^2. \tag{14.34}$$

Next, denote by $V = \text{var}(\gamma(U))$ and $C_{\gamma,F} = \text{Cov}\left(F^{-1}(U), \gamma(u)\right)$; then the expression of $b_\lambda = \text{std}(\gamma(U) + \lambda F^{-1}(U))$ becomes $b_\lambda^2 = V + 2\lambda C_{\gamma,F} + \lambda^2 \sigma_F^2$, and the square of equation (14.34) becomes

$$K^2(V + 2\lambda C_{\gamma,F} + \lambda^2 \sigma_F^2) = (\sigma C_{\gamma,F} + \lambda \sigma \sigma_F^2)^2.$$

This is a second-degree equation in λ, which can also be written as

$$\sigma_F^2 \lambda^2 + 2C_{\gamma,F}\lambda + \frac{K^2 V - \sigma^2 C_{\gamma,F}^2}{K^2 - \sigma^2 \sigma_F^2} \tag{14.35}$$

$$= \sigma_F^2 \lambda^2 + 2C_{\gamma,F}\lambda - \frac{1}{\sigma_F^2}\left(\Delta - C_{\gamma,F}^2\right) = 0,$$

or $$\sigma_F^2 \frac{\sigma^2 C_{\gamma,F}^2 - K^2 V}{K^2 - \sigma^2 \sigma_F^2} + C_{\gamma,F}^2 = \Delta. \tag{14.36}$$

Observe that if $\gamma(\cdot) \neq F^{-1}(\cdot)$, then $C_{\gamma,F} > 0$ as the covariance between two comonotonic variables. $K^2 - \sigma^2 \sigma_F^2 < 0$ by the definition of K as a covariance between two variables with respective variances σ^2 and σ_F^2 and that are not perfectly correlated, and $C_{\gamma,F}^2 - V\sigma_F^2 \leq 0$, by the definition of $C_{\gamma,F}$ as a covariance between two variables with respective variances V and σ_F^2. Thus, from the expression (14.36) we have that $\Delta > 0$. As a consequence, (14.35) then has two roots, of which only one, $\lambda = \frac{\sqrt{\Delta} - C_{\gamma,F}}{\sigma_F^2}$, is positive.   □

The following lemma is needed for the proof of Theorem 14.10.

Lemma 14.21   *Let $\lambda > 0$ and g be a distortion function. Then*

$$\inf_{h \in \mathcal{F}^{-1}(\mu,\sigma)} \langle \gamma - \lambda F^{-1}, h \rangle$$

is uniquely attained by

$$h_\lambda^\downarrow(u) = \mu + \sigma\left(\frac{a_\lambda - k_\lambda^\downarrow(u)}{b_\lambda}\right), \quad 0 < u < 1, \tag{14.37}$$

where $a_\lambda = E\left(k_\lambda^\downarrow(U)\right)$, $b_\lambda = \text{Std}\left(k_\lambda^\downarrow(U)\right)$, *and* $k_\lambda^\downarrow$ *is the projection of* $\gamma - \lambda F^{-1}$ *onto the set of non-increasing functions. Moreover,* $\text{Cor}(F^{-1}(U), h_\lambda^\downarrow(U))$ *is continuous in* λ, *for* $\lambda > 0$.

Proof For $\lambda > 0$, note that $k_\lambda^\downarrow$ is not constant, and $h_\lambda^\downarrow$, as defined in (14.37), fulfills $h_\lambda^\downarrow \in \mathcal{F}^{-1}(\mu, \sigma)$. Let $h \in \mathcal{F}^{-1}(\mu, \sigma)$; then $k(u) = b_\lambda (\mu - h(u)) / \sigma + a_\lambda$ is a non-increasing function. Moreover, it holds that $||k_\lambda^\downarrow||^2 = ||k||^2$. Thus, the following inequalities are equivalent:

$$||\gamma - \lambda F^{-1} - k_\lambda^\downarrow||^2 \leq ||\gamma - \lambda F^{-1} - k||^2$$
$$\Longleftrightarrow \quad \langle \gamma - \lambda F^{-1}, k_\lambda^\downarrow \rangle \geq \langle \gamma - \lambda F^{-1}, k \rangle$$
$$\Longleftrightarrow \quad \langle \gamma - \lambda F^{-1}, h_\lambda^\downarrow \rangle \leq \langle \gamma - \lambda F^{-1}, h \rangle.$$

Note that unless $k_\lambda^\downarrow = k$, the inequalities are strict, which implies that the solution is unique. Finally, the continuity of $\mathrm{Cor}(F^{-1}(U), h_\lambda^\downarrow(U)) = \mathrm{Cor}(F^{-1}(U), k_\lambda^\downarrow(U))$ with respect to λ follows from continuity of the isotonic projection $k_\lambda^\downarrow$, that is, Proposition 14.23, property 2. $\qquad\square$

Proof of Theorem 14.3

For concave distortion risk measures, (14.1b) is equivalent to

$$\sup_{h \in \mathcal{F}_\varepsilon^{-1}(\mu, \sigma)} \langle \gamma, h \rangle, \tag{14.38}$$

where γ is non-decreasing. For $\lambda \geq 0$, let h_λ denote the quantile function defined in (14.28). By construction, $h_\lambda \in \mathcal{F}^{-1}(\mu, \sigma)$. The Wasserstein distance between F^{-1} and h_λ is given by

$$d_W(F^{-1}, h_\lambda)^2 = E\left((F^{-1}(U) - h_\lambda(U))^2\right)$$
$$= \mu_F^2 + \sigma_F^2 + \mu^2 + \sigma^2 - 2\,\mathrm{Cov}\left(F^{-1}(U), h_\lambda(U)\right) - 2\mu\mu_F$$
$$= (\mu_F - \mu)^2 + (\sigma_F - \sigma)^2 + 2\sigma\sigma_F(1 - c_\lambda),$$

where $c_\lambda = \mathrm{Cor}(F^{-1}(U), h_\lambda(U))$. Note that by Lemma 14.18, $c_\lambda : [0, +\infty) \longrightarrow [-1, 1]$ is continuous with $c_0 = \mathrm{Cor}(F^{-1}(U), h_0(U)) = \mathrm{Cor}(F^{-1}(U), \gamma(U))$ and $\lim_{\lambda \nearrow +\infty} c_\lambda = 1$.

Case 1. Let $(\mu_F - \mu)^2 + (\sigma_F - \sigma)^2 < \varepsilon < (\mu_F - \mu)^2 + (\sigma_F - \sigma)^2 + 2\sigma\sigma_F(1 - c_0)$. We split this part into three steps. First, we show that the optimal quantile function has to be of the form h_λ for some $\lambda \geq 0$. Second, we show that the solution lies at the boundary of the Wasserstein ball. Third, we show uniqueness.

For the first step, let $h_1 \in \mathcal{M}_\varepsilon(\mu, \sigma)$ be a solution to problem (14.38) with $d_W(F^{-1}, h_1) = \sqrt{\varepsilon_1}$ and $(\mu_F - \mu)^2 + (\sigma_F - \sigma) < \varepsilon_1 \leq \varepsilon$. Then, by continuity of c_λ, there exists $\lambda_1 > 0$ and a corresponding $h_{\lambda_1} \in \mathcal{M}_\varepsilon(\mu, \sigma)$, such that $d_W(F^{-1}, h_{\lambda_1}) = \sqrt{\varepsilon_1}$. Moreover, it holds that

$$d_W(F^{-1}, h_1) = d_W(F^{-1}, h_{\lambda_1}) \quad \Longleftrightarrow \quad \langle F^{-1}, h_1 \rangle = \langle F^{-1}, h_{\lambda_1} \rangle.$$

Applying Lemma 14.18, we obtain that $\langle \gamma, h_1 \rangle \leq \langle \gamma, h_{\lambda_1} \rangle$; thus h_{λ_1} improves upon h_1, and we conclude the first step.

For the second step, let $\lambda \geq 0$ be such that h_λ satisfies $d_W(F^{-1}, h_\lambda) = \sqrt{\varepsilon}$. By construction, it holds that

$$d_W(F^{-1}, h_{\lambda_1}) \leq d_W(F^{-1}, h_\lambda) \iff \langle F^{-1}, h_{\lambda_1} \rangle \geq \langle F^{-1}, h_\lambda \rangle. \tag{14.39}$$

By Lemma 14.18 we have that $\langle \gamma + \lambda F^{-1}, h_\lambda \rangle \geq \langle \gamma + \lambda F^{-1}, h_{\lambda_1} \rangle$, which, together with (14.39), implies that

$$\langle \gamma, h_\lambda - h_{\lambda_1} \rangle \geq \lambda \langle F^{-1}, h_{\lambda_1} - h_\lambda \rangle \geq 0.$$

Therefore, we obtain that $\langle \gamma, h_\lambda \rangle \geq \langle \gamma, h_{\lambda_1} \rangle$ and hence that the solution lies at the boundary of the Wasserstein ball. Uniqueness follows from uniqueness of h_λ as the solution of Lemma 14.18. The worst case bound $H_g(h_\lambda)$ is then

$$
\begin{aligned}
H_g(h_\lambda) &= \int_0^1 h_\lambda(u) \gamma(u) \, du \\
&= \mu + \frac{\sigma}{b_\lambda} \left(\int_0^1 \gamma(u) \left(\gamma(u) + \lambda F^{-1}(u) \right) du - a_\lambda \right) \\
&= \mu + \sigma \operatorname{Std}(\gamma(U)) \operatorname{Cor}\left(\gamma(U), \gamma(U) + \lambda F^{-1}(U) \right).
\end{aligned}
$$

The explicit expression for λ then follows from Lemma 14.20 and is given in (14.30).

Case 2. Let $(\mu_F - \mu)^2 + (\sigma_F - \sigma)^2 + 2\sigma\sigma_F(1 - c_0) \leq \varepsilon$; then $h_0 \in \mathcal{F}_\varepsilon^{-1}(\mu, \sigma)$ and thus is admissible. Moreover, it is straightforward that for all $\lambda > 0$, $\langle \gamma, h_0 \rangle \geq \langle \gamma, h_\lambda \rangle$. Hence, h_0 solves problem (14.38).

Proof of Theorem 14.7

For $\lambda > 0$, define $h_\lambda^\uparrow$ as in (14.29), and note that $h_\lambda^\uparrow \in \mathcal{F}^{-1}(\mu, \sigma)$. The Wasserstein distance between F^{-1} and $h_\lambda^\uparrow$ is given by (see also Theorem 14.3)

$$d_W(F^{-1}, h_\lambda^\uparrow)^2 = (\mu_F - \mu)^2 + (\sigma_F - \sigma)^2 + 2\sigma\sigma_F(1 - c_\lambda),$$

where $c_\lambda = \operatorname{Cor}(F^{-1}(U), h_\lambda^\uparrow(U))$. By Lemma 14.19, $c_\lambda \colon (0, +\infty) \to [-1, 1]$ is continuous in $\lambda > 0$, with $\lim_{\lambda \nearrow +\infty} c_\lambda = 1$.

Case 1. Let $(\mu_F - \mu)^2 + (\sigma_F - \sigma)^2 < \varepsilon < (\mu_F - \mu)^2 + (\sigma_F - \sigma)^2 + 2\sigma\sigma_F(1 - c_0)$. Similar to the proof of Theorem 14.3, we split the proof into three steps: first, the worst case quantile function is of the form h_λ; second, the solution lies at the boundary of the Wasserstein ball; and third uniqueness. The first two steps follow along the lines of the proof of Theorem 14.3 case 1, replacing h_λ with $h_\lambda^\uparrow$ and using the properties of the isotonic projection $h_\lambda^\uparrow$ in the second step. Uniqueness follows from the uniqueness of $h_\lambda^\uparrow$ as the solution of (14.19) in Lemma 14.19.

Case 2. If $(\mu_F - \mu)^2 + (\sigma_F - \sigma)^2 + 2\sigma\sigma_F(1 - c_0) \leq \varepsilon$ and $\gamma^\uparrow$ is not constant, then Lemma 14.19 applies for all $\lambda \geq 0$. Let $h_0^\uparrow$ be as defined in (14.29) with $\lambda = 0$. Then $h_0^\uparrow \in \mathcal{F}_\varepsilon^{-1}(\mu, \sigma)$. Moreover, for all $h \in \mathcal{F}^{-1}(\mu, \sigma)$ it holds, applying Lemma 14.19 with $\lambda = 0$, that

$$\langle \gamma, h \rangle \leq \langle \gamma, h_0^\uparrow \rangle.$$

Hence, $h_0^\uparrow$ cannot be improved and thus yields the unique optimal solution. If $\gamma^\uparrow$ is constant, then we find that for all $h \in \mathcal{F}^{-1}(\mu, \sigma)$,

$$\langle \gamma, h \rangle \leq \langle \gamma^\uparrow, h \rangle = \mu,$$

where the first inequality follows from the properties of the isotonic projection (Proposition 14.24, property 3) and the equality follows from the fact that $\gamma^\uparrow = 1$ as $\langle \gamma, 1 \rangle = \langle \gamma^\uparrow, 1 \rangle$ (applying Proposition 14.24, property 3 with $k \equiv 1$ and $k \equiv -1$). Finally, we show that μ is the supremum. Consider the case where $\mu \geq 0$, and define $t_\alpha = \alpha \mu \gamma^\uparrow + (1 - \alpha) z_\alpha$, for $0 < \alpha < 1$, and where z_α is a non-negative non-decreasing function such that $E(z_\alpha(U)) = \mu$ and $\mathrm{Std}(z_\alpha(U)) = \sigma/(1 - \alpha)$. Hence, $t_\alpha \in \mathcal{M}^{-1}(\mu, \sigma)$, for $0 < \alpha < 1$. Moreover, for every $\varepsilon > 0$ small enough, there exists $0 < \alpha < 1$ such that

$$\langle \gamma, t_\alpha \rangle = \langle \gamma, \alpha \mu \gamma^\uparrow \rangle + \langle \gamma, (1 - \alpha) z_\alpha \rangle \geq \alpha \mu \geq \mu - \varepsilon.$$

For $\mu < 0$ the proof is similar and thus omitted.

Proof of Corollary 14.8
Corollary 14.8 follows from Theorem 14.7 (case 2). We only prove that equation (14.13) (when $\gamma^\uparrow$ is not constant) is equivalent to the worst case bound in Theorem 14.7 (case 2). From the characterization of the isotonic projection, that is applying Property 3 of Proposition 14.24 with $k = \pm 1$, we obtain that $E(\gamma^\uparrow(U)) = E(\gamma(U)) = 1$. Further, by Property 3 of Proposition 14.24, it holds that $E\left(\gamma^\uparrow(U)^2\right) = E\left(\gamma^\uparrow(U)\gamma(U)\right)$. Hence,

$$\mathrm{Cor}\left(\gamma(U), \gamma^\uparrow(U)\right) = \frac{\mathrm{Cov}\left(\gamma^\uparrow(U), \gamma(U)\right)}{\mathrm{Std}\left(\gamma^\uparrow(U)\right)\mathrm{Std}\left(\gamma(U)\right)}$$

$$= \frac{\mathrm{Var}\left(\gamma^\uparrow(U)\right)}{\mathrm{Std}\left(\gamma^\uparrow(U)\right)\mathrm{Std}\left(\gamma(U)\right)} = \frac{\mathrm{Std}\left(\gamma^\uparrow(U)\right)}{\mathrm{Std}\left(\gamma(U)\right)}.$$

Proof of Theorem 14.10
For $\lambda > 0$, define $h_\lambda^\downarrow$ as in (14.37), and observe that $h_\lambda^\downarrow \in \mathcal{F}^{-1}(\mu, \sigma)$. Then,

$$d_W(F^{-1}, h_\lambda^\downarrow)^2 = (\mu_F - \mu)^2 + (\sigma_F - \sigma)^2 + 2\sigma\sigma_F(1 - c_\lambda),$$

where $c_\lambda = \mathrm{Cor}(F^{-1}(U), h_\lambda^\downarrow(U))$. By Lemma 14.21, $c_\lambda : (0, +\infty) \to [-1, 1]$ is continuous in λ, $\lambda > 0$, with $\lim_{\lambda \nearrow +\infty} c_\lambda = 1$. The remainder of the proof follows analogously to the arguments of the proof of Theorem 14.7.

Proof of Proposition 14.14
By Theorem 14.7 case 1, we have to calculate the projection of $\frac{1}{\beta - \alpha}\mathbb{1}_{(\alpha, \beta]} + \lambda F^{-1}$ onto the space of non-decreasing functions. According to Lemma 5.1 by Brighi and Chipot (1994), the projection has the form

$$k_\lambda^\uparrow(u) = \left(\tfrac{1}{\beta - \alpha}\mathbb{1}_{(\alpha, \beta]}(u) + \lambda F^{-1}(u)\right)\mathbb{1}_{(\bigcup_{j \in \mathcal{J}} I_j)^c}(u) + \sum_{j \in \mathcal{J}} c_j \mathbb{1}_{I_j}(u),$$

$$0 < u < 1, \qquad (14.40)$$

where $\mathcal{J}$ is a countable index set, and I_j, $j \in \mathcal{J}$, are disjoint intervals.

We first show that $\mathcal{J}$ contains only one interval. For this, let k^* be an optimal solution, with representation given by (14.40). By definition k^* minimizes $\|\frac{1}{\beta-\alpha}\mathbb{1}_{(\alpha,\beta]} + \lambda F^{-1} - k\|_2^2$ over all non-decreasing functions k. Moreover,

$$\left\|\frac{1}{\beta-\alpha}\mathbb{1}_{(\alpha,\beta]} + \lambda F^{-1} - k^*\right\|_2^2 = \sum_{j\in\mathcal{J}} \int_{I_j} \left(\frac{1}{\beta-\alpha}\mathbb{1}_{(\alpha,\beta]}(u) + \lambda F^{-1}(u) - c_j\right)^2 du. \quad (14.41)$$

Case 1: If $I_j \cap (\alpha,\beta] = \phi$, then the jth summand in (14.41) can be improved by choosing $k^*(u) = \min\{\lambda F^{-1}(u), k^*(u_j^+)\}$, for all $u \in I_j$, and where u_j^+ denotes the right endpoint of I_j.

Case 2: If $I_j \subset [\alpha,\beta)$, then the jth summand in (14.41) can be improved by setting $k^*(u) = \min\{\frac{1}{\beta-\alpha} + \lambda F^{-1}(u), k^*(u_j^+)\}$ on I_j, and where u_j^+ denotes the right endpoint of I_j.

Therefore, $|\mathcal{J}| = 1$, and the projection k^* is constant on one interval $I = (w_0, w_1]$, with $\alpha \le w_0 \le \beta \le w_1$. Define $c = k^*(u)$, $u \in I$. Then, c crosses $\lambda F^{-1}(u)$ at $u = w_1$, and, moreover, c crosses $1/(\beta - \alpha) + \lambda F^{-1}(u)$ when $u = w_0$, provided the jump $1/(\beta - \alpha)$ is not too large. More precisely, for a given c, w_0 and w_1 follow from

$$\lambda F^{-1}(w_0) = \begin{cases} c - \frac{1}{\beta-\alpha}, & \text{if } \frac{1}{\beta-\alpha} \le \lambda \left(F^{-1}(w_1) - F^{-1}(\alpha)\right), \\ \lambda F^{-1}(\alpha), & \text{otherwise,} \end{cases}$$

$$\lambda F^{-1}(w_1) = c.$$

The optimal c, which minimises (14.41), fulfils

$$\arg\inf_{w_0,w_1,c} \int_{w_0}^{w_1} \left(\frac{1}{\beta-\alpha}\mathbb{1}_{(\alpha,\beta]}(u) + \lambda F^{-1}(u) - c\right)^2 du$$

$$= \arg\inf_{w_0,w_1,c} -2c\left(\frac{\beta-w_0}{\beta-\alpha} + \lambda\int_{w_0}^{\beta} F^{-1}(u)\,du\right) + c^2(w_1 - w_0).$$

Taking the derivative with respect to c, the optimal c is given by

$$c = \frac{1}{w_1 - w_0}\frac{\beta - w_0}{\beta - \alpha} + \frac{\lambda}{w_1 - w_0}\int_{w_0}^{w_1} F^{-1}(u)\,du.$$

Proof of Proposition 14.15

The proof follows analogously to the proof of Proposition 14.14, and is thus omitted.

14.6   Appendix: Isotonic Projection

The important role of isotonic projections for dealing with various statistical optimization problems, mainly in the frame of L^p– spaces, was earlier stressed in Barlow et al. (1972).

We first collect properties of the metric projection defined in (14.9). For this, we denote by $\mathcal{N} = \{\ell \mid \ell\colon (0,1) \to \mathbb{R}, \int_0^1 \ell(u)^2\,du < +\infty\}$ the space of square-integrable functions on $(0, 1)$ and consider the following metric projection from N to $\mathcal{K}$ as defined in (14.10):

$$P_{\mathcal{K}}(\ell) = \arg \inf_{k \in \mathcal{K}} \|\ell - k\|^2, \qquad \ell \in N.$$

A metric projection is called *isotone* if the projection preserves the order induced by $\mathcal{K}$ (Németh, 2003). The partial order $\leq_{\mathcal{K}}$ on N induced by $\mathcal{K}$ is given, for any $\ell_1, \ell_2 \in N$, by

$$\ell_1 \leq_{\mathcal{K}} \ell_2 \qquad \text{if} \quad \ell_2 - \ell_1 \in \mathcal{K}.$$

Proposition 14.22 (Theorem 3 of Németh (2003)) *The metric projection $P_{\mathcal{K}}$ is isotone and subadditive, that is, $P_{\mathcal{K}}$ fulfils for all $\ell_1, \ell_2 \in N$:*

1. **Isotone:** *If $\ell_1 \leq_{\mathcal{K}} \ell_2$, then $P_{\mathcal{K}}(\ell_1) \leq_{\mathcal{K}} P_{\mathcal{K}}(\ell_2)$; and*
2. **Subadditive:** *$P_{\mathcal{K}}(\ell_1 + \ell_2) \leq_{\mathcal{K}} P_{\mathcal{K}}(\ell_1) + P_{\mathcal{K}}(\ell_2)$.*

We deal with projections of functions of the type $\gamma + \lambda F^{-1}$, for fixed F and γ and $\lambda \geq 0$. Thus, we view the projection of $\gamma + \lambda F^{-1}$ as a function of λ and define $k_\lambda = P_{\mathcal{K}}(\gamma + \lambda F^{-1})$, $\lambda \geq 0$.

Proposition 14.23 *For fixed γ and F, the metric projection k_λ is an isotonic projection with the following properties:*

1. **Isotone:** *For $\lambda_1 \leq \lambda_2$, it holds that $k_{\lambda_1} \leq_{\mathcal{K}} k_{\lambda_2}$.*
2. **Continuous:** *If $\lambda_n \to \lambda$, then $\lim_{n \nearrow +\infty} \|k_{\lambda_n} - k_\lambda\|^2 = 0$.*

Proof Isotonicity of k_λ (Property 1) follows by noting that $\lambda_1 \leq \lambda_2$ implies $\gamma + \lambda_2 F^{-1} - (\gamma + \lambda_1 F^{-1}) = (\lambda_2 - \lambda_1) F^{-1} \in \mathcal{K}$. Thus, by isotonicity of $P_{\mathcal{K}}$ we obtain

$$k_{\lambda_1} = P_{\mathcal{K}}(\gamma + \lambda_1 F^{-1}) \leq_{\mathcal{K}} P_{\mathcal{K}}(\gamma + \lambda_2 F^{-1}) = k_{\lambda_2}.$$

To prove continuity of k_λ (Property 2) denote $\lambda_n = \lambda + e_n$, for a sequence e_n with $e_n \to 0$, as $n \to +\infty$. Note that $|e_n| F^{-1} - e_n F^{-1} \in \mathcal{K}$; thus, using first subadditivity and then isotonicity of $P_{\mathcal{K}}$ in the second inequality, we obtain

$$\begin{aligned}
\|k_{\lambda_n} - k_\lambda\|^2 &= \|P_{\mathcal{K}}(\gamma + (\lambda + e_n) F^{-1}) - P_{\mathcal{K}}(\gamma + \lambda F^{-1})\|^2 \\
&\leq \|k_\lambda + P_{\mathcal{K}}(e_n F^{-1}) - k_\lambda\|^2 \\
&\leq \|P_{\mathcal{K}}(|e_n| F^{-1})\|^2 \\
&= |e_n| \|F^{-1}\|^2 \to 0, \quad \text{for} \quad n \to +\infty. \qquad \square
\end{aligned}$$

The next proposition collects a few properties of the isotonic projection $P_{\mathcal{K}}$ and thus of k_λ. For this we denote by $\langle \cdot, \cdot \rangle$ the inner product on N, that is, $\langle \ell_1, \ell_2 \rangle = \int_0^1 \ell_1(u) \ell_2(u)\, \mathrm{d}u$, for $\ell_1, \ell_2 \in N$.

Proposition 14.24 *The isotonic projection $P_{\mathcal{K}}$ satisfies the following properties for all $\ell \in N$ and $k \in \mathcal{K}$:*

1. *$\langle \ell - P_{\mathcal{K}}(\ell), P_{\mathcal{K}}(\ell) - k \rangle \geq 0$;*
2. *$\langle \ell, P_{\mathcal{K}}(\ell) \rangle = \langle P_{\mathcal{K}}(\ell), P_{\mathcal{K}}(\ell) \rangle$;*

3. $\langle \ell, k \rangle \leq \langle P_{\mathcal{K}}(\ell), k \rangle.$

Proof The proof follows along the lines of the proof of Theorem 1.3.2 by Barlow et al. (1972).

Property 1. For $\ell \in \mathcal{N}$, $f \in \mathcal{K}$, and $w \in [0, 1]$ define the function $(1 - w) P_{\mathcal{K}}(\ell) + w f$. By definition of the isotonic projection, it holds that $(1 - w) P_{\mathcal{K}}(\ell) + w f \in \mathcal{K}$. Next, we consider the function

$$w \mapsto \langle \ell - (1 - w) P_{\mathcal{K}}(\ell) - w f, \ell - (1 - w) P_{\mathcal{K}}(\ell) - w f \rangle,$$

which obtains its minimum in $w = 0$, by the very definition of the isotonic projection of $P_{\mathcal{K}}(\ell)$. The first order condition at $w = 0$ implies that

$$2 \langle \ell - (1 - w) P_{\mathcal{K}}(\ell) - w f, P_K(\ell) - f \rangle |_{w=0}$$
$$= 2 \langle \ell - P_K(\ell), P_K(\ell) - f \rangle \geq 0.$$

Property 2. By the definition of an isotonic projection, it holds that for all $\ell, k \in \mathcal{N}$,

$$\langle P_{\mathcal{K}}(\ell) - \ell, P_{\mathcal{K}}(\ell) - \ell \rangle \leq \langle k - \ell, k - \ell \rangle.$$

Choosing $k := c\, P_{\mathcal{K}}(\ell)$ with $c > 0$, we obtain

$$2 (c - 1) \langle P_{\mathcal{K}}(\ell), \ell \rangle \leq (c^2 - 1) \langle P_{\mathcal{K}}(\ell), P_{\mathcal{K}}(\ell) \rangle.$$

Thus, for $0 < c_1 < 1$ and $c_2 > 1$, the above inequality becomes

$$\frac{1 - c_1^2}{2(1 - c_1)} \langle P_{\mathcal{K}}(\ell), P_{\mathcal{K}}(\ell) \rangle \leq \langle P_{\mathcal{K}}(\ell), \ell \rangle \leq \frac{c_2^2 - 1}{2(c_2 - 1)} \langle P_{\mathcal{K}}(\ell), P_{\mathcal{K}}(\ell) \rangle.$$

Taking limits with $c_1 \nearrow 1$ and $c_2 \searrow 1$ concludes the statement of the property.

Property 3. For $\ell \in \mathcal{N}$ and $k \in \mathcal{K}$, by first applying Property 2 and then Property 1, we obtain

$$\langle \ell - P_{\mathcal{K}}(\ell), P_{\mathcal{K}}(\ell) - k \rangle$$
$$= \langle \ell, P_{\mathcal{K}}(\ell) \rangle - \langle \ell, k \rangle - \langle P_{\mathcal{K}}(\ell), P_{\mathcal{K}}(\ell) \rangle + \langle k, P_{\mathcal{K}}(\ell) \rangle$$
$$= -\langle \ell, k \rangle + \langle k, P_{\mathcal{K}}(\ell) \rangle$$
$$\geq 0. \qquad \square$$

14.7 Appendix: Models for Insurance Portfolio Losses

Here we collect the alternative models used in Section 14.4 to compare them with the reference Pareto–Clayton model. An alternative model assumes that the portfolio loss S follows a distribution function detailed below. We further provide the first two moments, which are used to calculate the alternative models' parameters (by matching the first two moments of the alternative models to the reference Pareto–Clayton model).

1. **Lognormal model (LN)**: In this case, we express the portfolio loss S as $S = e^Z$, where Z is a normally distributed random variable with mean μ and variance σ^2. The first two moments of S are then $ES = e^{\mu + \frac{1}{2}\sigma^2}$ and $ES^2 = e^{\mu + 2\sigma^2}$, respectively.

2. **Gamma model (Γ)**: Under this model, the portfolio loss S is Gamma-distributed with parameters $\alpha, \beta > 0$, density

$$f_S(s) = \frac{\beta^\alpha s^{\alpha-1} e^{-\beta s}}{\Gamma(\alpha)}, \quad s > 0.$$

The first two moments are $ES = \alpha\beta$ and $ES^2 = (1 + \alpha)\alpha\beta^2$, where $\Gamma(\cdot)$ denotes the Gamma function.

3. **Weibull model (W)**: The portfolio loss S follows a Weibull distribution with parameters $\lambda, k \in (0, +\infty)$, distribution function

$$F_S(s) = 1 - e^{-(s/\lambda)^k}, \quad s > 0.$$

The first two moments are $ES = \lambda\Gamma(1 + \frac{1}{k})$ and $ES^2 = \lambda^2\Gamma(1 + \frac{2}{k})$, respectively.

4. **Inverse Gaussian model (IG)**: Under this model, the distribution of the portfolio loss S is inverse Gaussian with parameters $\mu, \lambda > 0$ and

$$F_S(s) = \Phi\left(\sqrt{\frac{\lambda}{s}\left(\frac{s}{\mu} - 1\right)}\right) + e^{\frac{2\lambda}{\mu}} \Phi\left(-\sqrt{\frac{\lambda}{s}\left(\frac{s}{\mu} + 1\right)}\right), \quad s > 0.$$

The first two moments are $ES = \mu$ and $ES^2 = \frac{\mu^3}{\lambda} + \mu^2$.

5. **Inverse Gamma model (IΓ)**: The portfolio loss S follows an inverse Gamma distribution with parameters $\alpha, \beta > 0$, density function

$$f_S(s) = \frac{\beta^\alpha (\frac{1}{s})^{\alpha+1} e^{-\frac{\beta}{s}}}{\Gamma(\alpha)}, \quad s > 0.$$

The first two moments are $ES = \frac{\beta}{\alpha-1}$ and $ES^2 = \frac{\beta^2}{(\alpha-1)(\alpha-2)}$.

6. **Inverse Weibull model (IW)**: Under this model, the distribution function of the portfolio loss S is inverse Weibull with parameters $\lambda, k > 0$ and density

$$f_S(s) = e^{-(\lambda/s)^k}, \quad s > 0.$$

The first two moments are $ES = \lambda\Gamma(1 - \frac{1}{k})$ and $ES^2 = \lambda^2\Gamma(1 - \frac{2}{k})$, respectively.

7. **Log-logistic model (LL)**: Under this model, the portfolio loss S follows a log-logistic distribution with parameters $\alpha, \beta > 0$ and quantile function

$$F_S^{-1}(s) = \alpha\left(\frac{p}{1 - p}\right)^{1/\beta}, \quad s > 0.$$

The first two moments are $ES = \frac{\alpha\pi/\beta}{\sin(\pi/\beta)}$ and $ES^2 = \frac{\alpha^2 2\pi/\beta}{\sin(2\pi/\beta)}$, respectively.

References

Aas, K., Czado, C., Frigessi, A., and Bakken, H. 2009. Pair-copula constructions of multiple dependence. *Insurance: Math. Econom.*, **44**(2), 182–198.

Abdous, B., Genest, C., and Rémillard, B. 2005. Dependence properties of meta-elliptical distributions. Pages 1–15 of: Duchesne, P., and Rémillard, B. (eds), *Statistical Modelling and Analysis for Complex Data Problems*. Springer.

Acciaio, B., Beiglböck, M., Penkner, F., and Schachermayer, W. 2016. A model-free version of the fundamental theorem of asset pricing and the super replication theorem. *Math. Finance*, **26**, 233–251.

Actuarial Association of Europe (AAE). 2017. *Comments template on discussion paper on the review of specific items in the Solvency II delegated regulation*. Technical report. Actuarial Association of Europe.

Albrecher, H., Constantinescu, C., and Loisel, S. 2011. Explicit ruin formulas for models with dependence among risks. *Insurance: Math. Econom.*, **48**(2), 265–270.

Ansari, J., and Rüschendorf, L. 2021. Ordering results for elliptical distributions with applications to risk bounds. *J. Multivariate Anal.*, **182**, Art.–Id. 104709.

Artzner, P., Delbaen, F., Eber, J.-M., and Heath, D. 1999. Coherent measures of risk. *Math. Finance*, **9**(3), 203–228.

Barlow, R. E., Bartholomew, D. J., Bremner, J. M., and Brunk, H. D. 1972. *Statistical Inference under Order Restrictions: The Theory and Application of Isotonic Regression*. Wiley.

Barrieu, P., and Scandolo, G. 2015. Assessing financial model risk. *Eur. J. Oper. Res.*, **242**(2), 546–556.

Basel Committee on Banking Supervision. 2010 (Oct.). *Developments in Modelling Risk Aggregation*. Bank for International Settlements.

Basu, S., and DasGupta, A. 1997. The mean, median, and mode of unimodal distribution: A characterization. *Theory Probab. Appl.*, **41**(2), 210–223.

Bäuerle, N., and Müller, A. 2006. Stochastic orders and risk measures: Consistency and bounds. *Insurance: Math. Econom.*, **38**, 132–148.

Beiglböck, M., Henry-Labordère, P., and Penkner, F. 2013. Model-independent bounds for option prices – a mass transport approach. *Finance Stoch.*, **17**(3), 477–501.

Benson, F. 1949. A note on the estimation of mean and standard deviation from quantiles. *J. R. Stat. Soc., Ser. B, Stat. Methodol.*, **11**, 91–100.

Bernard, C., and McLeish, D. 2016. Algorithms for finding copulas minimizing convex functions of sums. *Asia-Pacific J. Oper. Res.*, **33**(5), Art.–Id. 1650040.

Bernard, C., and Vanduffel, S. 2014. Mean-variance optimal portfolios in the presence of a benchmark with applications to fraud detection. *Eur. J. Oper. Res.*, **234**(2), 469–480.

Bernard, C., and Vanduffel, S. 2015. A new approach to assessing model risk in high dimensions. *J. Banking Finance*, **58**, 167–178.

Bernard, C., Bondarenko, O., and Vanduffel, S. 2018b. Rearrangement algorithm and maximum entropy. *Ann. Oper. Res.*, **261**(1–2), 107–134.

Bernard, C., Bondarenko, O., and Vanduffel, S. 2021. A model-free approach to multivariate option pricing. *Rev. Deriv. Res.*, **24**(2), 1–21.

Bernard, C., Boyle, P., and Vanduffel, S. 2014a. Explicit representation of cost-efficient strategies. *Finance*, **35**(2), 5–55.

Bernard, C., Pesenti, S. M., and Vanduffel, S. 2023. Robust distortion risk measures. *Mathematical Finance*, https://doi.org/10.1111/mafi.12414.

Bernard, C., Denuit, M., and Vanduffel, S. 2018a. Measuring portfolio risk under partial dependence information. *J. Risk Insurance*, **85**(3), 843–863.

Bernard, C., Jiang, X., and Vanduffel, S. 2012. A note on 'Improved Fréchet bounds and model-free pricing of multi-asset options' by Tankov (2011). *J. Appl. Probab.*, **49**(3), 866–875.

Bernard, C., Jiang, X., and Wang, R. 2014b. Risk aggregation with dependence uncertainty. *Insurance: Math. Econom.*, **54**, 93–108.

Bernard, C., Kazzi, R., and Vanduffel, S. 2020. Range value-at-risk bounds for uni-modal distributions under partial information. *Insurance: Math. Econom.*, **94**(1), 9–24.

Bernard, C., Liu, Y., MacGillivray, N., and Zhang, J. 2013. Bounds on capital requirements for bivariate risk with given marginals and partial information on the dependence. *Depend. Model.*, **1**, 37–53.

Bernard, C., Rüschendorf, L., and Vanduffel, S. 2017c. Value-at-risk bounds with variance constraints. *J. Risk Insurance*, **84**(3), 923–959.

Bernard, C., Rüschendorf, L., Vanduffel, S., and Wang, R. 2017b. Risk bounds for factor models. *Finance Stoch.*, **3**, 631–659.

Bernard, C., Rüschendorf, L., Vanduffel, S., and Yao, J. 2017a. How robust is the value-at-risk of credit risk portfolios? *Eur. J. Finance*, **23**(6), 507–534.

Bignozzi, V., Puccetti, G., and Rüschendorf, L. 2015. Reducing model risk via positive and negative dependence assumptions. *Insurance: Math. Econom.*, **61**, 17–26.

Boudt, K., Jakobsons, E., and Vanduffel, S. 2018. Block rearranging elements within matrix columns to minimize the variability of the row sums. *4OR*, **16**(1), 31–50.

Brighi, B., and Chipot, M. 1994. Approximated convex envelope of a function. *SIAM J. Numer. Anal.*, **31**(1), 128–148.

Burgert, C., and Rüschendorf, L. 2006. Consistent risk measures for portfolio vectors. *Insurance: Math. Econom.*, **38**, 289–297.

Carhart, M. M. 1997. On persistence in mutual fund performance. *J. Finance*, **52**(1), 57–82.

Chamberlain, G., and Rothschild, M. 1983. Arbitrage, factor structure, and mean-variance analysis on large asset markets. *Econometrica*, **51**, 1281–1304. http://doi.org/10.3386/w0996

Chernoff, H., and Reiter, S. 1954. *Selection of a distribution function to minimize an expectation subject to side conditions.* Technical Report. Stanford University: Applied Mathematics and Statistics Labs.

Chong, K.-M. 1974. Some extensions of a theorem of Hardy, Littlewood and Pólya and their applications. *Canad. J. Math.*, **26**, 1321–1340.

Chong, K.-M., and Rice, N. M. 1971. *Equimeasurable Rearrangements of Functions.* Queen's Papers in Pure and Applied Mathematics, vol. 28. Queen's University.

Choquet, G. 1954. Theory of capacities. *Ann. Inst. Fourier*, **5**, 131–295.

Coffman, E. G., and Yannakakis, M. 1984. Permuting elements within columns of a matrix in order to minimize maximum row sum. *Math. Oper. Res.*, **9**(3), 384–390.

Connor, G., and Korajczyk, R. A. 1993. A test for the number of factors in an approximate factor model. *J. Finance*, **48**(4), 1263–1291.

Cont, R., Deguest, R., and Scandolo, G. 2010. Robustness and sensitivity analysis of risk measurement procedures. *Quant. Finance*, **10**(6), 593–606.

Cope, E., Mignola, G., Antonini, G., and Ugoccioni, R. 2009. Challenges and pitfalls in measuring operational risk from loss data. *J. Oper. Risk*, **4**(4), 855–863.

Cornilly, D., Puccetti, G., Rüschendorf, L., and Vanduffel, S. 2022. Fair allocation of indivisible goods with minimum inequality or minimum envy. *Eur. J. Oper. Res.*, **297**(2), 741–752.

Cornilly, D., Rüschendorf, L., and Vanduffel, S. 2018. Upper bounds for strictly concave distortion risk measures on moment spaces. *Insurance: Math. Econom.*, **82**, 141–151.

Credit Suisse First Boston. 1997. *CreditRisk+: A credit risk management framework.* Technical report. Credit Suisse First Boston Bank.

Cuberos, A., Masiello, E., and Maume-Deschamps, V. 2015. High level quantile approximations of sums of risks. *Depend. Model.*, **3**(1), 141–158.

Cuesta-Albertos, J. A., Rüschendorf, L., and Tuero-Diaz, A. 1993. Optimal coupling of multivariate distributions and stochastic processes. *J. Multivariate Anal.*, **46**, 335–361.

Czado, C. 2010. Pair copula constructions of multivariate copulas. Pages 93–109 of: Jaworski, P., Durante, F., Härdle, W. K., and Rychlik, T. (eds), *Copula Theory and Its Applications.* Lect. Notes Stat., vol. 198. Springer.

Dacorogna, M. M., Elbahtouri, L., and Kratz, M. 2016. Explicit diversification benefit for dependent risks. *SCOR Papers*, **38**(1), 1–25.

das Gupta, S., Olkin, I., Savage, L. J., Eaton, M. L., Perlman, M., and Sobel, M. 1972. Inequalities on the probability content of convex regions for elliptically contoured distributions. Pages 241–265 of: Le Cam, L. M., Neyman, J., and Scott, E. L. (eds), *Proc. 6th Berkeley Sympos. Math. Statist. Probab.*, vol. 2. Berkeley, University of California Press.

Day, P. W. 1972. Rearrangement inequalities. *Canad. J. Math.*, **24**, 930–943.

de Schepper, A., and Heijnen, B. 2010. How to estimate the value at risk under incomplete information. *J. Comput. Appl. Math.*, **233**(9), 2213–2226.

de Vylder, F. E. 1982. Best upper bounds for integerals with respect to measures allowed to vary under conical and integral constraints. *Insurance: Math. Econom.*, **1**(2), 109–130.

de Vylder, F. E. 1996. *Advanced Risk Theory: A Self-contained Introduction.* Éd. de L'Univ. de Bruxelles.

Deelstra, G., Diallo, I., and Vanmaele, M. 2008. Bounds for Asian basket options. *J. Comput. Appl. Math.*, **218**(2), 215–228.

Denuit, M. 1999. The exponential premium calculation principle revisited. *Astin Bull.*, **29**(2), 215–226.

Denuit, M., Genest, C., and Marceau, É. 1999. Stochastic bounds on sums of dependent risks. *Insurance: Math. Econom.*, **25**(1), 85–104.

Denuit, M., Lefèvre, C., and Shaked, M. 1998. The *s*-convex orders among real random variables, with applications. *Math. Inequal. Appl.*, **1**(4), 585–613.

Dhaene, J., Laeven, R. J. A., Vanduffel, S., Darkiewicz, G., and Goovaerts, M. J. 2008. Can a coherent risk measure be too subadditive? *J. Risk Insurance*, **75**(2), 365–386.

Dhaene, J., Tsanakas, A., Valdez, E. A., and Vanduffel, S. 2012. Optimal capital allocation principles. *J. Risk Insurance*, **79**(1), 1–28.

Dhaene, J., Vanduffel, S., Goovaerts, M., Olieslagers, R., and Koch, R. 2003. On the computation of the capital multiplier in the Fortis credit economic capital model. *Belg. Actuar. Bull.*, **3**(1), 50–57.

Dhaene, J., Vanduffel, S., Goovaerts, M. J., Kaas, R., Tang, Q., and Vyncke, D. 2006. Risk measures and comonotonicity: A review. *Stoch. Models*, **22**(4), 573–606.

Dubey, S. D. 1970. Compound gamma, beta and *F* distributions. *Metrika*, **16**(1), 27–31.

Ekern, S. 1980. Increasing *n*-th degree risk. *Economics*, **6**(4), 329–333.

El Ghaoui, L., Oks, M., and Oustry, F. 2003. Worst-case value-at-risk and robust portfolio optimization: A conic programming approach. *Oper. Res.*, **51**(4), 543–556.

Embrechts, P., and Puccetti, G. 2006a. Aggregating risk capital, with an application to operational risk. *Geneva Risk Insur. Rev.*, **31**(2), 71–90.

Embrechts, P., and Puccetti, G. 2006b. Bounds for functions of dependent risks. *Finance Stoch.*, **10**(3), 341–352.

Embrechts, P., and Puccetti, G. 2010. Risk aggregation. Pages 111–126 of: Jaworski, P., Durante, F., Härdle, W. K., and Rychlik, T. (eds), *Copula Theory and Its Applications.* Springer.

Embrechts, P., Höing, A., and Juri, A. 2003. Using copulae to bound the value-at-risk for functions of dependent risks. *Finance Stoch.*, **7**(2), 145–167.

Embrechts, P., McNeil, A., and Straumann, D. 1999. Correlation: Pitfalls and alternatives. *Risk Mag.*, **12**, 69–71.

Embrechts, P., McNeil, A., and Straumann, D. 2002. Correlation and dependence in risk management: Properties and pitfalls. Pages 176–223 of: Dempster, M., and Moffatt, H. (eds), *Risk Management: Value-at-Risk and Beyond*. Cambridge, Cambridge University Press.

Embrechts, P., Puccetti, G., and Rüschendorf, L. 2013. Model uncertainty and VaR aggregation. *J. Banking Finance*, **37**(8), 2750–2764.

Embrechts, P., Puccetti, G., Rüschendorf, L., Wang, R., and Beleraj, A. 2014. An academic response to Basel 3.5. *Risks*, **2**(1), 25–48.

Embrechts, P., Wang, B., and Wang, R. 2015. Aggregation-robustness and model uncertainty of regulatory risk measures. *Finance Stoch.*, **19**(4), 763–790.

Engle, R. F., Ng, V. K., and Rothschild, M. 1990. Asset pricing with a factor-ARCH covariance structure: Empirical estimates for treasury bills. *J. Econometrics*, **45**(1), 213–237.

European Central Bank (ECB). 2017. *Guide for the targeted review of internal models (TRIM)*. Technical report. European Central Bank.

Everett III, H. 1963. Generalized Lagrange multiplier method for solving problems of optimum allocation of resources. *Oper. Res.*, **11**(3), 399–417.

Fama, E. F., and French, K. R. 1993. Common risk factors in the returns on stocks and bonds. *J. Financial Econom.*, **33**(1), 3–56.

Fan, K., and Lorentz, G. G. 1954. An integral inequality. *Amer. Math. Monthly*, **61**, 626–631.

Föllmer, H., and Schied, A. 2004. *Stochastic Finance. An Introduction in Discrete Time*. 2nd revised and extended edn. Berlin, de Gruyter.

Frank, M. J., and Schweizer, B. 1979. On the duality of generalized infimal and supremal convolutions. *Rend. Mat., VI. Ser.*, **12**, 1–23.

Fréchet, M. 1951. Sur les tableaux de corrélation dont les marges sont données. *Ann. l'Université de Lyon, Section A*, **14**, 53–77.

Gaffke, N., and Rüschendorf, L. 1981. On a class of extremal problems in statistics. *Math. Operationsforsch. Stat., Ser. Optimization*, **12**(1), 123–135.

Genest, C., Marceau, É., and Mesfioui, M. 2002. Upper stop-loss bounds for sums of possibly dependent risks with given means and variances. *Statist. Probab. Lett.*, **57**, 33–34.

Gordy, M. B. 2000. A comparative anatomy of credit risk models. *J. Banking Finance*, **24**(1), 119–149.

Gordy, M. B. 2003. A risk-factor model foundation for ratings-based bank capital rules. *J. Financial Intermed.*, **12**(3), 199–232.

Gumbel, E. J. 1954. The maxima of the mean largest value and of the range. *Ann. Math. Stat.*, **25**, 76–84.

Haaf, H., Reiss, O., and Schoenmakers, J. 2004. Numerically stable computation of CreditRisk$^+$. Pages 69–77 of: Gundlach, M., and Lehrbass, F. (eds), *CreditRisk$^+$ in the Banking Industry*. Springer.

Hardy, G. H., Littlewood, J. E., and Pólya, G. 1952. *Inequalities*. 2nd edn. Cambridge.

Haus, U.-U. 2015. Bounding stochastic dependence, complete mixability of matrices, and multidimensional bottleneck assignment problems. *Oper. Res. Lett.*, **43**, 74–79.

Hoeffding, W. 1940. *Maßstabinvariante Korrelationstheorie*. Ph.D.Thesis, Schr. Math. Inst. u. Inst. Angew. Math. Univ. Berlin 5, 181–233 (1940) and Berlin: Dissertation.

Hofert, M. 2020. Implementing the rearrangement algorithm: An example from computational risk management. *Risks*, **8**(2), 47.

Hofert, M., Memartoluie, A., Saunders, D., and Wirjanto, T. 2017. Improved algorithms for computing worst value-at-risk. *Stat. Risk Model.*, **34**(1–2), 13–31.

Hürlimann, W. 2002. Analytical bounds for two value-at-risk functionals. *Astin Bull.*, **32**(2), 235–265.

Ingersoll, J. E. 1984. Some results in the theory of arbitrage pricing. *J. Finance*, **39**(4), 1021–1039.

Joe, H. 1997. *Multivariate Models and Dependence Concepts*. Monographs on Statistics and Applied Probability, vol. 73. London, Chapman & Hall.

Jorion, P. 2006. *Value at Risk: The New Benchmark for Managing Financial Risk*. 3rd edn. New York, McGraw-Hill.

Jouini, E., Schachermayer, W., and Touzi, N. 2006. Law invariant risk measures have the Fatou property. Pages 49–71 of: Kusuoka, S., and Yamazaki, A. (eds), *Advances in Mathematical Economics*, vol. 9 Springer.

Kaas, R., and Goovaerts, M. J. 1986. Best bounds for positive distributions with fixed moments. *Insurance: Math. Econom.*, **5**(1), 87–92.

Kaas, R., Dhaene, J., and Goovaerts, M. J. 2000. Upper and lower bounds for sums of random variables. *Insurance: Math. Econom.*, **27**(2), 151–168.

Kantorovich, L. V. 1942. On the translocation of masses. *C. R. (Dokl.) Acad. Sci. URSS, Ser.*, **37**, 199–201.

Karlin, S. 1968. *Total Positivity*, vol. I. Stanford, Stanford University Press.

Karlin, S., and Novikoff, A. 1963. Generalized convex inequalities. *Pac. J. Math.*, **13**, 1251–1279.

Karlin, S., and Shapley, L. S. 1953. *Geometry of Moment Spaces*, vol. 12. American Mathematical Society.

Karlin, S., and Studden, W. J. 1966. *Tchebycheff Systems: With Applications in Analysis and Statistics*, vol. 15. Interscience Publishers.

Kellerer, H. G. 1988. Measure theoretic versions of linear programming. *Math. Z.*, **198**, 367–400.

Khintchine, A. Y. 1938. On unimodal distributions. *Izv. Nauchno-Issle., Inst. Mat. Mekhhaniki*, **2**(2), 1–7.

Kusuoka, S. 2001. On law invariant coherent risk measures. Pages 83–95 of: Kusuoka, S., et al. (eds), *Advances in Mathematical Economics*. Springer.

Lai, T. L., and Robbins, H. 1978. A class of dependent random variables and their maxima. *Z. Wahrscheinlichkeitstheor. Verw. Geb.*, **42**, 89–111.

Lai, T. L., and Robbins, M. 1976. Maximally dependent random variables. *Proc. Nat. Acad. Sci. USA*, **73**, 286–288.

Levy, H., and Kroll, Y. 1978. Ordering uncertain options with borrowing and lending. *J. Finance*, **33**(2), 553–574.

Lewbel, A. 1991. The rank of demand systems: Theory and nonparametric estimation. *Econometrica*, **59**, 711–730.

Li, J. Y.-M. 2018. Closed-form solutions for worst-case law invariant risk measures with application to robust portfolio optimization. *Oper. Res.*, **66**(6), 1533–1541.

Li, L., Shao, H., Wang, R., and Yang, J. 2018. Worst-case range value-at-risk with partial information. *SIAM J. Financial Math.*, **9**(1), 190–218.

Lo, A. W. 1987. Semi-parametric upper bounds for option prices and expected payoffs. *J. Financial Econom.*, **19**, 373–387.

Lorentz, G. G. 1953. An inequality for rearrangements. *Amer. Math. Monthly*, **60**, 176–179.

Lowenstein, R. 2008. Long-term capital management: It's a short term memory. *New York Times*, **6**(1), 1–3.

Lux, T., and Papapantoleon, A. 2017. Improved Fréchet–Hoeffding bounds on *d*-copulas and applications in model-free finance. *Ann. Appl. Probab.*, **27**(6), 3633–3671.

Lux, T., and Papapantoleon, A. 2019. Model-free bounds on value-at-risk using partial dependence information. *Insurance: Math. Econom.*, **86**, 73–83.

Lux, T., and Rüschendorf, L. 2018. VaR bounds with two-sided dependence information on the copula. *Math. Finance*, **29**, 967–1000.

Luxemburg, W. A. J. 1967. Rearrangement invariant Banach function spaces. *Queen's Papers Pure Appl. Math.*, **10**, 83–144.

Mainik, G., and Rüschendorf, L. 2010. On optimal portfolio diversification with respect to extreme risks. *Finance Stoch.*, **14**, 593–623.

Makarov, G. D. 1981. Estimates for the distribution function of a sum of two random variables when the marginal distributions are fixed. *Theory Probab. Appl.*, **26**, 803–806.

Marshall, A. W., Olkin, I., and Arnold, B. C. 2011. *Inequalities: Theory of Majorization and Its Applications*. 2nd edn. Springer Series in Statistics. Springer.

McNeil, A. J., Frey, R., and Embrechts, P. 2005. *Quantitative Risk Management*. Princeton Series in Finance. Princeton, Princeton University Press.

McNeil, A. J., Frey, R., and Embrechts, P. 2015. *Quantitative Risk Management: Concepts, Techniques and Tools*. Princeton University Press.

Meilijson, I., and Nadas, A. 1979. Convex majorization with an application to the length of critical paths. *J. Appl. Probab.*, **16**, 671–677.

Mesfioui, M., and Quessy, J.-F. 2005. Bounds on the value-at-risk for the sum of possibly dependent risks. *Insurance: Math. Econom.*, **37**(1), 135–151.

Moscadelli, M. 2004 (July). *The Modelling of Operational Risk: Experience with the Analysis of the Data Collected by the Basel Committee*. Available at http://ideas .repec.org/p/bdi/wptemi/td_517_04.html.

Moynihan, R., Schweizer, B., and Sklar, A. 1978. Inequalities among operations on probability distribution functions. Pages 133–149 of: Beckenbach, E. F. (ed), *Allgemeine Ungleichungen 1*. Intern. Ser. Num. Math., vol. 41.

Müller, A., and Scarsini, M. 2001. Stochastic comparison of random vectors with a common copula. *Math. Oper. Res.*, **26**, 723–740.

Müller, A., and Scarsini, M. 2005. Archimedean copulae and positive dependence. *J. Multivariate Anal.*, **93**(2), 434–445.

Müller, A., and Stoyan, D. 2002. *Comparison Methods for Stochastic Models and Risks*. Chichester, John Wiley & Sons Ltd.

Natarajan, K., Sim, M., and Uichanco, J. 2010. Tractable robust expected utility and risk models for portfolio optimization. *Math. Finance*, **20**(4), 695–731.

Nelsen, R. B. 2006. *An Introduction to Copulas. Properties and Applications*. 2nd edn. Lect. Notes Stat., vol. 139. Springer.

Nelsen, R. B., Quesada Molina, J. J., Rodríguez Lallena, J. A., and Úbeda Flores, M. 2004. Best-possible bounds on sets of bivariate distribution functions. *J. Multivariate Anal.*, **90**(2), 348–358.

Németh, A. B. 2003. Characterization of a Hilbert vector lattice by the metric projection onto its positive cone. *J. Approx. Theo.*, **123**(2), 295–299.

Nešlehová, J., Embrechts, P., and Chavez-Demoulin, V. 2006. Infinite-mean models and the LDA for operational risk. *J. Oper. Risk*, **1**, 3–25.

Oakes, D. 1989. Bivariate survival models induced by frailties. *J. Amer. Statist. Assoc.*, **84**(406), 487–493.

Office of the Superintendent of Financial Institutions (OSFI). 2014. *Quantitative Impact Study No. 4: General – Aggregation and Diversification – Supplementary Information*. Online document. www.osfi-bsif.gc.ca/Eng/Docs/qis4_ir_sup.pdf.

Pan, X., Qiu, G., and Hu, T. 2016. Stochastic orderings for elliptical random vectors. *J. Multivariate Anal.*, **148**, 83–88.

Panjer, H. H. 1981. Recursive evaluation of a family of compound distributions. *Astin Bull.*, **12**(1), 22–26.

Panjer, H. H. 2001. *Measurement of Risk, Solvency Requirements and Allocation of Capital Within Financial Conglomerates*. University of Waterloo, Institute of Insurance and Pension Research.

Prudential Regulation Authority (PRA). 2018. *Model risk management principles for stress testing*. Supervisory statement. Bank of England, Prudential Regulation Authority.

Puccetti, G. 2005. *Bounds for Functions of Univariate and Multivariate Risks*. Ph.D. Thesis, University of Pisa, Italy.

Puccetti, G., and Rüschendorf, L. 2012a. Bounds for joint portfolios of dependent risks. *Stat. Risk Model.*, **29**, 107–132.

Puccetti, G., and Rüschendorf, L. 2012b. Computation of sharp bounds on the distribution of a function of dependent risks. *J. Comput. Appl. Math.*, **236**, 1833–1840.

Puccetti, G., and Rüschendorf, L. 2013. Sharp bounds for sums of dependent risks. *J. Appl. Probab.*, **50**(1), 42–53.

Puccetti, G., and Rüschendorf, L. 2014. Asymptotic equivalence of conservative VaR- and ES-based capital charges. *J. Risk*, **16**(3), 3–22.

Puccetti, G., Rüschendorf, L., and Manko, D. 2016. VaR bounds for joint portfolios with dependence constraints. *Depend. Model.*, **4**(1), 368–381.

Puccetti, G., Rüschendorf, L., Small, D., and Vanduffel, S. 2017. Reduction of value-at-risk bounds via independence and variance information. *Scand. Actuar. J.*, **3**, 245–266.

Puccetti, G., Wang, B., and Wang, R. 2012. Advances in complete mixability. *J. Appl. Probab.*, **49**(2), 430–440.

Puccetti, G., Wang, B., and Wang, R. 2013. Complete mixability and asymptotic equivalence of worst-possible VaR and ES estimates. *Insurance: Math. Econom.*, **53**(2), 821–828.

Rachev, S. T., and Rüschendorf, L. 1994. Solution of some transportation problems with relaxed and additional constraints. *SIAM Contr. Optim.*, **32**, 673–689.

Rachev, S. T., and Rüschendorf, L. 1998a. *Mass Transportation Problems. Vol. I: Theory.* Springer.

Rachev, S. T., and Rüschendorf, L. 1998b. *Mass Transportation Problems. Vol. II: Applications.* Springer.

Ross, S. A. 1976. The arbitrage theory of capital asset pricing. *J. Econom. Theory*, **13**(3), 341–360.

Rüschendorf, L. 1980. Inequalities for the expectation of Δ-monotone functions. *Z. Wahrscheinlichkeitstheor. Verw. Geb.*, **54**, 341–354.

Rüschendorf, L. 1981a. Characterization of dependence concepts for the normal distribution. *Ann. Inst. Stat. Math.*, **33**, 347–359.

Rüschendorf, L. 1981b. Sharpness of Fréchet bounds. *Z. Wahrscheinlichkeitstheor. Verw. Geb.*, **57**, 293–302.

Rüschendorf, L. 1982. Random variables with maximum sums. *Adv. Appl. Probab.*, **14**, 623–632.

Rüschendorf, L. 1983a. On the multidimensional assignment problem. *Methods Oper. Res.*, **47**, 107–113.

Rüschendorf, L. 1983b. Solution of a statistical optimization problem by rearrangement methods. *Metrika*, **30**, 55–62.

Rüschendorf, L. 1984. On the minimum discrimination information theorem. *Stat. Dec., Suppl. issue 1*, **1**, 263–283.

Rüschendorf, L. 1985. The Wasserstein distance and approximation theorems. *Z. Wahrscheinlichkeitstheor. Verw. Geb.*, **70**, 117–129.

Rüschendorf, L. 1991. Bounds for distributions with multivariate marginals. Pages 285–310 of: Mosler, K., and Scarsini, M. (eds), *Stochastic Order and Decision under Risk*, vol. 19. IMS Lecture Notes.

Rüschendorf, L. 2004. Comparison of multivariate risks and positive dependence. *J. Appl. Probab.*, **41**, 391–406.

Rüschendorf, L. 2005. Stochastic ordering of risks, influence of dependence and a.s. constructions. Pages 19–56 of: Balakrishnan, N., Bairamov, I. G., and Gebizlioglu, O. L. (eds), *Advances on Models, Characterizations and Applications.* Chapman & Hall/CRC Press.

Rüschendorf, L. 2013. *Mathematical Risk Analysis. Dependence, Risk Bounds, Optimal Allocations and Portfolios.* Springer.

Rüschendorf, L. 2017a. Improved Hoeffding–Fréchet bounds and applications to VaR estimates. Pages 181–202 of: Úbeda Flores, M., et al. (eds), *Copulas and Dependence Models with Applications.* Springer.

Rüschendorf, L. 2017b. Risk bounds and partial dependence information. Pages 345–366 of: Ferger, D., González Manteiga, W., Schmidt, T., and Wang, J.-L. (eds), *From Statistics to Mathematical Finance.* Springer.

Rüschendorf, L., and Uckelmann, L. 2002. Variance minimization and random variables with constant sum. Pages 221–222 of: Cuadras, C. M., et al. (eds), *Distributions With Given Marginals and Statistical Modelling*. Springer.

Rüschendorf, L., and Witting, J. 2017. VaR bounds in models with partial dependence information on subgroups. *Depend. Model.*, **5**(1), 59–74.

Rustagi, J. S. 1976. *Variational Methods in Statistics*. Mathematics in Science and Engineering, vol. 121. Elsevier, Amsterdam.

Rustagi, J. S. 1957. On minimizing and maximizing a certain integral with statistical applications. *Ann. Math. Stat.*, **28**, 309–328.

Salmon, F. 2009. Recipe for disaster: The formula that killed Wall Street. *Wired Magazine*, **17**(3), Available online at www.wired.com/2009/02/wp–quant/.

Santos, A. A. P., Nogales, F. J., and Ruiz, E. 2013. Comparing univariate and multivariate models to forecast portfolio value-at-risk. *J. Financial Econom.*, **11**(2), 400–441.

Scaillet, O. 2005. A Kolmogorov–Smirnov type test for positive quadrant dependence. *Can. J. Stat.*, **33**(3), 415–427.

Shaked, M., and Shantikumar, J. G. 2007. *Stochastic Orders*. New York, Springer.

Sharpe, W. F. 1964. Capital asset prices: A theory of market equilibrium under conditions of risk. *J. Finance*, **19**(3), 425–442.

Sklar, A. 1973. Random variables, joint distribution functions, and copulas. *Kybernetika*, **9**, 449–460.

Snijders, T. A. B. 1984. Antithetic variates for Monte Carlo estimation of probabilities. *Stat. Neerl.*, **38**, 55–73.

Strassen, V. 1965. The existence of probability measures with given marginals. *Ann. Math. Stat.*, **36**, 423–439.

Tankov, P. 2011. Improved Fréchet bounds and model-free pricing of multi-asset options. *J. Appl. Probab.*, **48**, 389–403.

Tsanakas, A. 2009. To split or not to split: Capital allocation with convex risk measures. *Insurance: Math. Econom.*, **44**(2), 268–277.

Tsanakas, A., and Desli, E. 2005. Measurement and pricing of risk in insurance markets. *Risk Anal.*, **25**(6), 1653–1668.

Vandendorpe, A., Ho, N.-D., Vanduffel, S., and Van Dooren, P. 2008. On the parameterization of the CreditRisk$^+$ model for estimating credit portfolio risk. *Insurance: Math. Econom.*, **42**(2), 736–745.

Vanduffel, S., Shang, Z., Henrard, L., Dhaene, J., and Valdez, E. A. 2008. Analytic bounds and approximations for annuities and Asian options. *Insurance: Math. Econom.*, **42**(3), 1109–1117.

Vanmaele, M., Deelstra, G., Liinev, J., Dhaene, J., and Goovaerts, M. J. 2006. Bounds for the price of discrete arithmetic Asian options. *J. Comput. Appl. Math.*, **185**(1), 51–90.

Wang, B., and Wang, R. 2011. The complete mixability and convex minimization problems with monotone marginal densities. *J. Multivariate Anal.*, **102**, 1344–1360.

Wang, B., and Wang, R. 2015. Extreme negative dependence and risk aggregation. *J. Multivariate Anal.*, **136**, 12–25.

Wang, B., and Wang, R. 2016. Joint mixability. *Oper. Res.*, **41**(3), 808–826.

Wang, B., Peng, L., and Yang, J. 2013. Bounds for the sum of dependent risks and worst value-at-risk with monotone marginal densities. *Finance Stoch.*, **17**(2), 395–417.

Wang, R. 2014. Asymptotic bounds for the distribution of the sum of dependent random variables. *J. Appl. Probab.*, **51**(3), 780–798.

Wang, R. 2015. Current open questions in complete mixability. *Probab. Surv.*, **12**, 13–32.

Wang, S. 1996. Premium calculation by transforming the layer premium density. *Astin Bull.*, **26**(1), 71–92.

Wei, G., and Hu, T. 2002. Supermodular dependence ordering on a class of multivariate copulas. *Stat. Probab. Lett.*, **57**, 375–385.

Whitt, M. 1976. Bivariate distributions with given marginals. *Ann. Stat.*, **4**, 1280–1289.

Williamson, R. C., and Downs, T. 1990. Probabilistic arithmetic. I. Numerical methods for calculating convolutions and dependency bounds. *Internat. J. Approx. Reason*, **4**(2), 89–158.

Wirch, J. L., and Hardy, M. R. 1999. A synthesis of risk measures for capital adequacy. *Insurance: Math. Econom.*, **25**(3), 337–347.

Yeh, H.-C. 2007. The frailty and the Archimedean structure of the general multivariate Pareto distributions. *Bull. Inst. Math. Acad. Sin. (N.S.)*, **2**(3), 713–729.

Yin, C. 2021. Stochastic ordering of multivariate elliptical distributions. *J. Appl. Probab.*, **58**(2), 551–568.

Index